Investigating Biological Systems Using Modeling

Strategies and Software

Investigating Biological Systems Using Modeling

Strategies and Software

Meryl E. Wastney
Division of Neonatology
Georgetown University Medical Center
Washington, District of Columbia

Blossom H. Patterson
Division of Cancer Prevention
National Cancer Institute
Bethesda, Maryland

Oscar A. Linares
Biomodeling Unit
Oakwood Healthcare System Southgate Center
Southgate, Michigan

Peter C. Greif
Division of Basic Sciences
National Cancer Institute, National Institutes of Health
Bethesda, Maryland

Ray C. Boston
School of Veterinary Medicine, Clinical Studies
Kennett Square, Pennsylvania

ACADEMIC PRESS
San Diego London Boston New York Sydney Tokyo Toronto

Chapter 2, fig. 2.4: reprinted from ref. 6 with permission of The American Society for Biochemistry and Molecular Biology; fig. 2.5: reprinted from ref. 11 with permission of the Federation of American Societies for Experimental Biology; Chapter 10, figs. 10.9 and 10.10: reprinted from ref. 18 with permission of The American Physiological Society; Chapter 19, fig. 1: reprinted from ref. 2 with permission of Elsevier Science; Chapter 21, figs. 21.1–5; reprinted from ref. 10 with permission of Elsevier Science; Chapter 12, table 12.1: reprinted from Wastney ME, Wang XQ, and Boston RC (1998). Publishing, interpreting, and accessing models. *J. Franklin Inst.* **35B**:281–301; Chapter 17, table 17.1: reprinted from Wastney ME, *et al.* (1995). World-wide access to computer models of biological systems. *Information Services and Use* **15**:185–191; Chapter 19, tables 19.1–4: reprinted from ref. 2 with permission of Elsevier Science; Chapter 21, table 21.1: reprinted from ref. 10 with permission.

This book is printed on acid-free paper. ∞

Academic Press
A Division of Harcourt Brace & Company
525 B Street, Suite 1900, San Diego, California 92101-4495, USA
http://www.academicpress.com

Academic Press
24–28 Oval Road, London NW1 7DX
http://www.hbuk.co.uk/ap/

Library of Congress Cataloging-in-Publication Data

Investigating biological systems using modeling : strategies and software / Meryl E. Wastney . . . [et al.].
p. cm.
Includes bibliographical references and index.
ISBN 0-12-736740-3 (alk. paper)
1. Biological systems--Mathematical models. 2. Biological systems--Computer simulation. I. Wastney, Meryl E.
QH323.5.I5675 1998
570′.1′5118--dc21 98-40389
CIP

PRINTED IN THE UNITED STATES OF AMERICA
98 99 00 01 02 03 MM 9 8 7 6 5 4 3 2 1

Loren Zech, B.E. (Elect. Eng.), M.D.
1943–1997

This book is dedicated to the memory of Dr. Loren Zech who was a senior investigator at the National Institutes of Health, Bethesda, for over 20 years. Following the death of Dr. Mones Berman, the instigator of SAAM in 1982, Loren became the guiding force in the continued development of the software while maintaining his work as a clinician and active researcher in the areas of lipoprotein and trace element metabolism, endocrinology, and pharmacokinetics. He was the author of over 100 papers, most of which involved modeling. He was dedicated, enthusiastic, and knowledgeable in the areas of mathematics, engineering, computer science, and biology. Most of all, he was a patient teacher and a gentle guide. Each of the authors of this book benefitted from his wisdom and experience. He was a teacher, colleague, and friend to each of us, and we dedicate this book to his memory and to his lifelong love of modeling.

Contents

12 Model Summarization

13 Multiple Studies

14 Information in the Model

15 Errors in Compartmental Modeling

16 Testing Robustness: Sensitivity, Identifiability, and Stability

V EVALUATING AND USING PUBLISHED MODELS

17 Why Use a Published Model?

Preface

Ample evidence now exists that complex biological systems can only be progressively understood with the aid of mathematical models. Mathematical modeling requires an understanding of several disciplines: biology, computing, kinetics, mathematics, and statistics. While many texts cover these topics individually, previous textbooks do not integrate this information to enable an investigator without a strong mathematical background to apply modeling for hypothesis testing. This textbook is an attempt to meet this need. References are made to material covered in other textbooks.

The approach described in this textbook is often called the "SAAM (simulation, analysis, and modeling) approach" or "Berman approach" to mathematical modeling. Briefly, it involves the process of mathematical modeling on a computer as an aid to understanding biological system behavior [1]. This approach has guided the development of the WinSAAM (Windows SAAM) modeling software. The SAAM program and its conversational version, CONSAM, have been under development at the Laboratory of Experimental and Computational Biology, National Institutes of Health, since the late 1950s. WinSAAM evolved from the SAAM and CONSAM programs. It was designed for use by biologists. For example, compartmental models to describe experiments performed on biological systems can be set up and changed by simply specifying differential equation parameters and providing the initial conditions for the experiment. Models can also be solved by entering explicit equations. Parameter values can be changed during a solution to simulate variable experimental conditions. Data can be fitted by ordinary, generalized, or weighted least-squares regression techniques. Plotting and statistical measures of fit make it easy to compare and evaluate mathematical or compartmental models. WinSAAM incorporates Windows features for interapplication communication, making interacting through WinSAAM with other software as simple as Select|Copy|Paste.

Section I is a general introduction to the field of modeling biological systems. Section II describes modeling software and compares several packages. Section III explains the tools for modeling and illustrates them using examples. Section IV covers topics related to the design of experimental studies and the steps involved in modeling biological data. Finally, Section V describes how to evaluate and use published models.

The book is designed for students (all sections), experimentalists who have data to analyze (Sections II–IV) or who are planning kinetic studies (Sections IV and V), and for modelers as a resource (Section III). Models contain a plethora of information about a system, and the book is also designed for those who wish to access information inherent in models for designing experiments, for assessing the state of knowledge in a particular area, or for teaching complex biophysical principles (Section V). There is some repetition in the book as the description of topics focuses more on theoretical aspects in some sections and practical approaches in others.

The principles of developing models to interpret kinetic data apply across disciplines, and examples in the text are chosen from many fields, including chemistry, biochemistry, physiology, pharmacology, animal science, medicine, and agriculture. Modeling is part art and part science and is best understood by a hands-on approach. The CD supplied with this book includes the latest version of WinSAAM as well as the examples that have been used in the text. Readers are encouraged to use and adapt the models for their own areas of interest and to incorporate modeling as a routine tool for data analysis and in experimental design.

We acknowledge the contributions made by our families and colleagues during the writing of this book, and specifically Dr. Janet Novotny for proofing some of the chapters.

Without data models can't exist . . .
Without models data can't be used . . .

R. Boston, 1998

Reference

1. Berman M. (1963). The formulation and testing of models. *Ann. N.Y. Acad. Sci.* **108**, 182–194.

SECTION I

Introduction

1

WHAT IS MODELING ?

This chapter will provide an introduction to modeling by defining terms and discussing modeling philosophy. Modeling straddles the fields of biology and mathematics. Therefore, it could be termed 'Biomodeling'. However, it differs from, but is related to, the fields of Biomathematics, Biostatistics, and Bioengineering where mathematics, statistics, and engineering are applied to the study of biological systems. Biomodeling involves the use of these scientific tools, a knowledge of biology, intuition, imagination, and creativity to answer the question, what biological process could explain these data? Modeling is therefore part art and part science. The approaches, concepts, and tools can be taught to some extent, but the art of modeling can only be acquired through practice and experience.

I. Definitions

Models are simplified representations of systems. They can be physical such as scale models of airplanes, or abstract such as mathematical models. Mathematical models are used widely in the fields of engineering, physics, economics, business, and meteorology where models are routinely used in weather prediction. Models are useful for studying complex systems where many processes occur simultaneously. In the case of weather forecasting, such processes might include temperature changes, precipitation patterns and expected movement of high and low pressure systems. The processes and their interactions can be represented mathematically by a set of equations (or model). Then, by solving the equations simultaneously, the solution of the model will mimic the behavior of the system (e.g., the path of a hurricane). As evident from this example, models are developed because it may not be possible, or it may be too costly, to probe the real system. Mathematical models therefore predict the response or behavior of a system. They can be also be used to predict the response of a system prior to an experiment on the actual system.

Models are used in all fields of biology. Specific examples are: 1) to calculate nutrient intake for optimal growth, 2) to represent blood circulation, 3) to predict the pharmacological response to a drug, 4) to determine the rate of uptake of compounds by cells, and 5) to calculate enzyme kinetics.

Modeling is the process of developing a model or set of equations to simultaneously represent the structure and behavior of a system. Modeling biological systems differs fundamentally from modeling physical systems because the structure of physical systems is usually known, whereas, the structure of biological systems is generally not known. Models of biological systems are based on observations of the system (Fig 1.1). This process of determining the structure of a system based on its behavior is called the inverse problem. There are a number of limitations in modeling biological compared to physical systems; data are often incomplete due to the limitations on sampling sites; sampling times and number of studies that can be performed; data are imprecise; constraints related to the biology and the experimental techniques must be embedded in the analysis (3, 4).

Because the structure of a biological system is not known, a model developed to fit data from the system also represents a hypothesis of the system. Models therefore are not static, but evolve over time as studies provide new information that extend and refine the model. A good example is the development of models that describe lipoprotein kinetics (5).

Tip: The aim of modeling is to create a mathematical 'likeness' of a system so that the model behaves in the same way as the system.

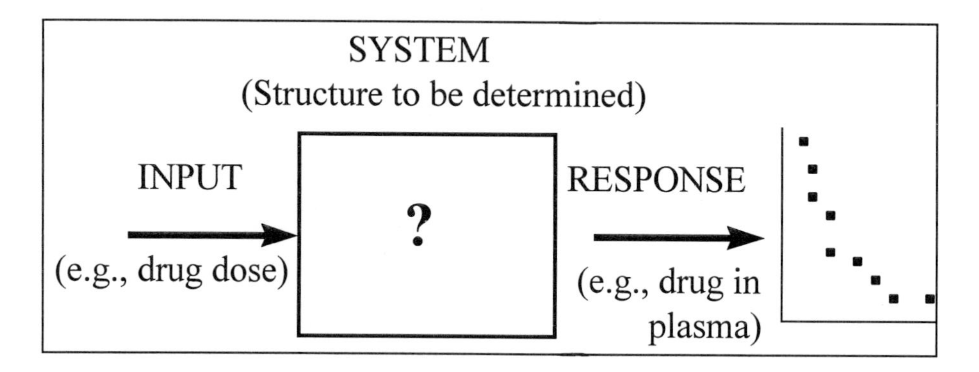

Fig 1.1 Biological modeling is the process of determining the structure of a system from its response.

II. Approaches to modeling

There are three general approaches to modeling biological systems. They relate to whether the model is defined before or after a study, whether the model parameters are related to the structure (physiology or chemistry) of a system, and the complexity of the model. The approach chosen by an investigator depends on the purpose for modeling the system.

1) *A priori* versus *post priori:*

A priori (or theoretical) models are developed based on existing information about a system. For example, a model for glucose metabolism could be based on *in vitro* studies of the individual enzyme reactions. An example of this approach is described by Garfinkel et al (11). By contrast, *post priori* (or empirical) models are based on new observations. With this approach, a model for glucose metabolism would be developed by injecting labeled glucose *in vivo*, measuring its disappearance from blood, and proposing a model to fit the data. An example of this approach is described by Foster et al. (10).

2) Descriptive versus mechanistic

Descriptive models consist of mathematical functions or equations that fit the observed data. Parameters in the equations generally have no relationship to entities in the system. An example is an equation that is the sum of exponentials. Although descriptive models are usually simpler than mechanistic models (described below), they are based on assumptions that need to be verified. Some limitations of these models are that they may not use all the information that is available in the data and they are often limited to the time range of the data.

As stated by Murray, a description of data is not an explanation (13). Mechanistic models consist of functions where the parameters are related to entities of the system under study. These models are based on the mechanisms of the system and can be used to explore and understand the structure and properties of a system. Mechanistic models have the disadvantage that the system structure needs to be specified. However, there are advantages in using these models, e.g., all the information in the data can be used, experiments can be proposed based on the findings of the model, and the models can be used to analyze data obtained from two or more different states which may serve to identify where differences in the system are occurring (6). This is important for comparing disease versus healthy states, or the effects of treated versus untreated states. The results of using descriptive versus mechanistic models have been described for calcium metabolism (12).

3) Large versus reduced

The goal of model development is to find the simplest model (i.e., fewest number of parameters) to fit the observed data. Within this framework, a system could be modeled using a 'minimal' model with a small number of parameters, or by a 'large' model with more parameters to represent additional biological complexity of the system. There is generally a tradeoff between mathematical rigor of a model and its biological accuracy. The minimal or reduced approach will often provide parameter values that are well-determined with low errors, but each parameter may represent the combination of several processes. By contrast, parameters in a large model generally relate to individual processes, but some may not be well-determined. A large model can sometimes be reduced to a simpler model by combining parts of the system for calculating specific parameters of the system (5).

> **Tip:** Terms are summarized in the glossary (Appendix 1).

III. Model Types

Models can be classified in several ways based on mathematical form (9). The form of a model chosen to investigate a system, its complexity, and how precisely it represents an actual system are all influenced by the purpose for which the model is developed. Some examples of different types of models applied to biological systems are found in Robson and Poppi (14).

1) 'Model-independent' versus model-dependent

Algebraic models are generally used to determine a specific parameter such as rate of clearance. They are often referred to as 'model-independent' because the value that is calculated does not depend on knowing the structure of the system.

However, some underlying model structure is assumed in these calculations. For example, absorption of a nutrient is sometimes calculated algebraically as the difference between intake and excretion of a labeled form of the nutrient in feces. This model assumes however, that the nutrient is not absorbed and then excreted back into the intestine. 'Model-dependent' implies that calculations about a system are related to the type of model used.

2) Deterministic versus stochastic

Deterministic models are specified explicitly and do not permit any random variability (e.g., the law of gravity) while stochastic models include an element of randomness and are based on probability. Stochastic models are used for determining rates of production, disposal and residence time of a compound (15). However, these parameters can also be determined from compartmental models (5, 8), (See Chapter 14). Most biological systems include features of both. For example, the existence of a pathway in a model (such as absorption from the gut) is deterministic in that the pathway exists in all subjects, but the rate of absorption differs among subjects, and is therefore, variable.

3) Linear versus Non-linear

A system is linear if any combination of inputs yields the same combination of outputs (7), i.e., the amount of a substance that is moved from one location or state to another is proportional to the amount of substance present. Such a system is said to follow first-order kinetics. Most biological systems are non-linear. This means that the movement is not proportional to the amount of substance present (6). A gentle perturbation of a nonlinear system however behaves linearly. Radioactive tracers are useful for studying biological systems because they have negligible mass, do not perturb the system, and therefore have linear kinetics (4). (Stable isotope tracers contribute mass and their use may perturb a system. Tracers are described in more detail in Chapter 3.)

4) Kinetic versus Dynamic

A kinetic model characterizes a system for a particular state while dynamic models describe changes in a system as it moves from one state to another. Kinetic models are generally linear while dynamic models include non-linearities. For example, a kinetic model could be used to describe glucose metabolism in a subject under normal conditions but a dynamic model would be required to describe the changes in glucose metabolism after the ingestion of a glucose load. The dynamic model would describe the glucose-induced release of insulin and the effect of insulin on returning blood glucose to normal levels.

5) Compartmental versus non-compartmental

Compartmental models represent a system by a series of ordinary differential equations. Non-compartmental models include all other types (partial differential equations, algebraic, stochastic). Compartmental models are useful for describing biological systems because these systems are often visualized in terms of pools or compartments. For example, a compound may be bound or free, inside the cell or outside, in plasma or in an extravascular space, and each state could be represented in a model by a separate compartment.

IV. Compartmental Models

Compartmental models have been widely applied in the study of biological systems, particularly in relation to analysis of isotope data (1, 2). Compartmental models assume that the material of interest is distributed throughout the system in discrete entities, called compartments. A compartment is considered to contain material that is homogeneous, or kinetically indistinguishable. A compartment may be defined physically (e.g., a specific body pool), or conceptually (e.g., all particles that turn over at a particular rate). For example, if a drug is lost from blood monoexponentially (Fig 1.2), metabolism of the drug can be represented by a one compartment model (Fig 1.3).

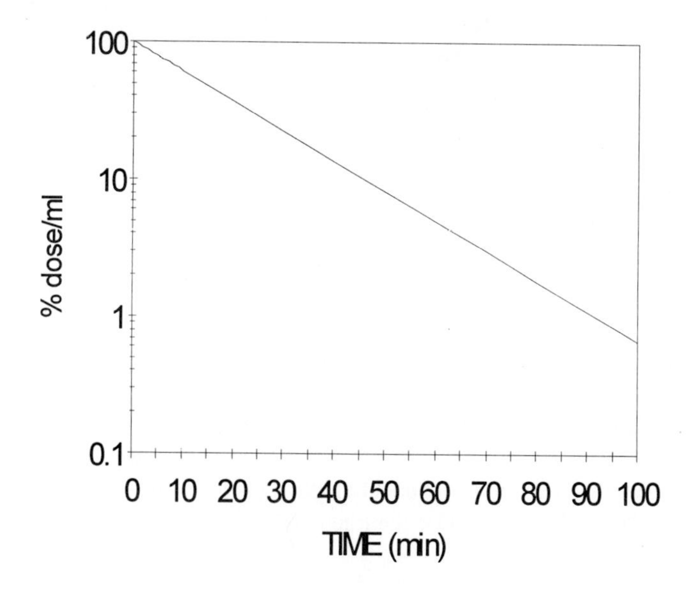

Fig 1.2. Drug disappearance from blood: monoexponential loss. Note that the Y axis is log scale, so that the exponential is given by a straight line.

To show the relationship between a monoexponential model and compartmental model, differentiate the equation for the exponential loss. It can be seen that the rate of loss is a constant fraction (k) of the amount present;

$$\begin{aligned} Y &= A e^{-kt} \\ \frac{dy}{dt} &= -kAe^{-kt} \\ &= -kY \end{aligned} \tag{1.1}$$

The model (Eq. 1.1) can be expressed using WinSAAM notation (see Appendix 2);

$$\frac{dF(1,t)}{dt} = -L(0,1)F(1) \tag{1.2}$$

where;

F(1,t) is material in compartment 1 at time t, and is often written as F(1), or Y in Eq. 1.1

L(0,1) is transfer of material from compartment 1 to the outside, or k in Eq 1.1.

Equation 1.2 is often written as,

$$F(1)' = -L(0,1) \bullet F(1) \tag{1.3}$$

and the model is shown graphically in Fig 1.3.

If the loss from a system is biexponential (Fig 1.4) the system can be described by two compartments (Fig 1.5).

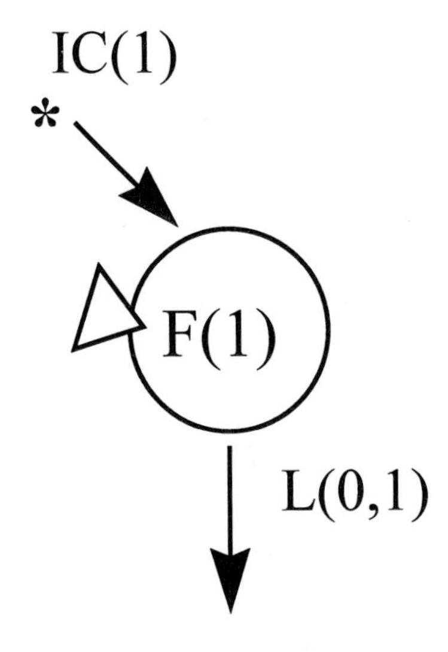

Fig 1.3 One compartment model for monoexponential loss. F(1) is compartment 1, IC(1) is the initial condition in compartment 1, and L(0,1) is the fractional rate of loss to the outside from compartment 1.

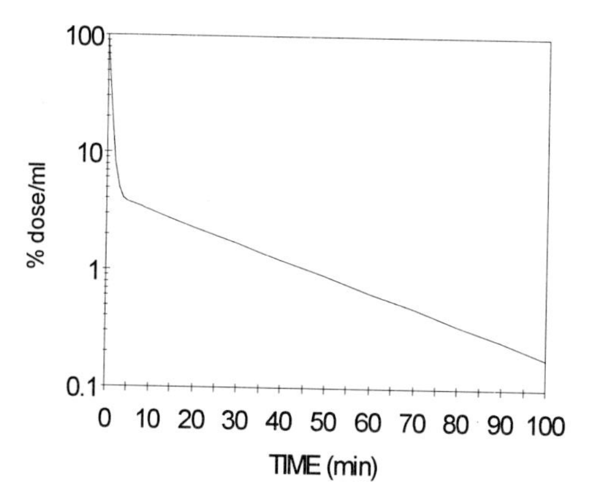

Fig 1.4 Drug disappearance from blood: biexponential loss.

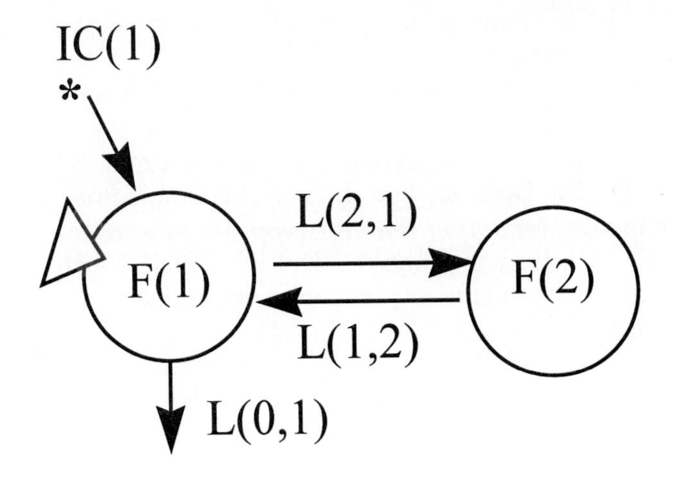

Fig 1.5 A two-compartment model for biexponential loss.

The system is described by the following two equations;

$$F(1)' = -L(0,1) \bullet F(1) - L(2,1) \bullet F(1) + L(1,2) \bullet F(2) \tag{1.4}$$
$$F(2)' = -L(1,2) \bullet F(2) + L(2,1) \bullet F(1)$$

In this case, the material is introduced into compartment 1 by bolus injection and is a compound that exchanges with compartment 2 (e.g., an IV drug input into plasma and taken up by the tissues). If we were interested in a compound that is synthesized by the system, such as glucose, the model would need to include this input, represented by U(i). The equations for the model become;

$$F(1)' = -L(0,1) \bullet F(1) - L(2,1) \bullet F(1) + L(1,2) \bullet F(2) + U(1) \tag{1.5}$$
$$F(2)' = -L(1,2) \bullet F(2) + L(2,1) \bullet F(1) + U(2)$$

This book will focus on mechanistic compartmental models, i.e., data-based modeling. This emphasis is chosen because this approach can be used to investigate the structure and function of a system.

REFERENCES

1. Anderson, D. H. 1983. Lecture Notes in Biomathematics; Compartmental modeling and tracer kinetics. Springer-Velag, Berlin.
2. Atkins, G. 1. 1969. Multicompartmental models in biological systems. Methuen, London.
3. Berman, M. 1969. Kinetic modeling in physiology. *FEBS Letters*. 2:S56-S57.
4. Berman, M. 1971. Compartmental modeling. *In* Advances in Medical Physics. J. Laughlin and E. Webster, editors, Boston.

5. Berman, M. 1979. Kinetic analysis of turnover data. *Progr. Biochem. Pharmacol.* 15:67-108.
6. Berman, M. 1982. Kinetic analysis and modeling: Theory and applications to lipoproteins. *In* Lipoprotein Kinetics and Modeling. M. Berman, S. M. Grundy, and B. V. Howard, editors. Academic Press, NY. 3-36.
7. Brownell, G. L., M. Berman, and J. S. Robertson. 1968. Nomenclature for Tracer Kinetics. *Int J Appl Radiat Isotop.* 19:249-262.
8. Covell, D. G., M. Bernam, and C. DeLisi. 1984. Mean residence time - Theoretical development, experimental determination, and practical use in tracer analysis. *Math Biosci.* 72:213-244.
9. Finklestein, L., and E. R. Carson. 1985. Mathematical Modelling of Dynamic Biological Systems. John Wiley and Sons, UK.
10. Foster, D. M., G. Hetenyi, Jr., and M. Berman. 1980. A model for carbon kinetics among plasma alanine, lactate, and glucose. *Am. J. Physiol.* 239:E30-E38.
11. Garfinkel, D., C. A. Kulikowski, V.-W. Soo, J. Maclay, and M. J. Achs. 1987. Modeling and artificial intelligence approaches to enzyme systems. *Fed Proc.* 46:2481-2484.
12. Jung, A., P. Bartholdi, B. Mermillod, J. Reeve, and R. Neer. 1978. Critical analysis of methods for analysing human calcium kinetics. *J. Theor. Biol.* 73:131-157.
13. Murray, J. D. 1993. Mathematical Biology. Springer-Velag, Berlin.
14. Robson, A. B., and D. B. Poppi. 1990. Modeling Digestion and Metabolism in Farm Animals. Lincoln University, Lincoln.
15. Shipley, R. A., and R. E. Clark. 1972. Tracer Methods for in vivo Kinetics: Theory and Applications. Academic Press, NY. 239.

2

THE STEPS IN BUILDING A MODEL

The goal of this chapter is to show how modeling fits into the overall experimental process. The steps involved in a modeling project are covered from identifying the problem to be studied, to publishing the results. Specifically, this chapter covers the formulation of an hypothesis, proposing a model and designing the experiments, testing the model against the data, refining and evaluating the model, and publishing the results. The individual steps in modeling are described in greater detail in subsequent chapters.

I. Relationship between hypothesis, data and model

The object of a scientific investigation is to improve understanding of a system by testing a hypothesis about that system. In the scientific process, a problem or scientific question is posed and stated as an hypothesis, or theoretical statement of the problem. The hypothesis is tested by designing an experiment and collecting data. For simple problems, data can be compared against an hypothesis intuitively, for some other problems, a statistical approach can be used. For complex biological systems, however, the hypothesis may need to be translated into a mathematical model and the model predictions compared to the observed data as a basis for rejecting or accepting an hypothesis (Fig 2.1).

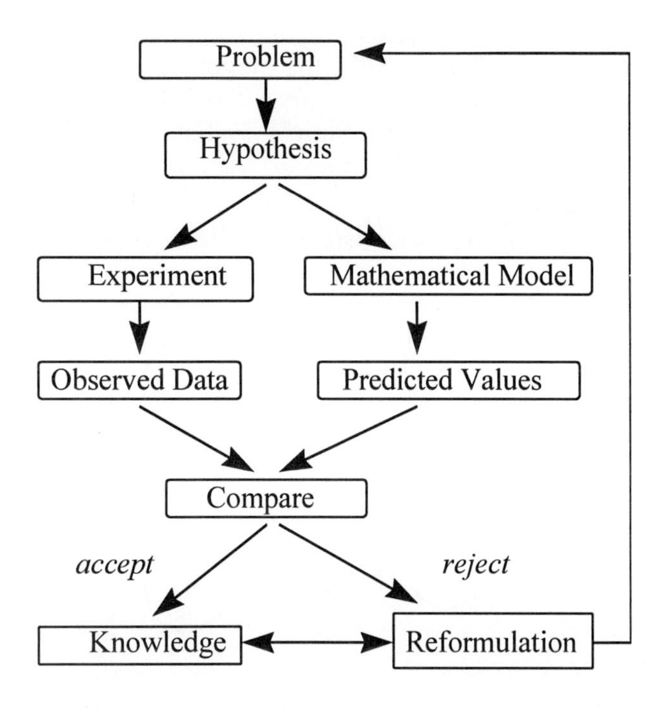

Fig 2.1 The modeling process.

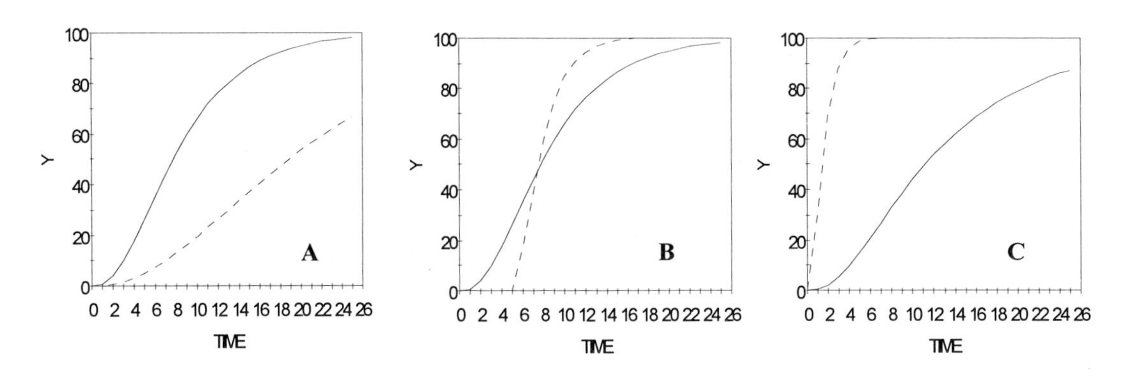

Fig 2.2 Release of compound Y from cells. Normal ________, Mutant -------------

For example, a cell line takes up a substance (X), converts it to another compound (Y), and then releases it. The hypothesis to be tested is that normal cells release the converted compound faster than do mutant cells. An experiment could be performed to measure the rate of release of Y. Several scenarios are possible (Fig 2.2). In the first scenario (A) release of Y is slower in the mutant cells (the hypothesis is accepted) while in the third scenario (C), the release is faster in the mutant cells and the hypothesis is rejected. However, in the second scenario (B) it is not clear whether the hypothesis should be accepted or rejected as the release by mutant cells is slower initially but then faster. One approach would be to set up a model and to use the model predictions to draw conclusions about the rate of release of Y.

A model for this system is shown in Fig 2.3. In this model several steps are involved in the uptake of X and release of Y. Although the hypothesis stated that release of Y was slower in mutant cells, the model suggests that the uptake and/or conversion pathways may also differ between the cell lines. This example shows that while a hypothesis relating to a simple system (e.g., processes involving a

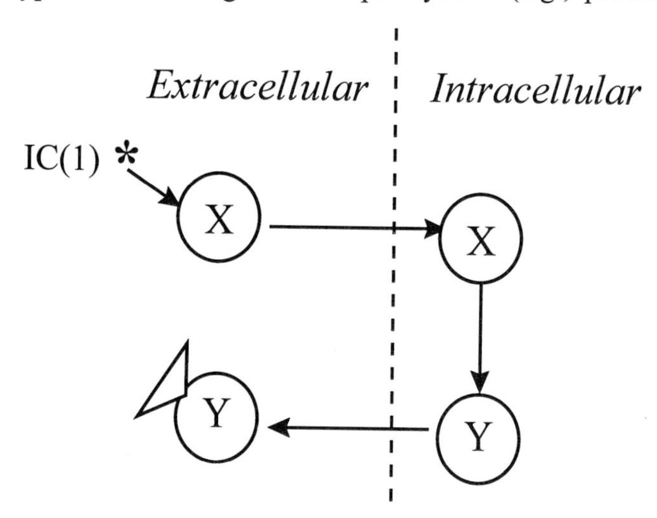

Fig 2.3 Model for cell uptake of compound X and release of compound Y.

single pathway) can be compared to data intuitively or by using simple statistics, when several processes are involved, a more complex mathematical model is required to compare in a detailed and rigorous way, experimental observations with hypotheses.

Models should be set up before experiments are performed on a system in order to optimize the experimental design with respect to the number of samples obtained, the timing of the samples, and the length of the study. In practice, however, data are often collected before the model is tested. As modeling becomes more accepted as part of biological investigations, models will be used, as described below, in the design of studies as well as the analysis of systems (1).

II. Define the System

The system under study may consist of a single molecule or, as in epidemiology, a whole population. The system may be open or closed. Closed systems do not have inputs from or losses to the outside and thus do not interact with their environment. One example of a closed system is an equilibrium reaction. Biological systems are open systems in that they have inputs and losses, although they may be reduced to closed systems as in *in vitro* studies. Once the system to be studied is defined, we set the boundaries for the investigation, and also identify the inputs and outputs, as well as the area for observation. A system is characterized by its processes, sub-systems, compounds, inputs and outputs where: processes are movement or changes in the system (e.g., absorption, metabolism, transport, oxidation), sub-systems are components of the whole system (e.g., red blood cells within blood) and compounds are the substances under study (e.g., drugs, nutrients, metabolites). An input is the addition of the compound of interest into the system. Losses include processes such as excretion or degradation. Before modeling a system it is important to know the purpose of the study, what is known about the biology of the system, and experimental details.

III. Identify the purpose

The purpose of the model usually dictates the form and detail of the model as well as the data that are required to develop and test it. Models are developed for various purposes (2). One purpose may be to calculate one or more parameters of interest, such as pool size, clearance rate, absorption, excretion, or transmission rates. If the purpose is to accurately determine a particular parameter, it is important that the model can be uniquely determined i.e., the parameters are unique (they are not highly correlated) and well determined statistically. Alternatively, the purpose may be to integrate a large amount of information on a system, perhaps to combine information obtained at the organ or cellular level with whole body level. For this type of model biological accuracy may be more important than mathematical rigor. A third purpose may be simulation, to predict responses to a perturbation and to identify differences in a system due to that perturbation. If the purpose is get a general idea on how the system would behave under a new condition, the accuracy of the model may not be important, however, if the purpose is to determine for example, a dosing regimen, the model accuracy may be critical. Finally, the purpose may be to derive the mechanistic principles underlying the behavior of a system. For this purpose, it may be necessary to obtain as much data from the system as possible (i.e., by sampling several sites and under several conditions). The purposes for modeling are described in more detail in Chapter 4.

IV. Locate and examine existing models

Models represent theories or hypotheses about how systems function. There are many models which will fit a particular set of data. To progress scientifically in understanding a system, it is important that new studies build upon information of earlier studies. The literature should be searched for models in the area under investigation. Differences between the models should be evaluated to choose the model most acceptable with current biological knowledge. Thus, a model should be tested against new data, rather than proposing a new model which only differs in small respects. If the current model does not fit the new data, then the model should be changed. Only on rare occasions should models be developed *de novo.*

V. Develop an initial model

If it is necessary to develop a new model there are two approaches: start with the simplest model (one-compartment) and add compartments and pathways as necessary to fit the data. The other approach is to develop the model including all known biological information, and to then simplify the model based on the data. An example of the first approach is shown in Fig 2.4 using a model for enzyme (E) kinetics (6). Three species were measured: the substrate (S), intermediate (N) and product (P). It became apparent that this model was inadequate to explain the data. It was necessary to add compartments to represent a binding step (ES), and a compartment for bound intermediate (EN).

An example of the second approach is shown in Fig 2.5 for ketone body kinetics in humans (11). The initial model included compartments for acetoacetate (A) and B-OH butyrate (B) in blood, tissues and liver. Conversion of A to B and vice versa, of the compounds however was too rapid to be resolved by the data and the model was therefore simplified by combining compartments for A and B in extravascular compartments.

Note that systems can be manipulated experimentally to simplify the modeling. For example, clamping is a technique used to maintain blood glucose at some prespecified level so that glucose production and utilization can be determined under controlled conditions. Some techniques appear to simplify the system but can result in erroneous conclusions. For example, comparing the amount of tracer in tissues at one time point after a single dose of tracer will not provide information on the relative distribution of tracer among tissues. It is necessary to measure the

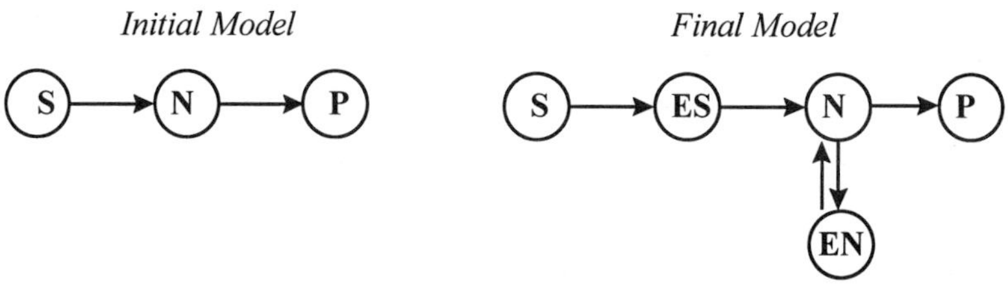

Fig 2.4 An initial and final model for enzyme kinetics.

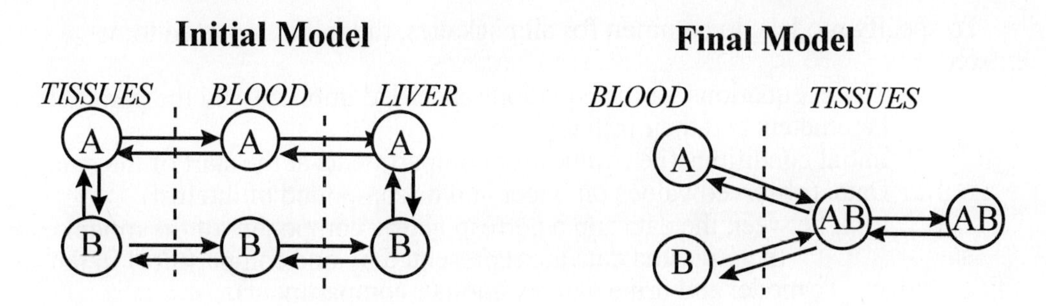

Fig 2.5 An initial and final model for ketone body metabolism in humans. A represents acetoacetate and B, β-OH butyrate.

amount of tracer in tissues over time (Fig 2.6). In this example, the ratio of tracer in serum, liver and muscle differs at 0.5, 1 and 5 hr. Sampling at only one time point provides no information about the uptake and turnover of the tracer by the various tissues.

VI. Set up model in form required by software

To compare the model predictions with the observed data it is necessary to convert the model into a format that can be solved by a modeling package. There are many software packages available (see Chapter 5). Each package has its strengths for solving particular types of models. Some packages are designed for simulation (e.g., ACSL (9)) while others simulate and fit data (e.g., SAAM (3), SAAMII (4), and WinSAAM (5) (see Chapter 6)). Some packages require that model equations be written explicitly (e.g., MLAB (8)) while in others, compartmental models can be set up by specifying the parameters of the model (e.g., WinSAAM), (5)) or graphically, by drawing the model (SAAMII (4) and STELLA (7)).

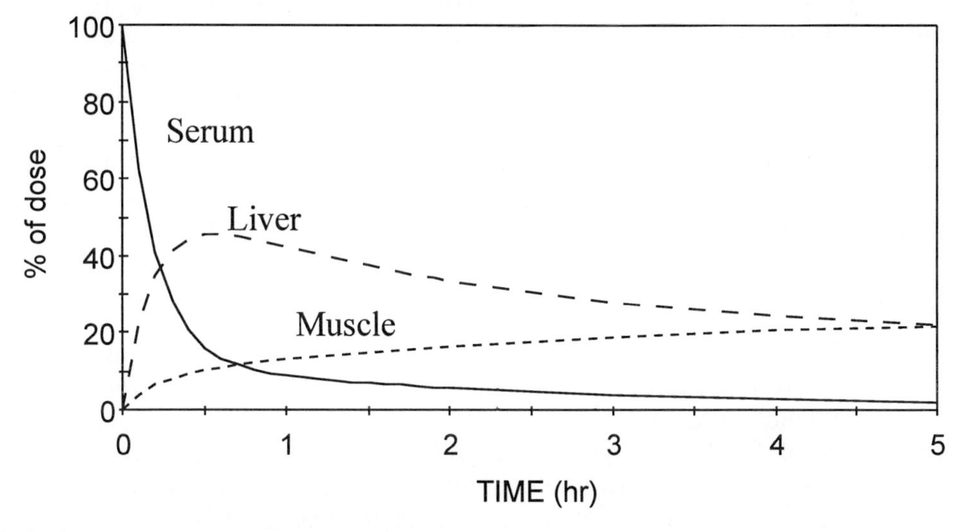

Fig 2.6 Distribution of a tracer among tissues.

To specify models, in common for all packages, the following need to be entered:

Model equations, or the equations specified implicitly by the parameters
Parameters and their initial values
Initial conditions (i.e., values for compartments at the start of the study)
Data (observed values on tracer and/or compound of interest)
Link between the data and a corresponding component of the model (e.g., plasma data are represented by one compartment in the model and urine data by another compartment).

In addition, the following may be required:

Changes in experimental conditions during the study
Conversion of units for observed data and calculated values
Functions (e.g., equations to calculate parameter values such as absorption, or to explain dependent relations between parameters in the model).

VII. Simulate experiments

Before performing an experimental study on the system, it is useful (although not always possible) to simulate the expected results using the model if a model exists. With this approach the experimental design can be optimized to ensure that sufficient data are being obtained from the necessary sites to identify the system, and that samples are obtained at times where the data will provide the most information about the system. For example, a sufficient number of samples should be obtained to define both the rapid and slow processes in a system.

VIII. Obtain data

The aim of the experiment is to obtain data to confirm or reject an hypothesis. The data need to be appropriate (i.e., obtained from the correct parts of the system), have minimal error, and for there to be a sufficient number of data points. It is necessary to have some knowledge of the error in the data when fitting the model to the data (see Chapter 15).

IX. Compare experimental data with model

Once the model has been set up and solved, the model-calculated values can be compared to the observed data. The comparison, initially, is performed best by graphical comparison. This allows the investigator to compare the shape of the predicted curve with the observed data. As the purpose in modeling is to develop a model that predicts all observed data, it is necessary to examine all portions of the curves. Small deviations may hold a large insight into processes occurring within the system. Patterson and Zech describe the development of a selenium model in humans, and how small but consistent fluctuations in the plasma tracer curve led them to postulate multiple plasma pools (10). It is useful for developing an understanding of the system to do the initial data fitting manually, i.e., select multiple values for a given parameter sequentially, examining the accompanying changes in fit. Through this process the investigator learns about the sensitivity of the system to certain processes, the structure of the system (whether additional compartments, delays, or pathways are required), and any abnormalities in the data. (See Chapter 10 for details and examples).

There are two common approaches to fitting models to data, the least squares approach where parameter values are adjusted to minimize the sum of squares of residuals about the data and the Maximum Likelihood approach. These are discussed in detail in Chapter 15. Once the model fit is close, the data can be fitted by iteration. The fit obtained will be influenced by the relative weights assigned to the data, and by how well the system is determined (i.e., whether there are sufficient data to calculate all parameters in the model).

X. Refine model

It is unlikely that the first model proposed for a system will fit all the data. It will be necessary to resolve differences between the observed data and model-calculated fits. Errors sometimes occur in the calculation of the data, and these can often be detected by careful examination of the initial fits to the data. Before changing the model, it is important to check the data (units, calculations, and the techniques used to obtain the data) to ensure the values are correct. For example, in fitting the ketone body model, two ketone bodies were converted rapidly, but the curves diverged later in the experiment. This was inconsistent, and was only resolved after it was discovered that the technique used to measure the ketone bodies also measured a third, slowly turning-over ketone body (11). Examples on refining models are shown in Chapter 11.

XI. Evaluate model

Once the model has been fitted to the data, it needs to be evaluated with respect to how well the parameters are determined (i.e., the standard deviations on the parameters) and the correlation between parameters. Although some parameters may not be uniquely determined, the model has heuristic value in terms of describing the structure of the system, relative pool sizes, and transport rates.

XII. Propose new studies

Once a model is consistent with available data, it can be tested under new conditions. This involves obtaining data under different conditions from those used to obtain data used to develop the model. The new conditions may be experimental perturbations (e.g., a new steady state) or a clinical condition (e.g., to compare normal subjects with patients).

XIII. Publish

The final step in a modeling project is to publish the model and to make it available for other investigators to use. The published form of the model should solve as described in the manuscript, and the experiments and justification for the final model should be described in detail. A set of representative data should be made available with the model. Publishing models will be discussed in more detail in Chapter 22.

REFERENCES

1. Bassingthwaighte, J. B. 1995. Toward modeling the human physionome. *Adv Exp Med Biol.*382:331-339.
2. Berman, M. 1971. Compartmental modeling. *In* Advances in Medical Physics. J. Laughlin and E. Webster, editors, Boston.
3. Berman, M., and M. F. Weiss. 1978. SAAM Manual. DHEW Publication No. NIH 78-180. US Printing Office, Washington, DC.
4. Foster, D. M., H. R. Barrett, B. M. Bell, W. F. Beltz, C. Cobelli, H. Golde, J. A. Jacquez, and R. D. Phair. 1994. SAAM II: simulation, analysis and modeling software. *BMES Bulletin.* 18:19-21.
5. Greif, P., M. Wastney, O. Linares, and R. Boston. 1998. Balancing needs, efficiency, and functionality in the provision of modeling software: a perspective of the NIH WinSAAM project. *Adv Exp Biol Med.* vol. 445.
6. Hensley, P., G. Nardone, J. G. Chirikjian, and M. E. Wastney. 1990. The time-resolved kinetics of superhelical DNA cleavage by BamHI restriction endonuclease. *J. Biol. Chem.* 265:15300-15307.
7. High Performance Systems, Inc. 1996. STELLA Technical Documentation, Hanover, NH.
8. Knott, G. 1992. MLAB Interactive computer program for mathematical modeling. *J. NIH Res.* 4:93.
9. Mitchell and Gauthier, A. 1986. Advanced Continuous Simulation Language (ACSL) User Guide/Reference Manual, Concord, MA.
10. Patterson, B. H., and L. A. Zech. 1992. Development of a model for selenite metabolism in humans. *J. Nutr.* 122:709-714.
11. Wastney, M. E., W. E. H. Hall, and M. Berman. 1984. Ketone body kinetics in humans: a mathematical model. *J Lipid Res.* 25:160-173.

3
THE DIFFERENCE BETWEEN BUILDING A MODEL AND USING A MODEL

In this chapter we discuss the basic concepts behind compartmental modeling. We begin by defining a compartment and a compartmental model, then discuss how compartmental models arise. We then turn to the tracer and its properties. Essential to the understanding of tracer kinetics is the concept of the response of a linear compartmental system to a small perturbation, such as that resulting from the introduction of a tracer into a system. The concept of steady state is then introduced along with its implications. The solution of the set of equations for a linear compartmental model is given; this more technical discussion can be skipped without loss of continuity. Finally, we briefly discuss mapping amongst similar compartmental models. These areas are discussed at greater length in Edelstein-Keshet and Jacquez (1,2).

I. What is a Compartment?

A compartment is any (conceptual) zone within a system in which particles of the same type are involved in the same process. A *zone* is a physiological region, or space, e.g. blood, rumen liquor, bone marrow, a physiological *system* such as the human body, the respiratory system or the circulatory system. *Particles* are entities of the same type, or chemical forms, e.g. prostaglandin, ceruloplasmin, alcohol dehydrogenase, involved in the *same process*, or transformation or movement, such as transport, absorption, elimination, or oxidation.

A compartment is a homogeneous, kinetically distinct space. It frequently has a physical counterpart, e.g., whole blood, red blood cells, or urine. A compartment has dimensions, or size, e.g. mass (Kg), volume (L), concentration (Mole/L), or pressure (P). At steady state, the particles that make up a compartment are invariant in that their distribution is not changing with time.

An input into a compartment is the introduction of the substance of interest or the synthesis of constituent particles. An output from a compartment can be

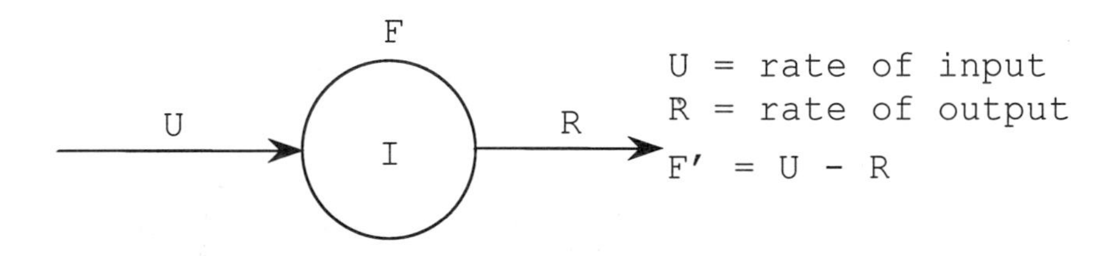

Fig 3.1 One Compartment Model

associated with the formation of other particles and forms from constituents of that compartment, or the transport of constituent particles out of the compartment. Consider a one-compartment model (Fig 3.1). If F represents the level of a substance in the compartment as a function of time, then we have, F', the rate of change of F with time. F' is given by the difference between U, the rate of input and R, the rate of output.

If the system is at steady state, then for our compartmental representation,

$$F' = 0 \text{ and } U = R \tag{3.1}$$

For alcohol metabolism we could demonstrate this structure as in Fig 3.2. Then, at steady state, acetaldehyde production just balances (i.e., equals) newly absorbed alcohol.

II. What is a linear compartmental model?

Compartmental models are made up of compartments, connections between the compartments describing the rate, direction, and manner (i.e., linear or non-linear) of exchange of 'particles' amongst the compartments, inputs (i.e., inputs from outside the particular system under study) and irreversible losses. In graphical representations of models, an arrow refers to a connection, lines joining two compartments represent linear processes, and a line joining a compartment and a transfers indicates a nonlinear process.

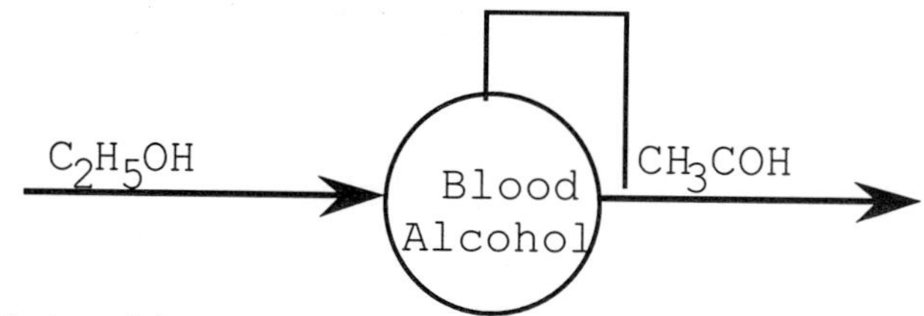

Fig 3.2 Blood alcohol model

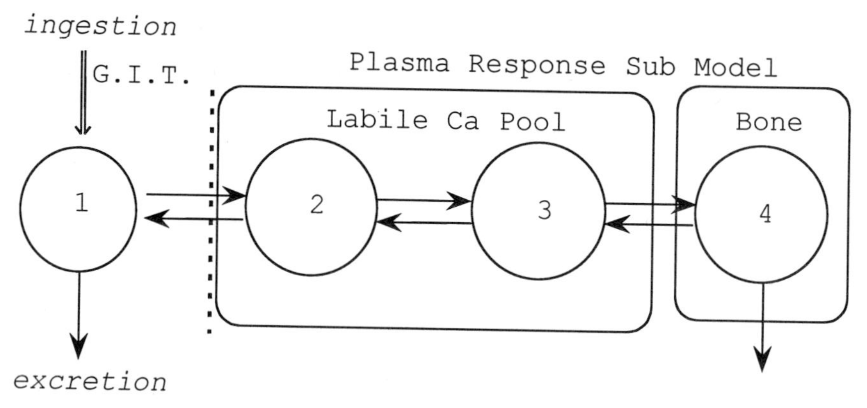

Fig 3.3 Model of calcium absorption and metabolism

For example, the action of alcohol dehydrogenase on blood alcohol is represented as a solid line joining the site of action and the relevant transfer (Fig 3.2).

In Fig 3.3, the dashed line represents an anatomical separation, and transfer to a compartment across this line is via absorption. "G.I.T." refers to the gastrointestinal tract, the double arrow represents a steady state input, and the single arrow from compartment 4 extending beyond the box represents irreversible loss from the system.

We now show the mathematical solution to a two-compartment model in detail. Consider the following 2-compartmental model (Fig 3.4).

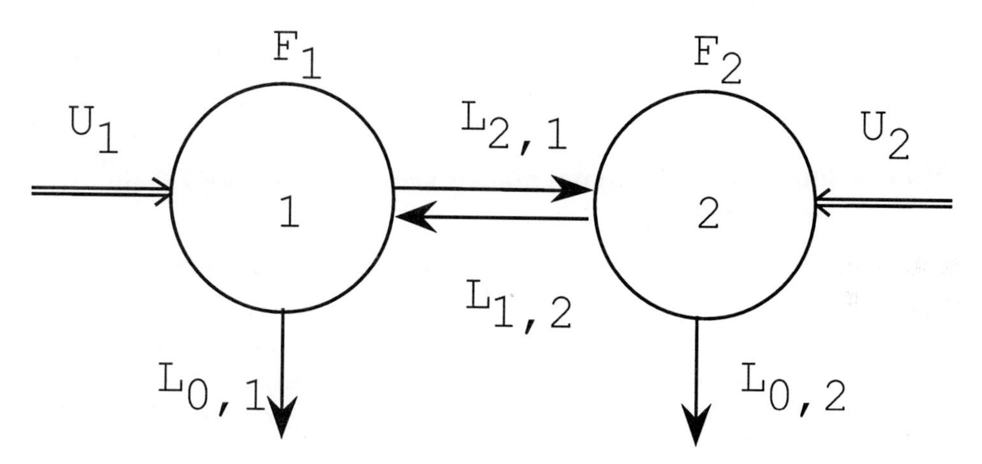

Fig 3.4 Two compartment model

F_j is the amount of particles in compartment j
L(j,i) is the fractional rate of transfer of particles to compartment j from compartment i
U_j is the exogenous input to compartment j

In the following, Li,j, LIJ and Lij are equivalent. From mass considerations, we can write our model:

Net Rate of Change = Rate In - Rate Out

or,

$$\begin{aligned} F1' &= -L11 \cdot F1 + L12 \cdot F2 + U1 \\ F2' &= L21 \cdot F1 - L22 \cdot F2 + U2 \end{aligned} \quad (3.2)$$

(3.2)

$$\begin{aligned} L11 &= L21 + L01 \\ L22 &= L12 + L02 \end{aligned} \quad (3.3)$$

i.e. at steady state ($F1' = F2' = 0$)

$$F1 = \frac{\begin{vmatrix} -U1 & L12 \\ -U2 & -L22 \end{vmatrix}}{\begin{vmatrix} -L11 & L12 \\ L21 & -L22 \end{vmatrix}} \qquad F2 = \frac{\begin{vmatrix} -L11 & -U1 \\ L21 & -U2 \end{vmatrix}}{\begin{vmatrix} -L11 & L12 \\ L21 & -L22 \end{vmatrix}}$$

where F1 is the steady state size of compartment 1.

III. How do linear compartmental models arise? Tracers and their properties

A linear compartmental model can describe any aspect of a genuinely linear system. Also, when a nonlinear system in steady state is perturbed to some small degree, a linear compartmental model can describe its response. These properties make linear compartmental modeling a powerful tool. If a system is described by;

$$F' = h(F) \tag{3.4}$$

then the linear model subject to these conditions is

$$f' = h'(F|_{F=F0})\, f \tag{3.5}$$

where f is the response of this system to a small perturbation and F_0 is the systems current steady state. In the absence of a tracer, the only aspects of a linear system that can be studied are the amounts and locations of those outputs accessible to measurement, based on controlled known inputs of a substance of interest into the system. To discover more about the spaces associated with the host, we need to perturb the system. Tracers, e.g. stable isotope tracers or radiotracers, are used to elucidate details of the transport or metabolism of host particles. They permit us to 'see inside the system', or rather to see echoes of their movement. For example, ^{45}Ca elucidates aspects of Ca movement in a living system. So as not to distort the system (i.e. move it away from its steady state) tracers are applied (injected, ingested) in small (or trace) amounts to plasma, a tissue of interest, or the gut. Tracers can usually be delivered quickly, for example as a bolus. A bolus can be readily represented in a model, as will be discussed later. Tracers move quickly in the space to which they are applied, and after a short while, are distributed within it in the same fashion as the tracee.

Tip: Suppose an amount of tracer, Q_0, is applied at time t_0. A 1 ml sample containing a uniform mixture of the tracer and tracee, is drawn shortly after t_0, from the compartment into which it was applied. If the 1 ml yields an amount of the tracer, Q_S, then the distribution space of the tracer and tracee is $\approx Q_0/Q_S$ ml.

When mixing is 'complete', the tracer moves with the host (tracee), and at the same rate as the host, in and out of the other compartments accessible to the host. Samples of the tracer (and tracee) can be collected at frequent points in time from accessible sites to which the tracer is transported (e.g., blood, gut, tissue). The tracer is distinguishable at these sites because of its other physical properties, e.g. radioactive decay or nuclear magnetic resonance pattern. It is important that the tracer decays or undergoes its own transformation (e.g., a protein denaturing) at a considerably slower rate than that relating to its metabolism, that is, the physical decay needs to be much slower than the biological decay. When adequate information is collected regarding the system response, the data collection ceases. A plot of a two-compartment model fit to data is shown in Fig 3.5. Data collection ceased in this study when the values were close to the limit of detection.

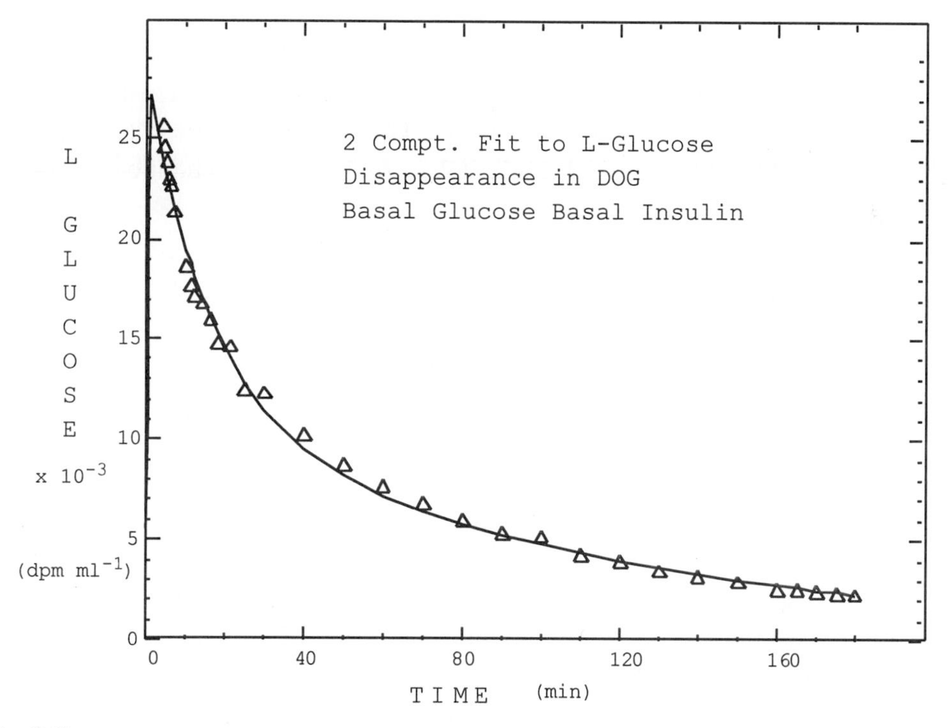

Fig.3.5 Two compartment fit to L-Glucose data

A good tracer must be indistinguishable from tracee to the 'system', be detectable for some pre-determined length of time, have adequate specific activity to study the system of interest, and denature more slowly than it is metabolized. The cost of obtaining and of disposing of tracers is now prompting investigators to use stable rather than radioactive isotope techniques to explore systems. Differences between these isotope forms are discussed further in Chapter 9.

IV. Steady state: definition and implications

A system is in steady state with respect to some particle, if the rate of input, and rate of output, balance for that particle. For a linear compartmental model

$$F_j' = -L_{jj}F_j + \sum_i L_{ji}F_i + Uj \tag{3.6}$$

at steady state $F_j' = 0$

or
$$U_j + \sum_i L_{ji}F_i = L_{jj}F_j \tag{3.7}$$

i.e.
$$F_j = (U_j + \sum_i (L_{(j,i)}F_i) / L_{(j,j)} \tag{3.8}$$

F_j is called a state variable of the model and the value of F_j at steady state is called the size of compartment j. When an input to a linear system is changed, the system moves from its current steady state. The response to the changed input is transitory, lasting until a new steady state, reflecting the changed input, is reached.

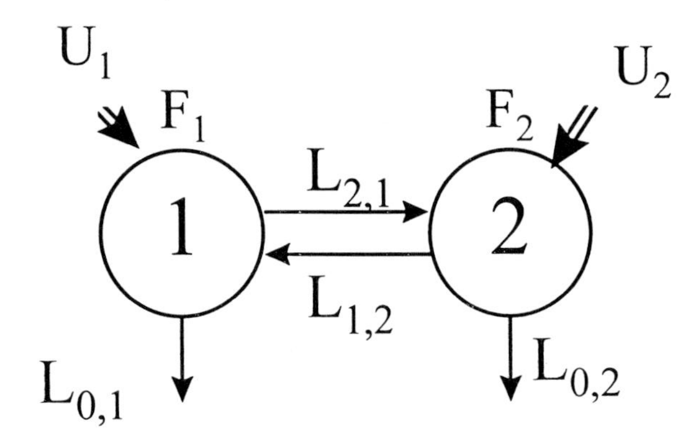

Fig 3.6 Two compartment model

For a compartment with no exogenous inputs i.e. $U_k = 0$ we find that

$$F_k = \sum_i L_{ki} F_i / L_{kk} \tag{3.9}$$

V. The response of a linear compartmental system to a small perturbation

For the model (Fig 3.6), we have (where the subscript notation has been suppressed);

$$\begin{aligned} F1' &= -L11 \cdot F1 + L12 \cdot F2 + U1 \\ F2' &= L21 \cdot F1 - L22 \cdot F2 + U2 \end{aligned} \tag{3.10}$$

$$\begin{aligned} L11 &= L21 + L01 \\ L22 &= L12 + L02 \end{aligned} \tag{3.11}$$

Expressed in matrix notation, we have:

$$\begin{bmatrix} F1' \\ F2' \end{bmatrix} = \begin{bmatrix} -L11 & L12 \\ L21 & -L22 \end{bmatrix} \begin{bmatrix} F1 \\ F2 \end{bmatrix} + \begin{bmatrix} U1 \\ U2 \end{bmatrix} \tag{3.12}$$

or

$$[F'] = [L][F] + [U] \tag{3.13}$$

where L is called the transfer matrix. We have indicated that to explore this system, we must perturb it. If a tracer is injected into the system, its time profile, f, can be denoted:

$$[F' + f'] = [L][F + f] + [U] \tag{3.14}$$

or using (3.13)

$$f' = L \cdot f \;. \tag{3.15}$$

The general solution to (3.15) is

$$f = Ae^{at} \tag{3.16}$$

and we can solve for A and a as follows:

Differentiating (3.16) we obtain

$$f' = a \cdot A \cdot e^{at} \tag{3.17}$$

and equating (3.15) and (3.17) we obtain

$$a \cdot A \cdot e^{at} = L \cdot Ae^{at} \tag{3.18}$$

or, since e^{at} is non-zero,

$$(L - a \cdot I) \cdot A = 0 \tag{3.19}$$

Equation (3.19) has a trivial solution at A = 0 which is of no interest. Thus we need to find solution(s) to (3.20).

$$(L - a \cdot I) = 0 \tag{3.20}$$

Values of 'a' satisfying (3.20) are referred to as the eigenvalues of 'L', and subject to certain regularity assumptions, are found using the determinant form of (3.20) viz:

$$|L - aI| = 0 \tag{3.21}$$

Once the a_j satisfying (3.21) are found, A_j satisfying

$$L \cdot A_j = a_j \cdot A_j \tag{3.22}$$

can be found resulting in a complete solution for (3.15).

The A_j are referred to as the eigenvectors of (3.15). For our simple two compartment system there will be two unique eigenvalues and two unique eigenvectors. This is not always the case. Occasionally situations arise in regard to the L matrix which limit the number of distinct eigenvalues. In this case, a pure exponential model is not an appropriate solution to the problem.

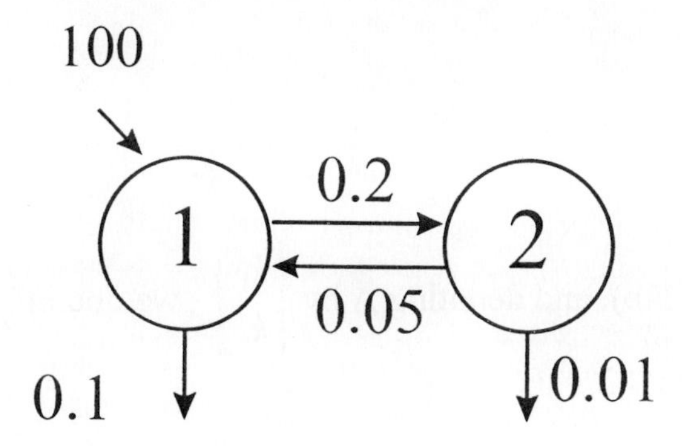

Fig 3.7 Two compartment model with bolus infusion

Example

Consider the response of the linear compartmental model below to a bolus of 100 units in compartment 1 (Fig 3.7). The system equations are;

$$f1' = -0.3f1 + 0.05f2 \qquad f1|_0 = 100$$

$$f2' = 0.2f1 - 0.06f2 \qquad f2|_0 = 0$$

and the response of the system is of the form:

$$f1 = c1 \cdot \exp(-a1 \cdot t) + c2 \cdot \exp(-a2 \cdot t)$$
$$f2 = c3 \cdot \exp(-a1 \cdot t) + c4 \cdot \exp(-a2 \cdot t)$$

1. Eigenvalues:

From (3.21)

$$\begin{vmatrix} -0.3-a & 0.05 \\ 0.2 & -0.06-a \end{vmatrix} = 0$$

or $\quad a^2 + 0.36a + 0.008 = 0$

or $\quad a = -.336, -.024$

2. Eigenvectors

We now find the non-zero null space vectors of A satisfying

$$L \cdot Aj = aj \cdot Aj \tag{3.23}$$

or more generally,

$$L \cdot A = a \cdot A \tag{3.24a}$$

$$\text{i.e. } [L - Ia]A = 0 \tag{3.24b}$$

Setting a = a1 in (3.24b), and denoting A by $\begin{bmatrix} k_1 \\ k_2 \end{bmatrix}$, we obtain

$$0.036 \text{ k1} + .05 \text{ k2} = 0 \tag{3.25}$$

i.e. k1 = -1.38 k2. Setting k2=-1 gives k1 = 1.38.

Similarly, setting a = a2 in 3.24b we obtain

$$-0.276 \text{ k1} + .05 \text{ k2} = 0 \tag{3.26}$$

By similar methods

k1 = .18 k2 whence, setting k2 = 1 gives k1 = 0.18

Thus our eigenvectors are $\begin{bmatrix} 1.38 \\ -1.0 \end{bmatrix}$ and $\begin{bmatrix} 0.18 \\ 1.0 \end{bmatrix}$

3. The exponential model

Our model (3.16) can now be refined to:

$$f_1 = 1.38c_1e^{-0.336t} + 0.18c_2e^{-0.024t}$$
$$f_2 = -c_1e^{-0.336t} + c_2e^{-0.024t}$$

From our initial conditions $[f_0] = \begin{bmatrix} 100 \\ 0 \end{bmatrix}$ we see that

$$1.38c_1 + 0.18c_2 = 100$$
$$c_1 = c_2$$

and c1 = 64.1.

Thus

$$f_1 = 88.46e^{-0.336t} + 11.54e^{-0.024t}$$
$$f_2 = -64.10e^{-0.336t} + 64.10e^{-0.024t}$$

Reversing this problem we note that the response of a linear compartmental model to a perturbation can be described as a sum of exponentials. The number of exponentials is equal to the number of (exchanging) compartments in the model. We can approach the two-compartment problem from the inverse point of view. If an experiment on a system yields a two-exponential response does that mean that only one two-compartment model describes the system? This figure (Fig 3.8) implies a model of the form (Fig 3.9).

We note that the experiment yields 3 pieces of information: A_{11}, a_1, and a_2. A_{12} is set by the experimental conditions. Accordingly our model may have at most three parameters. More than three parameters would lead to identification problems; fewer than three would lead to inconsistencies between the model and the observations. From the following expressions, we identify the model parameters:

$$
\begin{aligned}
f_1' &= -L_{11} f_1 + L_{12} f_2 \\
f_2' &= L_{21} f_1 - L_{22} f_2
\end{aligned}
\qquad (3.27)
$$

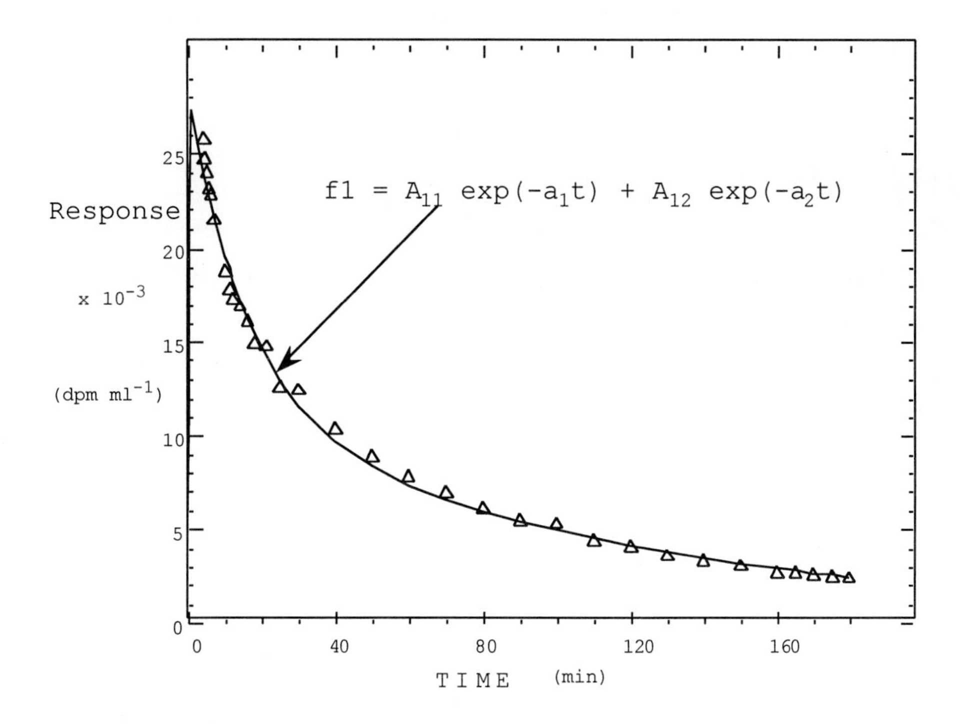

Fig 3.8 Data fitted with a biexponential model

From (3.24a), multiplying both sides of (3.24a) above by A^{-1}, and writing out terms in each of the matrices, we have:

$$L = AaA^{-1}$$

$$L = \begin{bmatrix} -L_{11} & L_{12} \\ L_{21} & -L_{22} \end{bmatrix}$$

and

$$a = \begin{bmatrix} a_1 & 0 \\ 0 & a_2 \end{bmatrix} \tag{3.28}$$

$$A = \begin{bmatrix} A_{11} & A_{12} \\ A_{21} & A_{22} \end{bmatrix}$$

From our initial conditions we have;

$A11 + A12 = 1$

$A21 + A22 = 0$

This can be seen from Fig 3.9. Setting f1=1 (the contents of compartment 1) and noting that at time t = 0, the terms $\exp(-a_1 t)$ and $\exp(-a_2 t)$ are both equal to 1, we see that f1 = A11 + A12 = 1. Similarly, noting that at time t = 0, compartment 2 is empty, (i.e., f2 = 0), f2 = A21 + A22 = 0.

Using (3.27) and (3.17),

$$\begin{aligned} L_{11} &= A_{11}a_1 + A_{12}a_2 \\ L_{22} &= A_{21}a_1 + A_{22}a_2 \end{aligned} \tag{3.29}$$

Collecting these equations and solving yields:

$$\begin{aligned} L_{12} &= A_{11}A_{12}(a_1 - a_2) / A_{22} \\ L_{21} &= A_{22}(a_1 - a_2) \\ L_{01} &= L_{11} - L_{12} \\ L_{02} &= L_{22} - L_{21} \end{aligned} \tag{3.30}$$

If only the response of compartment one is available, we only know the parameters (A11, A12, a1, and a2) of Equations (3.29) and (3.30). Thus it is not

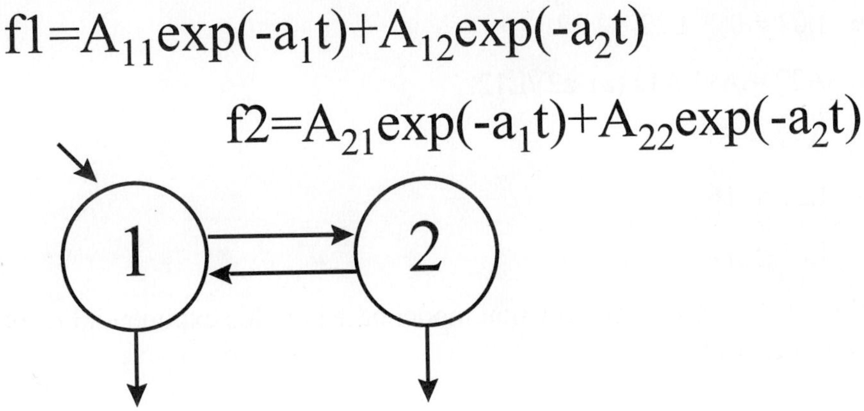

$$f1=A_{11}\exp(-a_1t)+A_{12}\exp(-a_2t)$$

$$f2=A_{21}\exp(-a_1t)+A_{22}\exp(-a_2t)$$

1 2

Fig 3.9 Schematic of a two compartment model with transient solutions

possible to determine all the model parameters of Equations 3.29 and 3.30. If we assume L(0,2) = 0 (Fig 3.10), we can determine the model parameters from the values at hand.

If we found that

$$f1 = 88.46 \exp(-.336t) + 11.54 \exp(-.024t)$$

i.e. A11 = .8846   a1 = .336

A12 = .1154   a2 = .024

we then have:

L11 = .3   L22 = .06

Note that

L11 + L22 = .36 and

a1 + a2 = .36

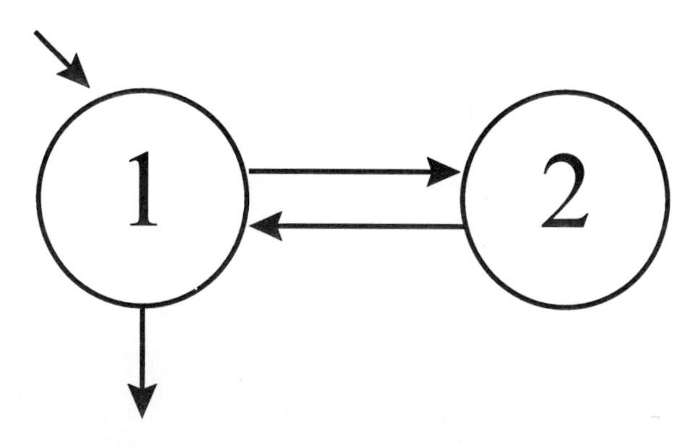

Fig 3.10 Two compartment model assuming one output is zero

Since L02 = 0, L22 = L12.

Thus A22 = A11·A12·(a1-a2)/L12.

Hence A22 = .513

and L21 = .16

L01 = .14.

Finally, our two-compartmental model based on this exponential response is shown in Fig 3.11.

and,

$$L = \begin{bmatrix} -0.3 & 0.06 \\ 0.16 & -0.06 \end{bmatrix}$$

the determinant of L is 0.008. The trace of L is 0.36.

Recall that our original two-compartmental model from which this exponential response was derived had two losses and the L matrix;

$$L = \begin{bmatrix} -0.3 & 0.05 \\ 0.2 & -0.06 \end{bmatrix}$$

the determinant of this L is 0.008 and its trace is 0.36.

We see that to conserve the eigenvalues of a linear compartmental model, both the trace and the determinant of the L matrix must be conserved. Notice that we have 'progressed' from a 4 parameter/process compartmental model to a 3 parameter/process model. When these steps are generalized the process is referred to as mapping. At the heart of mapping are similarity transformations.

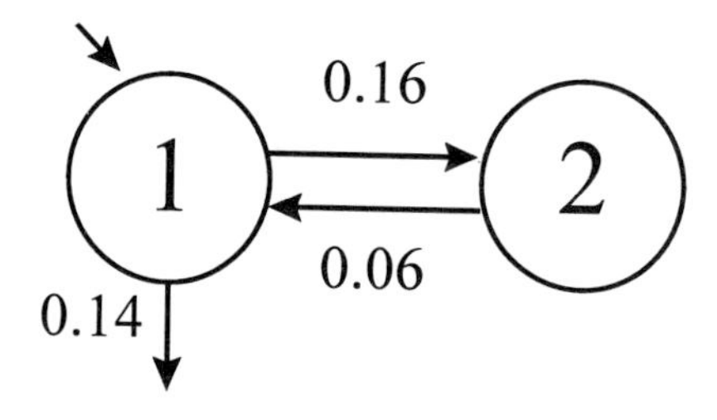

Fig 3.11 Solution of 2-compartment model based on exponential response in Fig 3.8.

REFERENCES

1. Edelstein-Keshet, L. 1988 Mathematical Models in Biology. Random House/Birkhauser Mathematics Series, NY
2. Jacquez, J. A. 1996 Compartmental Analysis in Biology and Medicine, 3rd Edition, BioMedware, Ann Arbor, MI.

4

WHY MODEL BIOLOGICAL SYSTEMS?

Modeling, once the domain of engineering, is being applied increasingly to the study of biological systems. The challenge of modeling biological systems is not to determine an arbitrary function to fit the data, but to use the modeling process to understand the system (2). This chapter addresses the issue of how modeling can be used to add to our understanding of biological systems.

Baldwin (1) divided purposes for modeling into generic and specific. Some generic purposes are described in this chapter including; to determine the structure of a system, to calculate parameters of interest, to integrate information on a system, to predict responses to a perturbation, to derive mechanistic principles underlying the behavior of a system, to identify differences under different conditions, and as an educational tool. Specific modeling purposes relate to a particular study and are described with examples later in the book.

I. Determine the 'structure' of a system

Modeling can be used to determine the structure of a system, where 'structure' refers to the relationships between various parts of a system. This relationship may be between species in an ecological system, a nutrient in blood and some particular tissue, or a metabolite distributed throughout a cell. A model can be used to determine the type of relationship as well as the sequence of events that occur between the various substances of interest. In some systems, this information may be known, whereas in others, the structure can only be inferred from the data.

II. Calculate parameters of interest

Models can be used to determine parameters of interest, such as pool sizes, clearance rates, or transport rates. If the purpose is to accurately determine a particular parameter, it is important that the model can be uniquely determined. However, the model must also be consistent with known biological information. For example, absorption can be determined by administering a tracer and measuring the tracer appearance in tissues (e.g., blood, urine, feces). If only feces are sampled, however, absorption may be overestimated if some tracer is absorbed and then resecreted into the gastrointestinal tract (9). Note a model may be well-determined but incorrect.

III. Integrate information on a system

The purpose of the model may be to determine the interaction of parts of a system. For example, digestion models for ruminants (1, 3, 5) are designed to show how the type of feed can affect bacterial levels and the nutrients that become available for absorption by the animal. The models integrate information on a number of subsystems (representing metabolism of individual nutrients) into an integrated form to represent digestion in the whole animal. Because of the interactions, dependencies and feedback of the processes, large complex systems such as these can only be understood using modeling. Other examples are to link metabolism in one tissue with metabolism in another, or to link metabolism of a nutrient in one form with the metabolite in a second form. Both these scenarios are demonstrated in a model for copper metabolism, where metabolism of different forms of copper in various tissues were investigated simultaneously by kinetic studies and modeling (4)

IV. Predict responses to a perturbation

When a model for a particular system of interest exists, it can be used to simulate and predict levels of drug in blood following various dosing regimes. Models can be used to simulate inputs into a system at various sites and to simulate short-term as well as long-term responses. Experiments can be simulated rapidly using a model to predict likely scenarios before undertaking expensive experimental data collection. While models will never replace experiments, they can be used to avoid experiments where insufficient or inappropriate data will be collected for testing a certain hypothesis.

V. Derive the mechanistic principles underlying the behavior of a system

Large systems and systems with dynamic properties can only be understood with the use of models. Because many processes in large systems occur simultaneously and dynamic systems are accompanied by complex interactions between processes, it is unlikely that responses to perturbations could be predicted based solely on intuition, i.e., without a model. Examples of complex dynamic systems are the circulation model of Guyton et al. (7) and the metabolic model of Garfinkel et al. (6).

VI. Identify differences under different conditions

An important use for models is to identify sites of change in a system when studied under different conditions. In some cases, the altered condition may result in large changes in the kinetic curves and several pathways in a system, or it may result in a subtle change in the data caused by a large change in only one parameter. A model helps to identify which parameters change between the conditions and the degree of change. The conditions may be an untreated vs. treated state, a healthy subject vs. a diseased subject, or a normal vs. high intake of a nutrient. Specifically, when zinc intake in humans was increased by ten-fold, the changes in kinetics were explained by parameter changes representing five sites in the body (8). These sites of regulation could only be identified with the use of modeling.

VII. Education

Models form valuable tools for teaching the process of scientific inquiry and for challenging students to think creatively and quantitatively about a system. Models can be used to demonstrate properties of a system, teach principles (such as feed-back loops, saturation kinetics, etc.), test theories, and design studies.

With the increased speed and convenience of computers, availability of modeling software, and access to the Internet, models can now be used as tools by students of all ages. Researchers who are developing and publishing models need to be cognizant of this. They need to make their models understandable and accessible to the biological community at large as well as to students. One way is to make working versions of models available through a facility such as a model library (see Chapter 22). In this way, the models can be used as educational tools, as well as their use for understanding systems and advancing scientific knowledge.

REFERENCES

1. Baldwin, R. L. 1995. Modeling Ruminant Digestion and Metabolism. Chapman & Hall, London. 578.
2. Berman, M. 1963. The formulation and testing of models. *Annal N Y Acad Sci.* 108:182-194.
3. Dijkstra, J., H. D. S. C. Neal, D. E. Beever, and J. France. 1992. Simulation of nutrient digestion, absorption and outflow in the rumen: Model description. *J. Nutr.* 122:2239-2256.
4. Dunn, M. A. 1995. Historical overview of copper kinetics. *In* Kinetic Models of Trace Element and Mineral Metabolism During Development. K. N. Siva Subramanian and M. E. Wastney, editors. CRC Press, Boca Raton, FL. 171-186.
5. France, J., H. M. Thornley, and D. E. Beever. 1982. A mathematical model of the rumen. *J Agric Sci Camb.* 99:343-353.
6. Garfinkel, D. 1984. Modeling of inherently complex biological systems: Problems, strategies, and methods. *Math Biosci.* 72:131-139.
7. Guyton, A. C., T. G. Coleman, R. D. J. Manning, and J. E. Hall. 1984. Some problems and solutions for modeling overall cardiovascular regulation. *Math Biosci.* 72:141-155.
8. Wastney, M. E., R. L. Aamodt , W. F. Rumble, and R. I. Henkin. 1986. Kinetic analysis of zinc metabolism and its regulation in normal humans. *Am J Physiol.* 251:R398-R408.
9. Wastney, M. E., and R. I. Henkin. 1989. Calculation of zinc absorption in humans using tracers by fecal monitoring and a compartmental approach. *J Nutr.* 119:1438-1443.

SECTION II

Modeling Software

5
REVIEW OF SOFTWARE

I. Introduction

Development of software for modeling is exciting because it enables investigators to model complex systems with increasing ease. However, software maintenance, the testing, debugging, and documentation of these efforts, is a time-consuming task that accounts for up to 60% of the work associated with the supply of software. This maintenance is necessary, however, for the software to run under new computer architectures, operating environments, processing paradigms, computational procedures, and to solve new problems being addressed in the application. This chapter will discuss several modeling packages, ACSL (Automated Continuous Simulation Language) (Mitchell & Gauthier, 1993), MLAB (Civilized Software, 1996), Scientist (MicroMath, 1995), and WinSAAM (Windows version of Simulation, Analysis, And Modeling) (Greif et al., 1998). They have all been developed over a number of years and have been revised extensively to meet the needs of the client community. Each program will be discussed in terms of its background, and then they will be reviewed and contrasted in terms of style of operation, documentation, model entry, data entry, model solution procedures, data fitting, control files, integrators and optimizers, output, and then in terms of their major strengths and weaknesses.

II. Background to Product

1. ACSL

ACSL originated in the early '70's and seems to have identified military simulation application as its primary mission. It initially bore close resemblance to the CSMP (Continuous Simulation Modeling Program) system and more than likely identified its need from the application available there (CSMP virtually vanished in the '80's). ACSL, like all simulation software at the time of its origin, was mainframe based and, at least initially, saw application amongst the large-scale (of order of 100 state variables) modelers. Unlike the other software reviewed here ACSL possessed no direct data manipulation and data fitting capabilities but rather joined with other software (e.g. SIMUSOLVE) to extend its capabilities into this area.

ACSL has been used for developing models for the sheep and dairy cow rumen, the sheep metabolic system, the beef cow, steer and the diary cow digestive system (Thornley, France, Baldwin, Dijkstra, Daenfer, and Gill). Based on the work primarily of Thornley, France, and Baldwin, routine, yet technically accurate, complex metabolic systems have been implemented in computer models. They used entities, called transactions (stripped down pathways), to represent key metabolic processes such as multi-substrate Michaelis Menten systems, inhibition systems, allosteric systems, sigmoidal responses, and hormonal effects. The predictions from their models have been confirmed by actual observations.

2. MLAB

MLAB originated in the late '60's as a modeling system called Modelaide by Dr. Richard Shrager. It bore close resemblance to SAAM at that time and was primarily intended for the same audience. Upon Gary Knott's 'return' to the Division of Computing Research and Technology at the National Institutes of Health (where Shrager was located) dramatic enhancements to Modelaide, now MLAB, were accomplished as more statistical power was added to the SAIL (Stanford Artificial Intelligence Language) program. In the late '80's Knott started a company, Civilized Software Inc., with the intention of migrating the code of MLAB from SAIL to a more common language (C) and commercializing the overall effort to develop and maintain the system. The following demonstrations are included in the MLAB Applications Manual; chemical kinetics, nerve axon conducting, pharmacological systems, survival models, and the glucose-insulin minimal model. There have been a number of recent biomedical publications in endocrinology using MLAB, specifically the glucose kinetics and glucose dynamics.

3. Scientist

The Scientist software seems to have evolved from a number of programs developed by Fox in the 80's for fitting kinetic models to pharmacological data using PC's. Amongst this suite were two extremely efficient programs, Minsq, and Deqsol. Minsq was a data fitting program enabling the fitting of explicit functions to data (for example JMP) and Deqsol was a differential equation solver enabling the user to examine the relative influence of rate constant values on product profiles. The first version of Scientist surfaced in the early '90's, and a Windows version in 1993. Early versions of Scientist were complete in regard to their overall functionality and the software is used widely in pharmacology (Wagner,1993). Libraries of models can be purchased from MicroMath to run with Scientist and amongst these are chemical and pharmacological model collections. The Scientist manual includes demonstrations involving chemical kinetics, compartmental models, chemical equilibria, fitting titration data, and Arrhenius plots.

4. WinSAAM

WinSAAM, the Windows version of SAAM, was released in 1998. SAAM is the oldest of the collection of modeling software to be reviewed here. Dr. Mones Berman began developing it in the 50's as a collection of scientific subroutines. During the 60's it was upgraded, ported to an array of mainframe computers, and released as SAAM25. This was the first attempt to integrate the tools of biokinetics into a single modeling package. Features of SAAM which made it popular were that it was not necessary to specify differential equations for a kinetic system, parameters could be fitted to turnover data, steady state data could be incorporated into the fitting process, and nonlinear systems could be readily represented and fitted to data. In the 70's an interactive version of SAAM, Consam was developed at Berman's laboratory and in the 80's this version was ported to the personal computer. It was estimated that by the late 80's there were approximately 3,000 Consam users around the world (Collins, 1992). SAAM was initially developed for the direct fitting of models to tracer data. Models have been fitted to iodine, calcium, magnesium, iron, glucose, lipoprotein, lithium, zinc, copper, phosphorous and selenium tracer data. SAAM has also been used for the investigation of lipid metabolism, for the development of models for VLDL, IDL, LDL, and HDL metabolism and also glucose-insulin kinetics.

5. Other

Although only four packages will be reviewed in this chapter, there are many other packages available. These include NONMEM (for fitting NONlinear Mixed Effects, statistical regression-type Models), (NONMEM Project Group, 1994), SAAMII (Foster et al., 1994), and STELLA (High Performance Systems, UT). Some of these, and other packages were reviewed by Collins (1992) and Van Milgen et al., (1996).

III. Style of Operation

The versions of the software discussed in this chapter are; ACSL: Level 10D, Release 2.1, August '93, MLAB: Revision January '97, Scientist: Version 2.01, July '95, WinSAAM: Beta Release 0.1.22, January '98. Later versions of the software may be available.

1. ACSL

To run a model and investigate a system using ACSL involves the following steps:

1. The model is written out in the ACSL syntax. This is similar to FORTRAN code but allows freedom in regard to equation order. We note

that France's group has developed some good practices for expressing models in ACSL syntax leading to very readable and maintainable models (9,10).

2. Using an editor (e.g. NotePad) the ACSL model is entered into the computer, and saved as a 'CSL' file. As we get to know the 'rewarding' paths for investigating our ACSL model we can incorporate these into a command ('CMD') file for direct processing of the model
3. Using ACSL the model is translated from ACSL syntax into FORTRAN
4. Using an appropriate compiler (at this stage PowerStation FORTRAN, or Visual FORTRAN, respectively by MicroSoft, and Digital are preferred) the translated code is compiled
5. Using a system linker the compiled code is linked with the necessary compiler library routines and an executable program (called a 'prx' file in ACSL jargon) is generated
6. The model can be run repeatedly by the user, using an efficient run time command repertoire, to explore critical features of the system and the model under various conditions.

The dialog box for communicating between ACSL modeler and ACSL, for running the model for various settings of the model parameters is shown in Fig 5.1. The changeable features of the ACSL model are specified as 'constants' in the model.

```
ACSL - FRANCE_S
File  Edit  Setup  Simulation  Analysis  Linear  Help
Automatic startup disabled.
Unable to open file: france_s.cmd
d ??
          BN 0.                    BR 0.                    PN 0.
          PR 0.                    AI 5.5555E+33            CI 5.5555E+33
          MI 5.5555E+33            NI 5.5555E+33            VI 5.5555E+33
          DA 5.5555E+33            DC 5.5555E+33            DM 5.5555E+33
          DN 5.5555E+33            DV 5.5555E+33            FC 5.5555E+33
          FN 5.5555E+33            GM 5.5555E+33            KC 5.5555E+33
          KN 5.5555E+33            LM 5.5555E+33            MM 5.5555E+33
          MU 5.5555E+33            RC 5.5555E+33            RN 5.5555E+33
          SN 5.5555E+33            SV 5.5555E+33            UA 5.5555E+33
          UC 5.5555E+33            UN 5.5555E+33            VO 5.5555E+33
start
d cint
        CINT 0.10000000
d tstp
        TSTP 6.00000000
pre ??
start
Start
```

Fig 5.1 Communication between the ACSL modeler and ACSL.

2. MLAB

The mode of operation of MLAB is quite different from that of ACSL. There need be no input file for an MLAB run, the model need not be specified in advance, there is no compilation, linking, or related processing chores. The model is simply entered, often involving just a couple of lines of MLAB instructions, and the processing results displayed. MLAB is a hybrid program, it is partly a statistical package, partly a differential equation solver, and partly a symbolic solver as well. At the heart of most processing tasks in MLAB is the matrix manipulation or matrix management consideration. To maintain a modeling session and to enable reproduction of the results achieved in conjunction with this, it is necessary to compile the salient instructions into a file and simply execute the stored instructions. MLAB supports a file management tool for this purpose and we will demonstrate its use, and further of the power of MLAB later in this chapter.

The screen shot (Fig 5.2) illustrates the tightness of MLAB analysis and the significance of matrices to its operation *per se*. Here we specify a single differential equation model as well its initial conditions, and allied parameter value, and we solve the model for t (time possibly) running from 1=0 to 10 in unit steps. The results are then printed and plotted (plot not shown here).

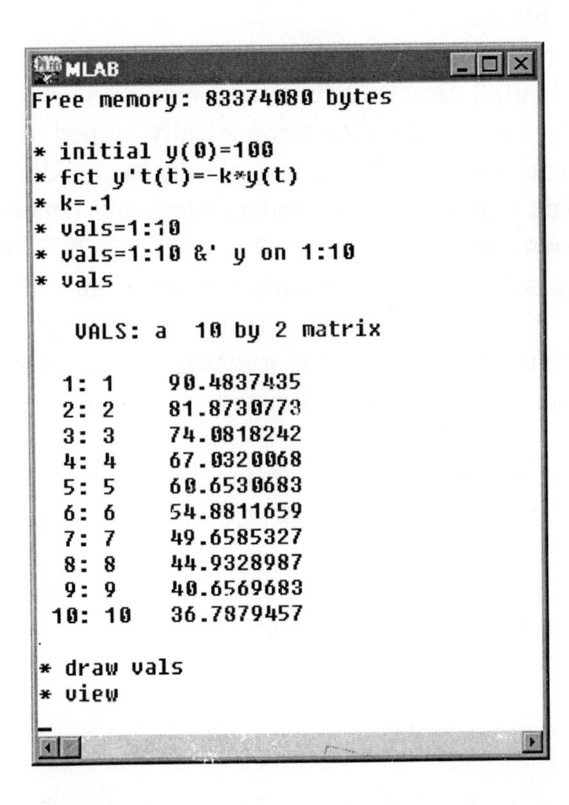

Fig 5.2. Interactive communication with MLAB for solving a one-compartment model.

3. Scientist

A third extreme variation on modeling software is Scientist, where commands are never entered and linkages to compilation or assemblage tools are not required. The model is specified in a 'text' window using a Pascal style of syntax. The modeling entities (independent variables, dependent variables, and parameters) can all be defined in conjunction with the investigation and their specification impacts the subsequent processing analysis paths. For example, by specifying parameters in the header section of the model, they are automatically entered into the parameter subsection of the model, for possible fitting to data. The independent variables, typically 't' only in straightforward kinetics, define the solution range, and solution points. Specification of the dependent variable set causes their calculated values to be entered into a spreadsheet for perusal, plotting, or fitting to data. An array of post-solution tools is available for protracted investigation of the model. One might advance a data fitting and model analysis session from scratch using Scientist is as follows:

1. Using a file/new model is entered into the model window with special attention to naming the dependent variables, as these will be eventually associated with observed values for data fitting
2. The model is successively compiled and debugged until a trouble-free compilation is achieved
3. A new parameter spreadsheet window is opened and assignments of adjustability and initial values made to the parameters. Adjustable parameters will subsequently have their values automatically refined during data fitting
4. The data spreadsheet is opened and checks are made to ensure that the names of the data items match the independent and dependent variable sets from the model window
5. The model is solved and the pertinent solution and observation columns inspected to see if it is likely that the parameters could possibly be refined to enable a consistent fit to the data. If a dependent variable is identified as Y then the solutions, in Scientist, associated with Y are automatically named Y_CALC
6. Data weighting is assigned and the fitting process initiated
7. Model-fitting statistics can be inspected after a model has been fitted, by returning to the Parameters window.

All tasks performed within Scientist are accessible from either menu bar buttons or selectable from drop down menu lists. The screen shot shows a typical model window (Fig 5.3).

4. WinSAAM

WinSAAM is a new Windows-based version of the Consam modeling software and a major issue of its design was the requirement that any current Consam user would be immediately able to use WinSAAM. WinSAAM was

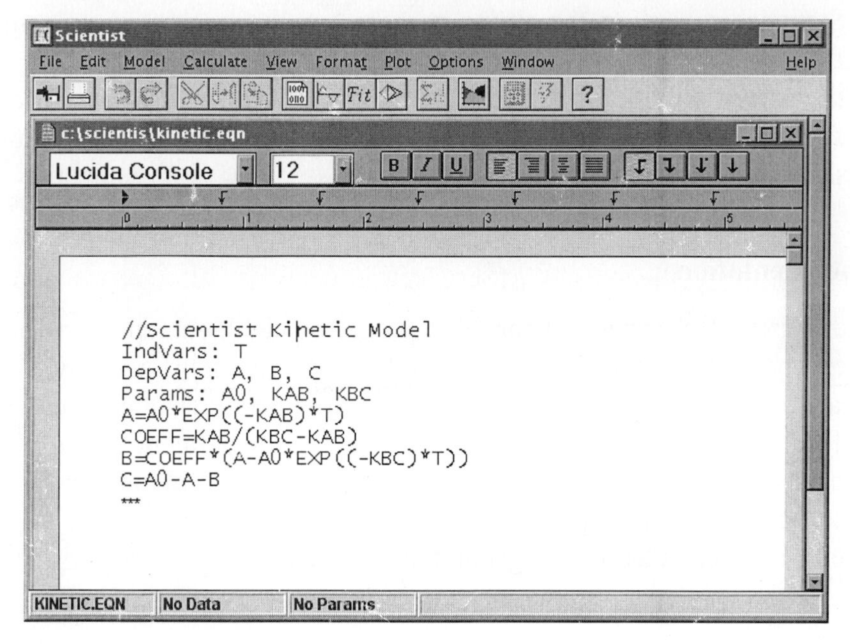

Fig 5.3. Scientist model window.

developed to 1) ensure the future viability of the SAAM modeling philosophy once DOS is phased out, 2) provide SAAM modelers with access to Windows-related tools and environmental advantages, and 3) integrate SAAM with routine statistical tools such as SAS, STATA, and SPSS, and productivity tools such as WORD, PowerPoint, and EXCEL in the Windows setting.

Like Scientist, WinSAAM comprises a set of windows, or working environments, where the investigator can advance aspects of the investigation. At the 'center' of the system is the terminal window, and here the user is able to enter models, save models, and retrieve past models. All communication with the computational kernel also takes place in the terminal window, via the 'common' Consam commands. From the terminal window we can access a graphic editor, a spreadsheet output analyzer, a data charting system, a batch output analyzer, and a logging facility. The strategy a modeler might explore to use WinSAAM in the development of a new model to describe a set of data could be as follows:

1. The model and data are entered, either using text-aligned fields, tab-aligned fields, or via a local spreadsheet
2. The model is compiled and corrected for coding errors
3. The model is solved using either the automatically selected computational procedure or one selected by the investigator
4. The solution is compared with the data by using either tables of numbers or plots, and decisions are made concerning appropriate model alterations
5. If the model is likely to yield improved estimates of adjustable parameters the model is fitted to the data

6. Statistics regarding the fitting process are displayed and judgements are made concerning the estimation process.

In the screen collage of screen shots (Fig 5.4) we show a section of; 1) the terminal window, 2) the charting window, 3) the text input window, and 4) the spreadsheet output window.

IV. Documentation

The degree of user support for these programs via documentation varies quite considerably. For example, Scientist offers a single user guide while MLAB offers an extensive array of manuals (e.g. Application Manual, and Graphics Manual). This should not in any way be construed as negligent by the MicroMath team because their software is so easy to use that a single manual is indeed quite adequate. MLAB, on the other hand is so vast and complex that even 4 substantial documents is barely adequate to get all the ideas across.

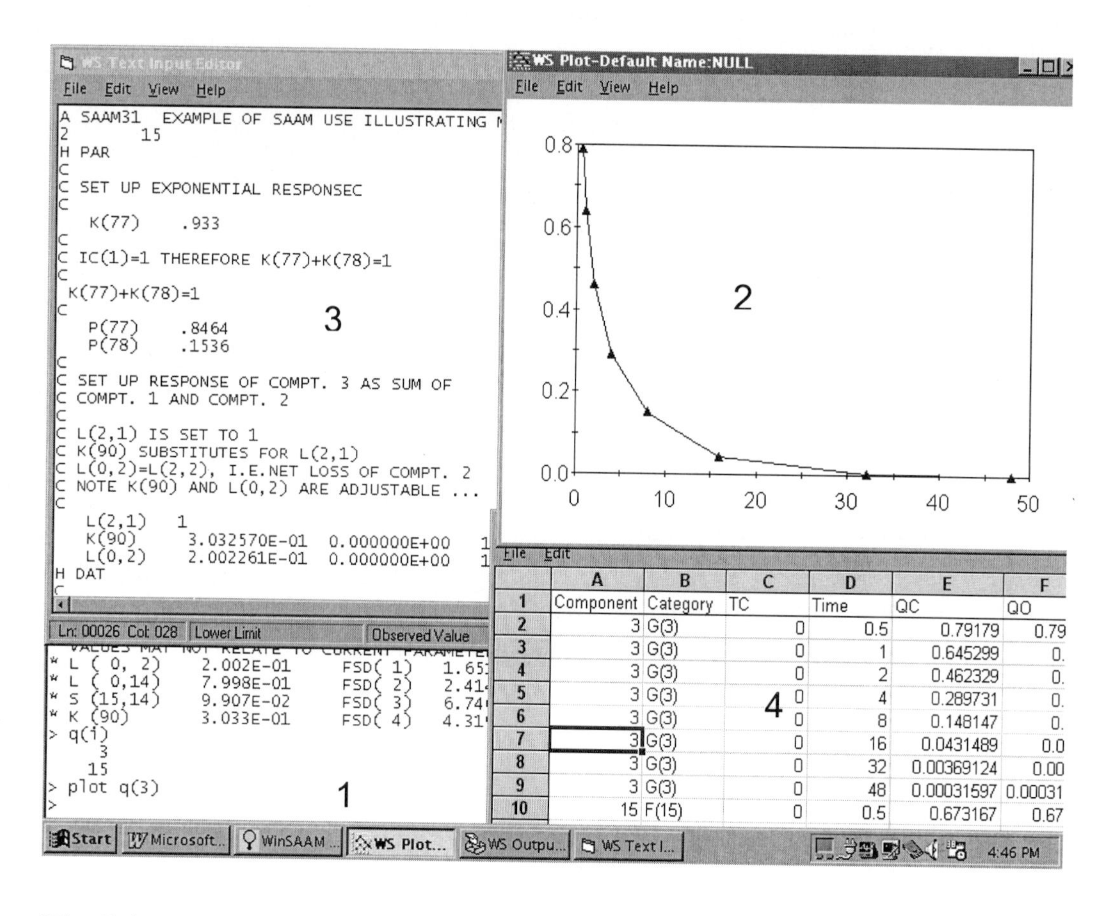

Fig 5.4. WinSAAM 1) terminal window, 2) Plot window, 3) Editor window and 4) output spreadsheet.

1. ACSL

ACSL provides: a Beginners Guide, a ACSL/PC for Windows Guide, and the ACSL Reference Manual. Each document is well organized and enables the user to resolve most problems. The writing style of each of these manuals is clear and accurate though a user with experience in traditional batch style computing will feel more comfortable than one who has only come to ACSL via Windows. The examples included in the Beginner's Guide and the Reference Manual are extremely helpful.

2. MLAB

The four manuals accompanying MLAB are; the User's Guide, the Applications Manual, the Reference Manual, and the Graphics Manual. Each of these documents is detailed and accurate however, they are difficult to understand. Because the manuals lack tutorials and other simple illustrations to clarify ideas and to present the MLAB functionality in a way that shows how the package meets a diverse set of needs it currently requires a tremendous effort to get ‘up to productive speed' with MLAB. As one who has invested the time to master a good deal of the MLAB capabilities I would urge that the serious investigator prepare to invest the effort mastering this product as the effort will be well worth it.

3. Scientist

The single manual accompanying Scientist is a Reference Manual/User's Guide. Because of the simplicity of Scientist this document is adequate however the allocation of space the writer gives to Scientist's operation and use is inappropriate. For example, 100 pages describe Equation Editor, a tool that bears minimal relationship mathematical modeling, another 100 pages are divided between a Text Window description and a Worksheet Window description, which are also tools that should be self-explanatory when it comes to truly applying the program to explore systems. The Calculation Chapter covers just sixty pages.

4. WinSAAM

Accompanying WinSAAM are the formidable SAAM User's Manual and the Consam User's Guide. The Consam User's Guide presents a detailed account of the commands supported by WinSAAM including illustrations for each command, but this documentation is not adequate for a user to gain a feel for the utility of this software. WinSAAM has a newly written on-line help facility.

V. Entering a Model

To illustrate the differences of these systems in regard to entering models we will enter a simple two-compartmental model (Fig 5.5) into each.

1. ACSL

The following 'CSL' file sets up this model in ACSL. Note the following:

1. The 'initial' section sets up the solution procedure and solution details, as well as the parameters and their initial values
2. The differential equations are specified in the 'derivative' sub section of the 'dynamic' section. The derivative sub section here represents the smoothly changeable aspects of the system whereas entries on the 'dynamic' section not in the 'derivative' sub section refer to model outputs and adjustments taking place at multiples of the communication interval, i.e. there is a potential for this intervention every 0.1 time unit (see CINT)
3. The solution runs until t=128 and then terminates as governed by the statement in the outermost section of the 'dynamic' section

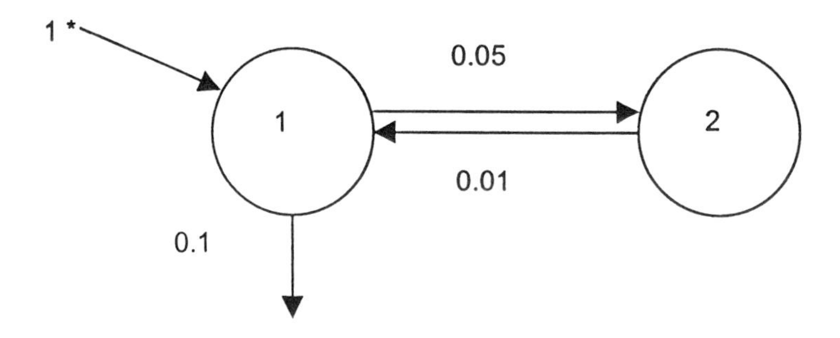

Fig 5.5 Two-compartment open model used to illustrate packages. A pulse bolus of 1 unit is injected into compartment 1, the fractional rate of exchange is 0.05 per unit time for transfer from compartment 1 to compartment 2, 0.01 per unit time for the reverse transfer, and 0.1 per unit time irreversible loss from compartment 1.

```
PROGRAM Two Compartmental System
! Demonstration of setting up and solving a compartmental
! system using ACSL
INITIAL

!     ACSL variables:
      Variable  Time = 0.
      IALG = 5                  ! R-K 4th order
      CINTERVAL CINT = .1        ! Output interval
      NSTEPS   NSTP = 1        ! Calculation interval
      MAXTERVAL MAXT = .0025    ! Maximum integration step
      CONSTANT  TSTP = 128.0    ! Termintaion time
! Model constants
constant  L01  = 0.1
constant  L21  = 0.05
constant  L12  = 0.01

       L11  = L01 + L21
       L22  = L12
! Initial Values for State variables
constant  F1i = 1
constant  F2i = 0
END !INITIAL
DYNAMIC
DERIVATIVE
      dF1dt = -L11 * F1 + L12 * F2
      dF2dt =  L21 * F1 - L22 * F2
! Integration Section
      F1 = INTEG(dF1dt, F1i)
      F2 = INTEG(dF2dt, F2i)
END     ! Derivative section
      TERMT (Time .GE. TSTP)
END     ! Dynamic section
END     ! Program
```

2. MLAB

One way of setting up this same system in MLAB would be as follows:

```
fct f1't(t)=-f1(t)*(l01 + l21) + f2(t)*l12
fct f2't(t)= f1(t)*l21        - f2(t)*l12
initial f1(0)=1
initial f2(0)=0
l01=.1 ;l21=.05 ; l12=.01
m=integrate(f1't,f2't,0:128!140)
```

The 'fct' indicates that a functional definition follows, and the ' implies a differential function. 'fct' is an abbreviation for 'function'. Initial, which can also be abbreviated 'init' initializes the differential equations, and much of the other syntactic specification of the two-compartment system is similar to other scientific programs. The Integrate operation causes the two differential equations to be solved simultaneously from t=0 to t=128 in 140 equal steps and it leads to the assignment of a 140 by 5 matrix to m. The first column of m will include the t values, and the second and successive columns will contain the following solutions at the specific t values of column 1, f1(t), f1'(t), f2(t), f2'(t). Once the solution is obtained in 'm' it can be extracted and explored as follows:

```
time = m col 1
fval1= m col 2
fval2 = m col 4
```

We have copied columns 1, 2, and 4 to the vectors time, fval1, and fval2. If we wanted to retain these solutions compactly back in a single matrix we could use:

```
mvals = time &' favl1 &' fvals
```

In this case the &' operation causes each vector to be adjoined to the preceding one leading to the creation of a 3 column matrix as indicated

3. Scientist

The model is created in the Model Window of Scientist using the following code:

```
// Scientist demonstration of Two Compartmental System
IndVars: T
Params: L01, L21, L12, K1, K2
DepVars:F1_Observed, F2_Observed
// Establish the Transfer Coefficients
L01 = 0.1
L21 = 0.05
```

```
L12 = 0.01
//
L11 = L01 + L21
L22 = L12
// Establish the Differential Equations
F1' = -L11 * F1 + L12 * F2
F2' =  L21 * F1 -  L22 * F2
// Setup the Observables
K1 = 1
K2 = 2
F1_Observed=F1 * K1
F2_Observed=F2 * K2
// Define the Initial Conditions and the Integration Range
T = 0
F1 = 1
F2 = 0
0 < T < 128
***
```

We have extended the model just a little to illustrate the formation of explicitly dependent variables as opposed to implicitly dependent variables. The model header section specifies the parameters, the independent variables, and the dependent variables. Only objects specified here emerge subsequently in Windows of the software, and hence only such objects can be factored, explicitly, into the data fitting process. The parameters specified here will be available for value assignment, or refinement by data fitting by their automatic mapping to the Parameters Window table following model compilation. The Independent Variables will surface either in conjunction with model solution, where we must provide both intervals and ranges for them before fitting can commence, or in conjunction with model fitting where we have the option of creating data weights based upon functions of the independent variables. The Dependent variables are solved for and their values automatically entered into the solutions Spreadsheet Window as soon as a solution is available. Care needs to be exercised to ensure that the correct associations of data and solutions are in effect for data fitting, especially when the dependent variable names do not initially relate to the data item names. This situation can be remedied by direct reassignment of variable names in the spreadsheet window.

Although we have specified L12, L21, and L01 with the Parameters header if they had not been needed for adjustment considerations they could simply have had values assigned and have then been treated as constants. Note that this is indeed how F1 and F2 as dependent variables have been treated, and we will not see the values for these variables in the solutions Spreadsheet.

Initial conditions and solution ranges have been set up in the last section of the model, and all that remains to complete the specifications for a solution is to indicate the 't' points within the solution range for which we require solutions. Scientist will guess that 20 solutions are needed and just prior to initiating a solution run we will have the opportunity of resetting the value.

4. WinSAAM

Somewhat more formally than the other software, we present the model in the syntax of WinSAAM. Here we see that there are two sections to the model, a parameters section under 'H PAR' and a solutions requesting section under 'H DAT'. Much of the material specified here is self-explanatory although it may need to be pointed out that the lines in the data and solutions section of the input call for 128 solutions for each of compartment '1' and '2' at uniform points in time.

```
A SAAM31
H PAR
  L(0,1)  .1
  L(2,1)  .05
  L(1,2)  .01
  IC(1)  1
  IC(2)=0
H DAT
101
      0
2     1              128
102
      0
2     1              128
```

Whereas ACSL, Scientist and WinSAAM are all 'compile' based systems, MLAB is more interpretative. As each command line is entered into MLAB it is interpreted and handled appropriately. All entries into MLAB can be viewed as commands. The other systems take in text files, translate the text into appropriate data structures and then move to a state where the user can advance the analysis. Of course ACSL requires more intermediate steps before the model is to state where it can be explored but much of the philosophy is similar to WinSAAM's and Scientist's.

VI. Entering Data

Data is accepted for a number of purposes. Firstly, as a basis for simply examining the degree of consistency between our model and the data to which it should relate. Secondly, to provide a basis for refining parameter values. Thirdly, to determine the precision (or conversely, uncertainty) of the less well known parameters. Finally, to measure of the diversity of parameter values for our model amongst individuals within the study population, or perhaps a series of populations.

1. ACSL

Because the version of ACLS reviewed does not support data fitting it we will not review any aspect of data entry for ACSL

2. MLAB

Data is entered into MLAB matrices using a variety of methods:

- list: enables the user to enter a list of values directly into a column of a matrix.
- kread: enables keyboard entries directly into a matrix.
- read: enables file-based values to be read into the columns of a matrix.

For example the sequence below would set up 'x' appropriately as a data set using the first approach.

```
x=1:7
x=x &' list(1.2,2.3,3.2,4.1,4.4,5.2,5.5)
```

Data sets are associated with functions by the 'fit' statement (see below)

3. Scientist

Scientist allows data entry into a spreadsheet either via the keyboard directly, or from a host of compatible spreadsheets. Once data is thus available it can be recalled through the 'OPEN | SPREADSHEET' sequence at an appropriate step in the modeling session. Data is associated with dependent variables (see functions with MLAB) using the 'spreadsheet' window in conjunction with the dependent variables specification section of the 'model' window.

4. WinSAAM

WinSAAM incorporates its data section directly into the file that also specifies the model. This means that associating a model with the data is trivial but it also implies an inappropriately tight connection of a model to the data. For scientific investigation to advance effectively, data may need to be associated with many models with each model exposing some unique feature of the system we are trying to describe with the aid of the (generic) model. It is planned to uncouple data and models in WinSAAM inputs in future versions of WinSAAM.

Data can be entered into WinSAAM using any combination of 5 means, from:

1. the terminal window using special tabulation facilities there
2. the text input window using standard pre-configured tabulation
3. the text input window by inserting text from a tab-delimited file
4. the text input window by pasting tab-delimited data, typically copied or cut from a spreadsheet.
5. the spreadsheet window using standard spreadsheet data entry techniques

VII. Model Solution Procedure

Producing model solutions means the steps needed to advance from a coded representation of the model to a state where model solutions are available for display and perhaps, appraisal.

1. ACSL

As we have indicated, building an executable version of the model in ACSL syntax involves model translation, model compilation, and object code linking. Once these steps are correctly negotiated, and we might mention that there can be 'breakages' at any of these stages, the ACSL user is exposed to an extremely powerful set of tools in the 'runtime' window allowing many features of the model to be explored. Indeed, apart from WinSAAM, ACSL was the only other modeling system providing direct access to the steady-state solutions.

A screen shot (Fig 5.6) shows some processing tasks following the creation of an executable version of the demonstration problem (Fig 5.5). Whereas these tasks have been accessed from a 'command' file they could just as easily have been entered directly from the ACSL runtime window command box. To see solutions for certain of the state variables four aspects of the ACSL analysis need to be specified. Firstly, a communication interval needs to be given an

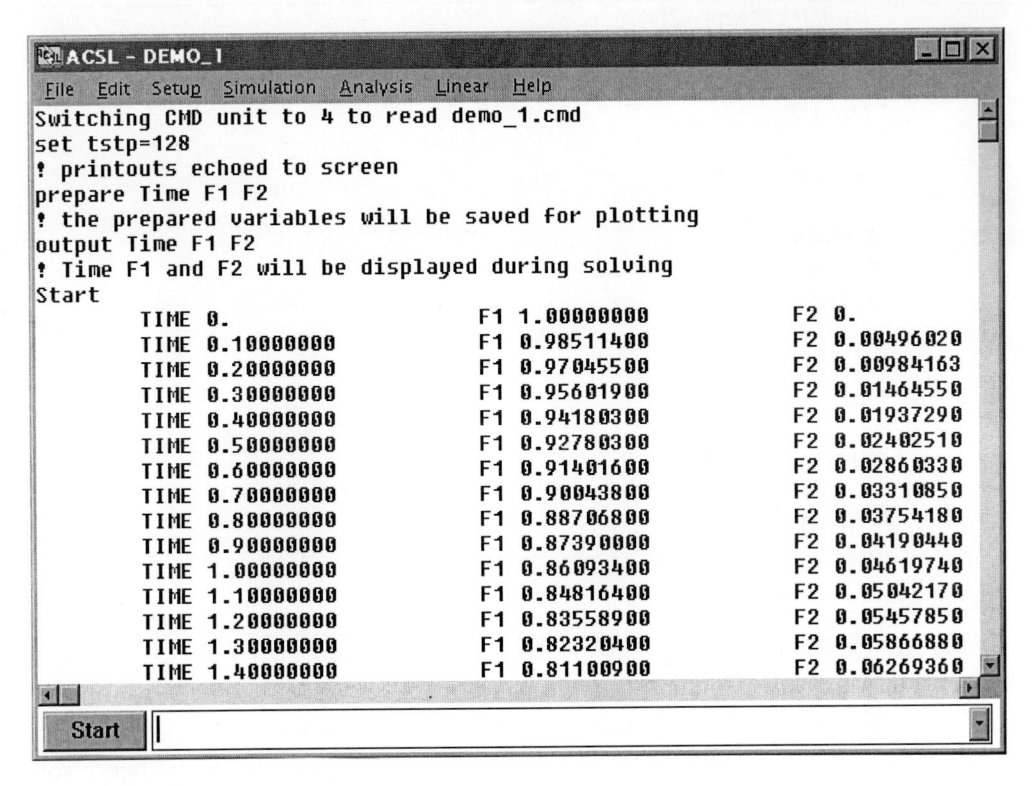

Fig 5.6. ACSL steady-state solution.

appropriate value (e.g. 'cint = 0.05'), secondly, the state variables need to be 'prepare'd for output (e.g. 'prepare Time F1 F2'), thirdly, the state variables to be observed need to be specified in an output statement (e.g. 'output Time F1 F2) and, finally, the job needs to be started ('start'). We will observe values for these variables at the intervening interval until time reaches the stopping time (see above).

2. MLAB

Solutions are obtained from MLAB models either via the explicit entry into a column of a matrix or via typing the solutions implicitly. There is no 'start' or 'solve' type commands for MLAB simply the generation of solutions either to a matrix or to the console (user window) in conjunction with another command. Initially this can be a disconcerting feature of MLAB because the user may be looking for solution access machinery however, after practice the consistency and compactness of MLAB's approach can become quite appealing.

```
* fct f(t)=k1*exp(-k2*t)
* k1=100; k2=.1

* f on 1:3

   : a  3 by 1 matrix

 1: 90.4837418
 2: 81.8730753
 3: 74.0818221

* m=f on 1:3
* m

  M: a  3 by 1 matrix

 1: 90.4837418
 2: 81.8730753
 3: 74.0818221
```

This dialog demonstrates how simple it is using MLAB to create an exponential function, produce some solutions for it, display the solutions, and insert the solutions into a column of a matrix. Of course only in a contrived situation would 't' be at fixed points and in any case we certainly need some flexibility in establishing solution points. The simple demonstration below shows how some control may be asserted here. First we set the points we seek as

```
* time=list(1,2,4,8,16,32)
* f on time

   : a  6 by 1 matrix

 1: 90.4837418
 2: 81.8730753
 3: 67.0320046
 4: 44.9328964
 5: 20.1896518
 6: 4.0762204

* m=time &' f on time
* m

  M: a  6 by 2 matrix

 1: 1    90.4837418
 2: 2    81.8730753
 3: 4    67.0320046
 4: 8    44.9328964
 5: 16   20.1896518
 6: 32   4.0762204
```

solution to 'f' as elements of a list which we transfer to the vector 'time'. Then we print solutions to 'f' for these time points. Finally we insert the time points and the solutions into the first two columns of a matrix 'm'.

3. Scientist

Generating solutions to a successfully compiled Scientist model, in principle, is no more complex than selecting 'simulation' from the 'calculate' drop down menu list. Then a confirmation of solution range and solution intervals is sought and the solution procedure is activated. We say 'in principle' because in fact we do have the option of setting certain aspects of the solution procedure prior to initiating the solution step. Fig 5.7 displays a table by Scientist just prior to initiating a solution and seeking confirmation of aspects of the solution to be obtained.

We see that the independent variable, 'T', of the model is reported (this is simply picked up from the 'independent variables' section of the model) and upper and lower solution points for the independent variable are also displayed. Again, the values here are those transported from the $0 < t < 128$ term in the model. In Scientist the default number of solution points is 20 as displayed here. Immediately the solution is available to a solution spreadsheet that is displayed in Fig 5.8. Note that part of the variable name for the calculated values is in fact concealed. This seems to be a shortcoming of Scientist as there is neither a way of widening the column to reveal the entire variable name nor automatically removing the '_calc' extension that Scientist gives when it creates the name for the new variable. But with the overall simplicity and 'fluidity' of Scientist this is a trivial objection.

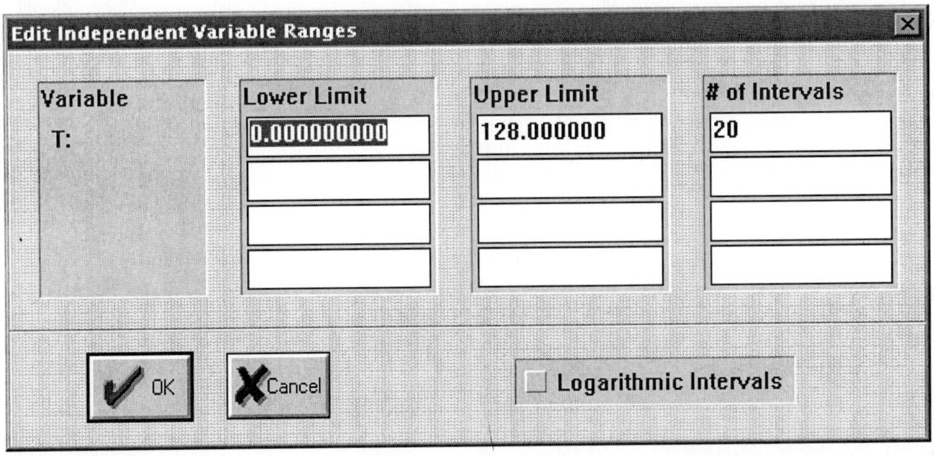

Fig 5.7. Scientist dialog box for solution request.

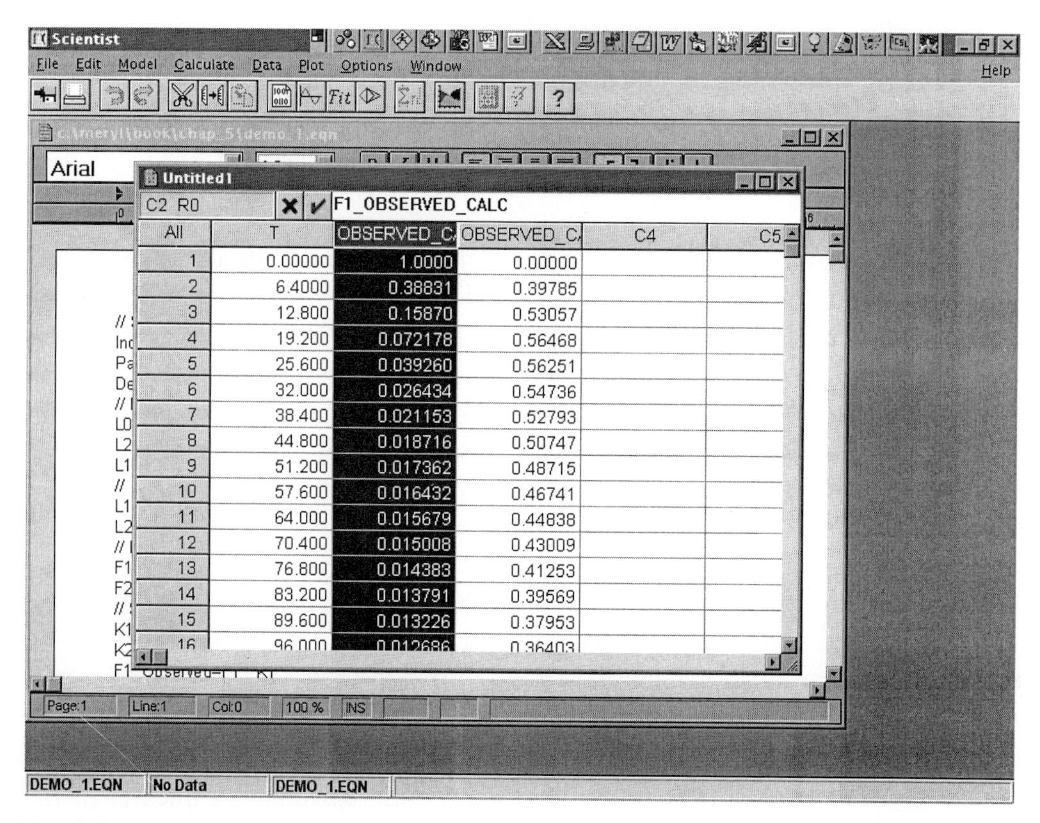

All	T	OBSERVED_C	OBSERVED_C	C4	C5
1	0.00000	1.0000	0.00000		
2	6.4000	0.38831	0.39785		
3	12.800	0.15870	0.53057		
4	19.200	0.072178	0.56468		
5	25.600	0.039260	0.56251		
6	32.000	0.026434	0.54736		
7	38.400	0.021153	0.52793		
8	44.800	0.018716	0.50747		
9	51.200	0.017362	0.48715		
10	57.600	0.016432	0.46741		
11	64.000	0.015679	0.44838		
12	70.400	0.015008	0.43009		
13	76.800	0.014383	0.41253		
14	83.200	0.013791	0.39569		
15	89.600	0.013226	0.37953		
16	96.000	0.012686	0.36403		

Fig 5.8. Scientist solution spreadsheet.

	A	B	C	D	E	F	G	H	I	J
1	Component	Category	TC	Time	QC	QO	SD	FSD	Theta	Weight
2	1	F(1)	0	0	1	0	0	0	0	
3	1	F(1)	0	1	0.860933	0	0	0	0	
4	1	F(1)	0	2	0.741633	0	0	0	0	
5	1	F(1)	0	3	0.639286	0	0	0	0	
6	1	F(1)	0	4	0.551482	0	0	0	0	
7	1	F(1)	0	5	0.47615	0	0	0	0	
8	1	F(1)	0	6	0.411517	0	0	0	0	
9	1	F(1)	0	7	0.35606	0	0	0	0	
10	1	F(1)	0	8	0.308473	0	0	0	0	
11	1	F(1)	0	9	0.267638	0	0	0	0	
12	1	F(1)	0	10	0.232593	0	0	0	0	
13	1	F(1)	0	11	0.202514	0	0	0	0	
14	1	F(1)	0	12	0.176696	0	0	0	0	
15	1	F(1)	0	13	0.154532	0	0	0	0	
16	1	F(1)	0	14	0.135502	0	0	0	0	
17	1	F(1)	0	15	0.119161	0	0	0	0	
18	1	F(1)	0	16	0.105126	0	0	0	0	

Fig 5.9. WinSAAM solution spreadsheet.

4. WinSAAM

Once an input file is compiled, using the 'DECK' command, solutions to system components become available by 'SOLV' (solving) the system equations. It should be reiterated here that solutions can only be accessed for those model constructs whose identifier is either explicitly or implicitly included as a category item on the component definition line. This is somewhat similar to ACSL where items needing to be monitored in conjunction with a model solution must first be 'prepared' and also reminiscent of Scientist where only explicitly defined 'dependent variables' can have their solutions accessed. Once a solution is obtained with WinSAAM, the solutions can be 'PRINted' or 'PLOTted' or displayed as a spreadsheet like the Scientist solutions. The spreadsheet screen shot (Fig 5.9) has been captured from a WinSAAM modeling session using the 2-compartment demonstration model discussed above.

Users familiar with earlier SAAM versions will see that the tabular layout of SAAM's very comprehensive printed output has been entirely preserved in this spreadsheet form. In addition, for cases where textual rather than tabular renditions of this output is needed, WinSAAM supports a batch output screen and here all the WinSAAM output associated with a processing run can be collected for direct incorporation into a report. The screen shot below captures an aspect of the batch output text window from WinSAAM (Fig 5.10).

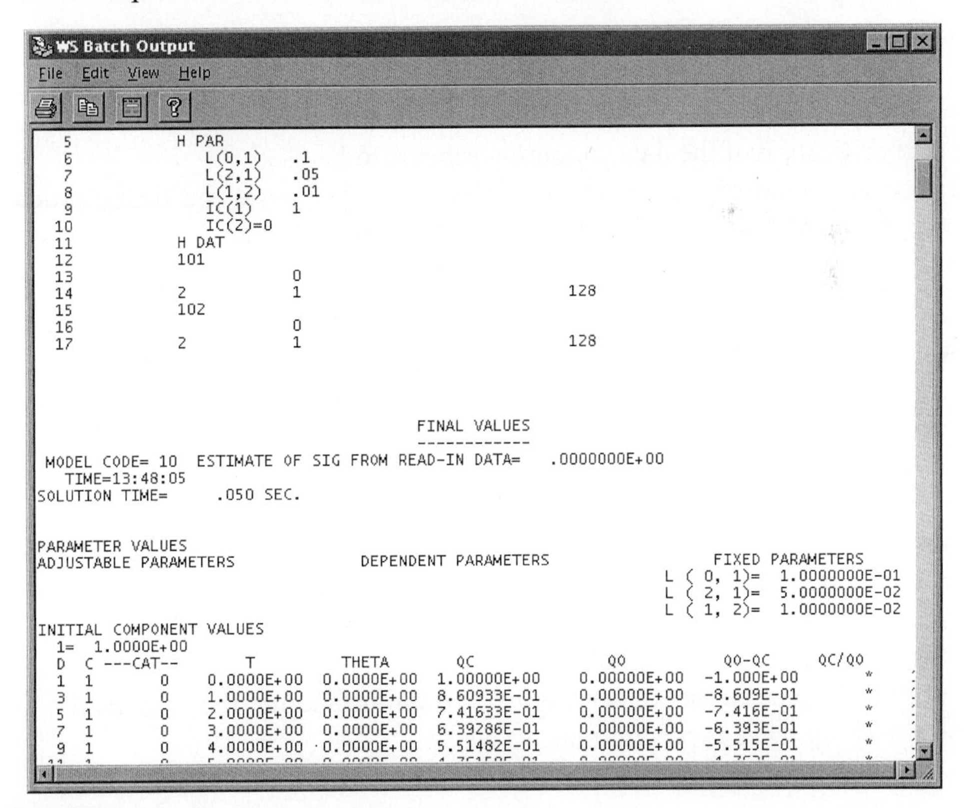

WS Batch Output
File Edit View Help
5 H PAR
6 L(0,1) .1
7 L(2,1) .05
8 L(1,2) .01
9 IC(1) 1
10 IC(2)=0
11 H DAT
12 101
13 0
14 2 1 128
15 102
16 0
17 2 1 128
FINAL VALUES
MODEL CODE= 10 ESTIMATE OF SIG FROM READ-IN DATA= .0000000E+00
TIME=13:48:05
SOLUTION TIME= .050 SEC.
PARAMETER VALUES
ADJUSTABLE PARAMETERS
DEPENDENT PARAMETERS
FIXED PARAMETERS
L (0, 1)= 1.0000000E-01
L (2, 1)= 5.0000000E-02
L (1, 2)= 1.0000000E-02
INITIAL COMPONENT VALUES
1= 1.0000E+00
D C ---CAT-- T THETA QC QO QO-QC QC/QO
1 1 0 0.0000E+00 0.0000E+00 1.00000E+00 0.00000E+00 -1.000E+00 *
3 1 0 1.0000E+00 0.0000E+00 8.60933E-01 0.00000E+00 -8.609E-01 *
5 1 0 2.0000E+00 0.0000E+00 7.41633E-01 0.00000E+00 -7.416E-01 *
7 1 0 3.0000E+00 0.0000E+00 6.39286E-01 0.00000E+00 -6.393E-01 *
9 1 0 4.0000E+00 0.0000E+00 5.51482E-01 0.00000E+00 -5.515E-01 *

Fig 5.10 WinSAAM batch- or text-output window.

VIII. Fitting Data

In a general sense data fitting implies flexing of a model enabling the parameters that we have least confidence in to have their values resolved by 'the' data. Data for the fitting process emanates from an experiment carefully designed for this parameter estimation purpose. Data fitting has become an integral part of modeling and it implies:

1. selection of the data to be fit.
2. selection of the function to fit to the data
3. selection of the parameters of the function whose values are to be adjusted in conjunction with the fitting process
4. specification of initial values for nonlinearly adjustable parameters
5. specification of a search domain for finding new values for the parameters
6. specification of an uncertainty surrounding the initial parameter values if known
7. specification of the data fitting objective
8. selection of the data fitting algorithm
9. specification of acceptance criteria in regard to the fitting process
10. specification of the data weighting scheme to be used
11. specification of any constraints which may influence the fitting process, i.e. in addition to search domains for parameter values

Modeling software that supports data fitting in general offers the user control over some of the above fitting specifications. More likely though a good proportion will be set within the program. For example, whereas the user may like to explore amongst least squares, maximum likelihood, and minimum chi-square as alternate objective functions, most packages will only offer least squares, indeed least squares with heavily defaulted fit acceptance criteria in place. It may not be appropriate to allow users to select amongst these data fitting settings, because some users may not appreciate the statistical ramifications of the settings, or the computational burden they impose on the fitting process *per se* as they arbitrarily explore the 'fit setting list'. Notwithstanding, they are available to the user desiring these expert options.

1. ACSL

The version of ACSL reviewed here contained no data fitting capability. Indeed, at the time of acquisition of the ACSL software data fitting in conjunction with ACSL use was available, albeit at a considerable extra cost, only by additionally acquiring the SIMUSOLVE data fitting software. Since writing this chapter, ACSL has now added a data fitting capability to the software.

2. MLAB

Data fitting is supported in MLAB using the 'fit' command. Here a series of clauses enable the user to specify, firstly the adjustable parameters, then sequence by sequence the function, the data set, the weighting scheme, and the constraints. For example, in the segment below we use MLAB to fit a simple, first-order differential equation to a synthetic set of data.

```
* data=list(99,96,92,88,75,63)
* data=1:6 &' data
* fct f't(t)=-k1*f(t)
* initial f(0)=ic1
* k1=.1;ic1=100
* fit(k1,ic1), f to data
final parameter values
    value              error            dependency    parameter
  0.07943094846     0.0154539453       0.7687754445  K1
   111.9142192      6.00553806         0.7687754445  IC1
3 iterations
CONVERGED
best weighted sum of squares = 1.189356e+02
weighted root mean square error = 5.452880e+00
weighted deviation fraction = 4.096866e-02
R squared = 8.757853e-01
* data=data &' f on 1:6
* data
  DATA: a  6 by 3 matrix

  1: 1  99  103.368651
  2: 2  96  95.4756062
  3: 3  92  88.1852604
  4: 4  88  81.4515924
  5: 5  75  75.2320952
  6: 6  63  69.4875075
```

To use empirical weighting we could have modified the above analysis as follows:

```
* for i=1:6 do {we[i]=1/data[i]^2}
* data=1:6 &' data
* fct f't(t)=-k1*f(t)
* initial f(0)=ic1
* k1=.1; ic1=100
* fit(k1,ic1), f to data with weight we
final parameter values
     value              error              dependency   parameter
  0.08765876283     0.01542350587        0.7936240264   K1
    114.2970718     6.924956303          0.7936240264   IC1
3 iterations
CONVERGED
best weighted sum of squares = 1.811352e-02
weighted root mean square error = 6.729325e-02
weighted deviation fraction = 4.629313e-02
R squared = 8.660312e-01
* data=data &' f on 1:6
* data
  DATA: a  6 by 3 matrix

  1: 1  99  104.70451
  2: 2  96  95.9170186
  3: 3  92  87.8670314
  4: 4  88  80.4926521
  5: 5  75  73.7371793
  6: 6  63  67.5486701
```

Note that the weights have not been normalized although they could have been using the 'rowsum' function. There are some interesting and important facts regarding MLAB data fitting that we need to be aware of to successfully fit data with MLAB. Firstly, data are specified in a two-column matrix where the first column is the set of values for the independent variable, and the second column is the observations at the points coinciding with the value of the independent variable. Secondly, the function to be fit is specified as the first element of each data-fitting clause. Finally each function to be fit is specified, i.e. the components of each data fitting clause are syntactically the same. Data fitting is discussed further in the section on control files.

3. Scientist

The only issues in data-fitting with Scientist are 1) ensuring that the data set and the calculation set names match and 2) ensuring that the data weights are set correctly, if default ($1/(obs)^2$) weighting is not wanted. A flexible approach to weight selection is available that permits either automated weight generation, where the weight determination is made automatically once the weighting scheme is selected, or a nominated user generated data column can be set to provide the weighting basis.

Once data fitting is complete comprehensive statistics can be made available in a 'statistics' text window. The sample below represents part of the statistics generated upon fitting the indicated three-compartment system to data for the three compartments. The model:

```
IndVars: T
DepVars: A, B, C
Params: A0, KAB, KBC
A'=-KAB*A
B'=KAB*A-KBC*B
C'=KBC*B
// Initial conditions
T=0.0
A=A0
B=0.0
C=0.0
```

A section of the statistics output relating to fitting the model to data:

```
*** MicroMath Scientist Statistics Report ***

Model File Name :   c:\scientis\kinediff.eqn
Data File Name :         c:\scientis\kinetic.mmd
Param File Name :   Untitled2

Goodness-of-fit statistics for data set: c:\scientis\kinetic.mmd
```

```
Data Column Name:        A
                              Weighted      Unweighted
Sum of squared observations : 0.263394639  0.263394639
Sum of squared deviations :        5.63483420E-5     5.63483420E-5
Standard deviation of data :  0.00265396736     0.00265396736
R-squared :                   0.999786069  0.999786069
Coefficient of determination : 0.999188922  0.999188922
Correlation :                 0.999774251  0.999774251

....

                              Weighted      Unweighted
Sum of squared observations : 0.527099465  0.527099465
Sum of squared deviations :        9.41489094E-5     9.41489094E-5
Standard deviation of data :  0.00177152391     0.00177152391
R-squared :                   0.999821383  0.999821383
Coefficient of determination : 0.999525734  0.999525734
Correlation :                 0.999785977  0.999785977
Model Selection Criterion :        7.47192340    7.47192340

Confidence Intervals:

Parameter Name :                   A0
Estimate Value =              0.298379612
Standard Deviation =               0.000762252818
95% Range (Univar) =               0.296822884  0.299936340
95% Range (S-Plane) =              0.296122670  0.300636554

Parameter Name :         KAB

Variance-Covariance Matrix:
    5.81029358E-7
    1.11675679E-10     9.09374458E-9
    -1.11569598E-7     -1.76124010E-8     2.77928023E-7

Correlation Matrix:
    1.00000000
    0.00153634301      1.00000000
    -0.277639122  -0.350332977  1.00000000

Rigorous Limits:

Parameter Name :         A0
Lower 95% univariate limit =   0.296823085
Upper 95% univariate limit =   0.299936491
Lower 95% support plane limit =     0.296123204
Upper 95% support plane limit =     0.300636893

Parameter Name :         KAB
Lower 95% univariate limit =   0.0198635752
Upper 95% univariate limit =   0.0202527504
Lower 95% support plane limit =     0.0197772326
Upper 95% support plane limit =     0.0203414719

Parameter Name :         KBC
Lower 95% univariate limit =   NAN
Upper 95% univariate limit =   0.0519280051
Lower 95% support plane limit =     0.0493143727
Upper 95% support plane limit =     0.0524358333
```

4. WinSAAM

The optimizer for data fitting using WinSAAM is invoked using the 'ITERate' command. Fitting the model to the data may require several invocations of this command. Indeed, only when the program responds that it is not possible to further improve the fit is it considered that the best possible combination of parameter values has been achieved. Output available following iterations includes: the percent reduction in the sum of squares, adjustment details of the parameter whose values are most altered in the fitting process, and the degree to which raw predicted parameter adjustments could be applied to the adjustable parameters. In our section below on major strengths we will return to the topic of data fitting using WinSAAM.

IX. Control Files

Control files (also called command files) cause the software to execute a sequence of processing steps. Specifically, a sequence of instructions or commands is assembled into an appropriately named file, and then the program carries out the commands in the files in strict sequence. When would control files be useful? Firstly, when we can't tolerate any command entry errors, e.g., when instructing a group in the use of the program, or instructing a group with the use of a particular strategy in using the software. Here we can assume that all instructions are both correct and sensible, or at least appropriate to the situation (demonstration of erroneous situations shouldn't be ruled out here), and that in conjunction with a 'more-like' utility they are simply automatically invoked in sequence. A second use for control files is when we wish to retain a strategy that we have evolved for analysis for subsequent invocations. When developed in conjunction with a software logging capability, it may be necessary to 'comment' out the command responses ensuring that only the instruction lines in the log sequence are interpreted by the program.

1. ACSL

The ACSL program supports very flexible control file based processing. Below we present a control file (called a 'CMD' file in ACSL syntax) to process the two compartmental model we have introduced earlier. To have control files automatically start up, upon selecting the run command from the project pull down menu list in the ACSL builder window, the control file must have the same name as the simulation (CSL) file name but have the extension, or file type, CMD. Of course there can be other control file assignments for a particular simulation run but these need to be selected from the runtime option in the project pull down menu list. Once the command file entry is selected, a file selection dialog box is opened permitting selection of the appropriate control file (command file in the syntax of ACSL):

```
set title='A Two Compartmental Model'
set hvdprn=.true.
set tstp=128
! printouts echoed to screen
prepare Time F1 F2
! the prepared variables will be saved for plotting
output Time F1 F2
! Time F1 and F2 will be displayed during solving
Start
plot fl F2/Xaxis=Time
print Time F1 F2
! Time H Mmean Cs and M will be printed
plot F1/Xaxis=F2
analyze/jacobian/trim/vectors=.TRUE./EIGEN
disp F1 F2
set L01=.09
set L21=.04
set L12=.009
disp L01 L21 L12
Start
analyze/jacobian/trim/vectors=.TRUE./EIGEN
disp F1 F2
plot F1 F2/Xaxis=Time
print F1 F2 Time
! quit
```

Here the 'set' commands each change an assignment of the values of the model constants. The 'prepare' command assembles solution-time values for the indicated variables, the plot commands produce graphic plots, and the 'analyze' command request causes the system jacobian to be generated permitting determination of the system eigenvalues and eigenvectors. The salient output generated by the 'analyze' is as follows, where we see the jacobian and then its eigenvalues;

```
Matrix elements - rows across, columns down
    1       2
 1 -0.1500000  0.0100002
 2  0.0500001 -0.0100001

Jacobian evaluated. Condition number is 0.23737000
Complex eigenvalues in ascending order
    REAL
 1 -0.00651531
 2 -0.15348500
```

We present a sample plot generated from this control sequence (Fig 5.11).

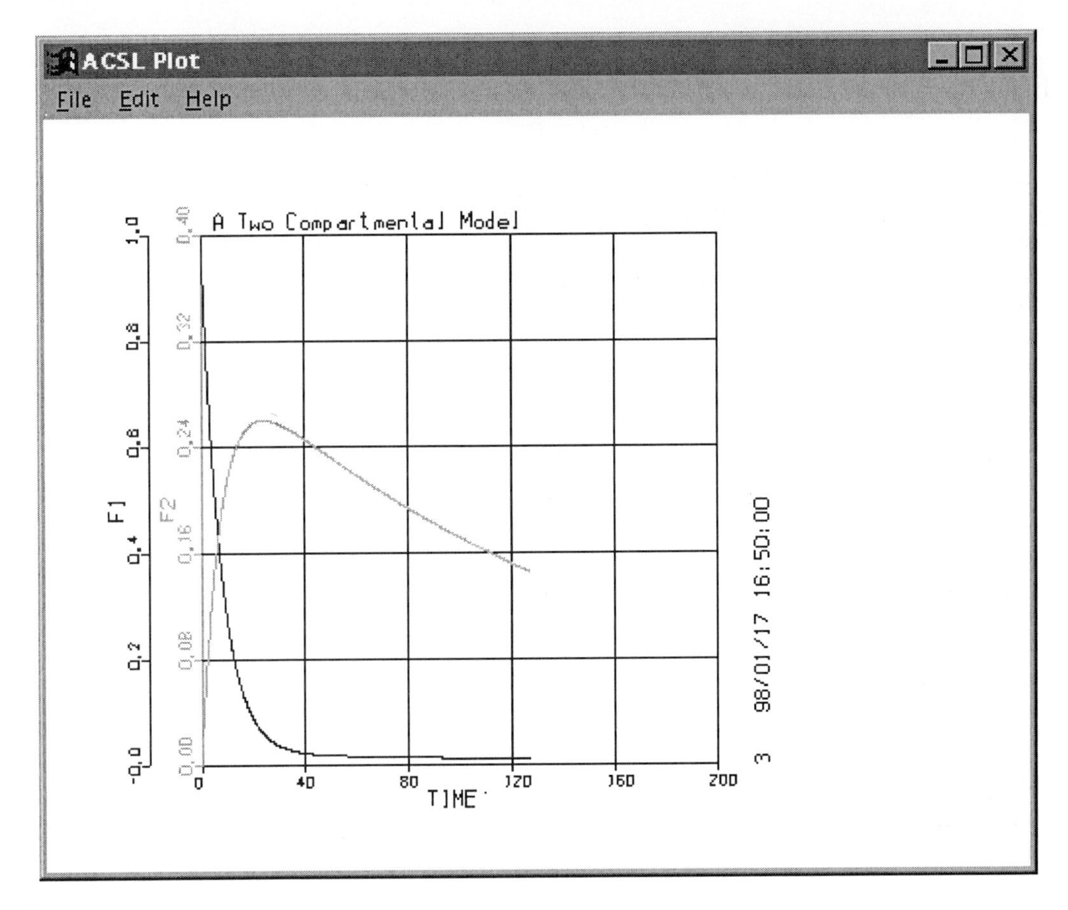

Fig 5.11 ACSL plot of fitted model.

2. MLAB

MLAB's control files are called 'do' files in the same vein as control files for the statistical package STATA. Almost any MLAB command is acceptable as an instruction to be included into a 'do' file. The do file below performs the following:

1. it defines a response function, say, r(t)
2. it defines a transfer function, say, h(t)
3. values of h(t) are generated for t=1 to 15 in unit steps and stored in the vector hh
4. a similar set of values are generated for r(t) and likewise stored in rr
5. hh is deconvolved with rr to see what form of input i(t), say, could have given rise to rr. The output from the deconvolution is stored in ii.
6. an exponential function, i(t), is fitted to the data ii with the amplitude and slope parameters, a1 and k1, allowed to adjust. Of course, k1 should emerge to be 0.2.
7. the fitting process yields a k1 value of 0.2 as needed

The 'do' file is invoked by the command do "dofile.do" where the actual 'do' file name is inserted into this position in the command. The 'do' file:

```
fct r(t)=100*(exp(-.1*t)-exp(-.2*t))
fct h(t)=exp(-.1*t)
hh=h on 1:15
rr=r on 1:15
ii=deconv(hh,rr,3)
fct i(t)=a1*exp(-k1*t)
a1=1; k1=.1
idata=1:15 &' ii
fit(a1,k1), i to idata
```

Output produced by the 'do'file

```
* echodo=3
* do "demo_4.do"
* fct r(t)=100*(exp(-.1*t)-exp(-.2*t))
Redefining R
* fct h(t)=exp(-.1*t)
Redefining H
* hh=h on 1:15
* rr=r on 1:15
* ii=deconv(hh,rr,3)
* fct i(t)=a1*exp(-k1*t)
Redefining I
* a1=1; k1=.1
* idata=1:15 &' ii
* fit(a1,k1), i to idata
final parameter values
    value           error               dependency   parameter
  11.62318401    4.941230029e-15        0.615514047   A1
        0.2    1.113246459e-16        0.615514047   K1
7 iterations
CONVERGED
best weighted sum of squares = 2.475174e-28
weighted root mean square error = 4.363462e-15
weighted deviation fraction = 6.987832e-16
R squared = 1.000000e+00
```

3. Scientist

The Scientist software has no control procedure capabilities or options.

4. WinSAAM

WinSAAM supports a limited form of control file. Indeed the operation is restricted to sequences of instructions in a file of any acceptable name located in the WinSAAM program directory. We have actually used this tool as a vehicle for verifying the processing accuracy of WinSAAM where, by storing the answers to problems in the files accessed in the control file, we are able to run test problems easily and immediately confirm the accuracy of the various integrators. To invoke to control file, tester, here we would use the command sequence:

```
file
input
tester
```

The output would appear in a setting dictated by the actual contents of the tester command file. A section of the tester command file is as follows:

```
deck
t1.tst
file
outp
t1.out
deck
solv
prin q(1)
file
deck
t2.tst
```

A small section of the output file, t1.out, generated by executing tester below shows the excellent agreement between the model code '2' solutions from WinSAAM and the precise values tabulated in the observation entry section of the test problem, in this case t1.tst.

D	C	---CAT--	T	THETA	QC	QO	QO-QC	QC/QO	WT
1	1	0	1.0000E-02	0.0000E+00	9.90050E-01	9.90050E-01	0.000E+00	1.000	1.00E+00
2	1	0	5.0000E-02	0.0000E+00	9.51229E-01	9.51229E-01	-5.960E-08	1.000	1.00E+00
3	1	0	1.0000E-01	0.0000E+00	9.04837E-01	9.04837E-01	-2.384E-07	1.000	1.00E+00
4	1	0	5.0000E-01	0.0000E+00	6.06531E-01	6.06531E-01	-5.960E-08	1.000	1.00E+00
5	1	0	1.0000E+00	0.0000E+00	3.67879E-01	3.67879E-01	-2.980E-08	1.000	1.00E+00
6	1	0	2.0000E+00	0.0000E+00	1.35335E-01	1.35335E-01	-1.490E-08	1.000	1.00E+00
7	1	0	4.0000E+00	0.0000E+00	1.83156E-02	1.83156E-02	-1.863E-09	1.000	1.00E+00
8	1	0	6.0000E+00	0.0000E+00	2.47875E-03	2.47875E-03	-6.985E-10	1.000	1.00E+00
9	1	0	8.0000E+00	0.0000E+00	3.35463E-04	3.35463E-04	-5.821E-11	1.000	1.00E+00
10	1	0	1.0000E+01	0.0000E+00	4.53999E-05	4.53999E-05	-3.638E-12	1.000	1.00E+00

An earlier version of the program, Consam, contained a macro operation that included commands for solving, plotting, printing and saving results. A future release of WinSAAM will have this macro operation of Consam fully restored.

X. Integrators and Optimizers

The utility and efficiency of modeling software tends to relate to its provision for addressing the widest possible range of modeling problems. With regard to problem solving, for example, this usually means the ease with which both stiff and non-stiff problems are managed. The overheads of a stiff solver for its application to non-stiff problems render it uncompetitive in comparison to common non-stiff solvers. Here techniques such as Adams-Bashforth and Runge-Kutta for such problems will prove both efficient and accurate. Nevertheless stiff solvers are definitely needed for stiff problems .

A similar situation exists for optimizers. Here we occasionally encounter the situation where poorly defined heavily parameterized minimization surfaces need to be fit to data of lower quality. Local minima, ridges, and general surface roughness, impose that we take pains to ensure that any minimum we find is lower than previous ones. Under such circumstances we may do well to have on hand gradient-free algorithms to assist in the analysis of any gradient-based minima.

Table 5.1 shows the array of integrators and optimization algorithms provided by the modeling software reviewed here.

XI. Output

We define output as the generation of potentially exportable objects, e.g. lexical files or graphic files, for printing directly, or for incorporation into other software. The situation arises especially in the preparation of reports, papers, and presentations where the results generated by modeling software require further editing prior to their final deployment. Under such circumstances it is most convenient when the modeling software offers a direct path between its output and the intended target.

1. ACSL

ACSL offers the following output forms:

1. log files output containing all user interaction with the ACSL system during a modeling session
2. print files which contain the 'print'ed prepared variables determined during a run.
3. screen shots of the output generated and displayed in the ACSL run-time modeling window.
4. plots produced in conjunction with plot commands and displayed in separate plot windows.

Table 5.1 Integrators and optimization algorithms provided by several modeling software packages

Software	Integrators	How Accessed	Optimizers	How Accessed
ACSL	Adams-Moulton Gear Runge-Kutta 1 Runge-Kutta 2 Runge-Kutta 4 Runge-Kutta-Fehlberg 2-3 Runge-Kutta- Fehlberg 4-5	IALG=1 IALG=2 IALG=3 IALG=4 IALG=5 IALG=8 IALG=9	---	Specified in CSL Stream
MLAB	Gear 3 Gear 2-3 Adams Adams/Gear 2	method = 3 method = 2 method = 1 method = 0	Marquardt-Levenberg	Command
Scientist	Euler Runge-Kutta 4 Runge Kutta-Fehlberg 4-5 Bulirsch-Stoer Adams Adams-Moulton BDF (Episode)	Dialog box entry cells selectable for the desired integrator	Modified Gauss-Newton Marquardt-Levenberg Steepest Descent Simplex	Dialog box entry cells selectable for the desired optimizer.
WinSAAM	Runge-Kutta 2 Runge-Kutta 4/5 Newton Method Convolution Adams-Moulton Gear	mode = 1 mode = 21 mode = 2 mode = 10 mode = 22 mode = 23	Modified Gauss-Newton	Command

Below we present samples of the log file output and the screen output generated in conjunction with a control file-coordinated analysis of our two-pool model presented above. The plot output has been shown previously (Fig 5.11).
Log segment:

```
ACSL Runtime Exec PC Version Level 10D  98/01/18 13:23:43     Page  1

Switching CMD unit to 4 to read demo_1.cmd
set title='A Two Compartmental Model'
set hvdprn=.true.
set tstp=128
! printouts echoed to screen
prepare Time F1 F2
! the prepared variables will be saved for plotting
output Time F1 F2
! Time F1 and F2 will be displayed during solving
Start
      TIME 0.              F1 1.00000000       F2 0.
      TIME 0.10000000          F1 0.98511400       F2 0.00496020
      TIME 0.20000000          F1 0.97045500       F2 0.00984163
      TIME 0.30000000          F1 0.95601900       F2 0.01464550
      TIME 0.40000000          F1 0.94180300       F2 0.01937290
      TIME 0.50000000          F1 0.92780300       F2 0.02402510
      TIME 0.60000000          F1 0.91401600       F2 0.02860330
      TIME 0.70000000          F1 0.90043800       F2 0.03310850
```

Screen output:

```
Switching CMD unit to 4 to read demo_1.cmd
set tstp=128
! printouts echoed to screen
prepare Time F1 F2
! the prepared variables will be saved for plotting
output Time F1 F2
! Time F1 and F2 will be displayed during solving
Start
      TIME 0.              F1 1.00000000       F2 0.
      TIME 0.10000000          F1 0.98511400       F2 0.00496020
      TIME 0.20000000          F1 0.97045500       F2 0.00984163
      TIME 0.30000000          F1 0.95601900       F2 0.01464550
      TIME 0.40000000          F1 0.94180300       F2 0.01937290
      TIME 0.50000000          F1 0.92780300       F2 0.02402510
      TIME 0.60000000          F1 0.91401600       F2 0.02860330
```

2. MLAB

MLAB supports a very similar array of output utilities to ACSL. Specifically, log files, and print files can be easily generated and accessed for subsequent incorporation into documents, and in addition, graphics can either be saved as files for later access or, again as with ACSL they can be captured from the graphics window for direct incorporation into another application.

We have already demonstrated the incorporation of log file contents into word processing documents, however, we present a plot generated by MLAB of some results generated in conjunction with the analysis of our two-compartment model using MLAB (Fig 5.12).

The salient component of the MLAB interaction leading to the generation of this plot is as follows (captured in the MLAB log file):

```
* fct f1't(t)=-f1(t)*(l01 + l21) + f2(t)*l12
* fct f2't(t)= f1(t)*l21       - f2(t)*l12
* initial f1(0)=1
* initial f2(0)=0
*
* l01=.1 ;l21=.05 ; l12=.01
*
* m=integrate(f1't,f2't,0:128!140)
*
* draw m col (1,2) color red lt (1,0,0,0,0,.0075,0)
* draw m col (1,4) color yellow lt (.01,.0075,.013,0,0,.0075,0)
* view
end of MLAB.LOG
```

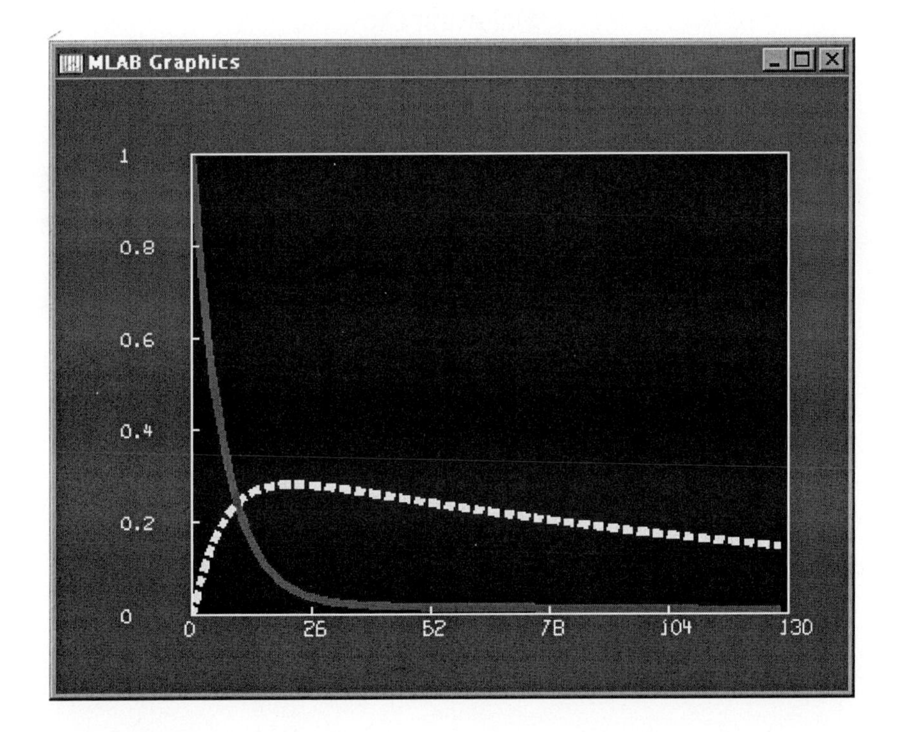

Fig 5.12. MLAB plot of fitted model.

3. Scientist

Scientist output is predominantly screen oriented, however, again via capture tools and an array of exporters graphics and spreadsheets can be exported in a host of alternate file formats. For example spreadsheet output can be exported to almost every imaginable spreadsheet system such as Lotus, Excel, and Quattro. The model window can be saved as ASCII text, or as Rich Text File Format (RTF). Of course, each of these as well as the parameters window can be managed within their own Scientist format.

The Statistics output text file which can be generated in conjunction with a data fitting session (see above) can also be saved as a text or RTF file for incorporation into another document or direct printing.

The figure below (Fig 5.13) was generated in conjunction with the analysis of the sample two-compartment problem using Scientist and reflects the profile of each of the responses over the first 128 or so time units of the solution. The figure was initially incorporated into PowerPoint and then copied from PowerPoint into WORD for Windows 97.

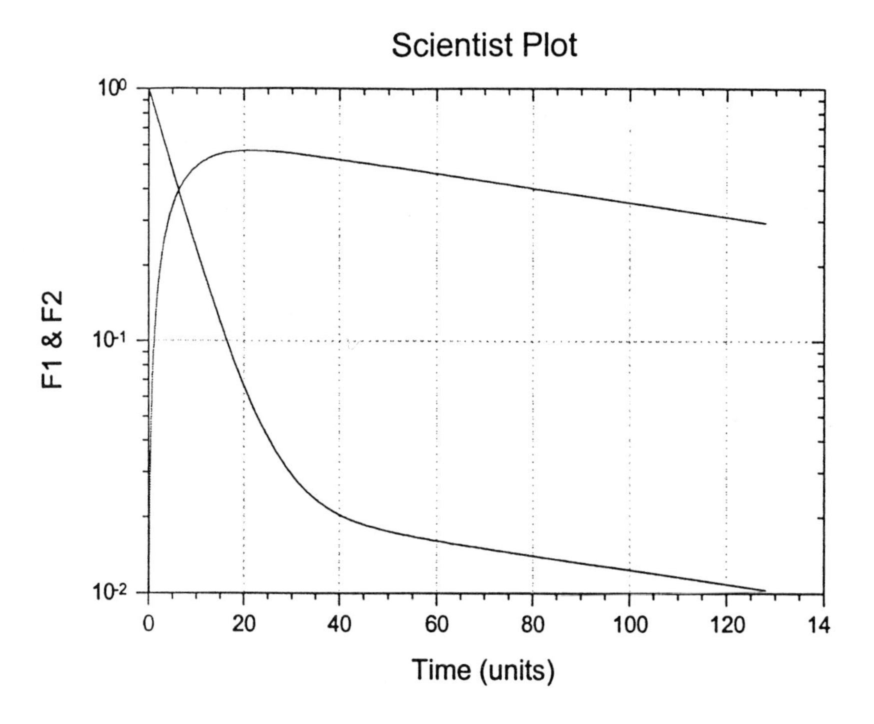

Fig 5.13 Scientist plot of fitted data.

1.4 WinSAAM

A major design concern in the development of WinSAAM was the provision for the widest possible flexibility regarding the exchange of WinSAAM output items with other software. Specifically:

1. All text windows including the terminal window, the logging window, the batch output window, and the text input window have their text as a selectable attribute for copying
2. The spreadsheet output window can be saved in tabbed text format, Excel 4 & 5 file formats, and formula 1 file format.
3. The plots can be saved as windows metafile, enhanced metafile, jpeg, and bitmap file formats as well as being able to be copied to the clipboard for direct incorporation into OLE (Object Linking and Embedding) ready software.

The WinSAAM plot (Fig 5.14) relates to graphic output from the familiar two-compartment sample problem.

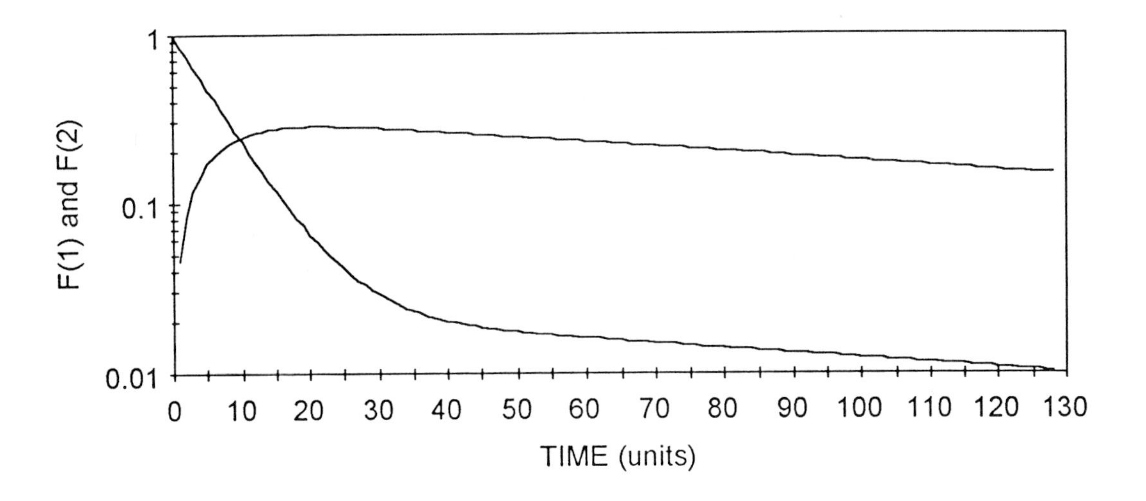

Fig 5.14 WinSAAM plot of fitted data.

A powerful feature of WinSAAM is its ability to transfer output results directly to statistical software for further analysis. In the figure below we show the capture of some WinSAAM processing results in the statistical package STATA. This data transfer called for less than six button presses. In the plot (Fig 5.15) we present a STATA rendition of the WinSAAM calculated values demonstrating options existing here for exploring alternate graphing systems for displaying model solutions.

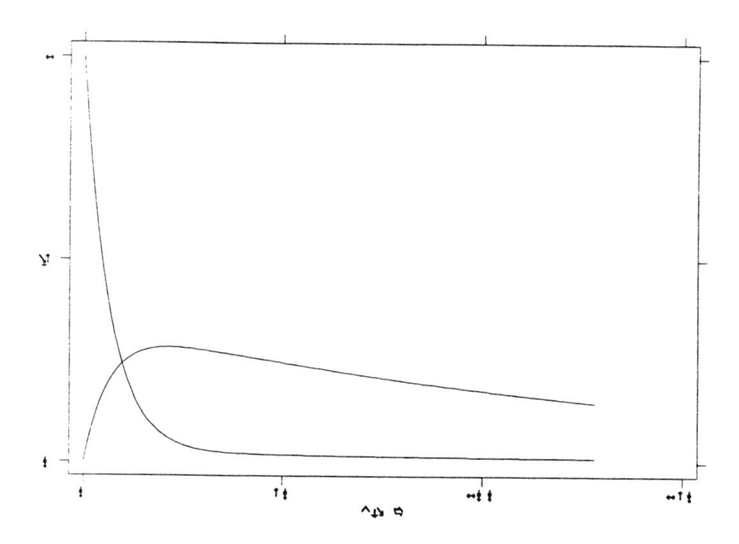

Fig 5.15 WinSAAM data transferred to STATA 5 via interapplication communication and then plotted.

XII. Major Strengths

1. WinSAAM

Perhaps the greatest strength of WinSAAM is the ease with which complex models can be inserted, explored and fitted to data. To illustrate the strength of WinSAAM in this regard we present an analysis of some data included in the MLAB User's guide. Here an experiment was reportedly undertaken to examine the metabolism of a drug. The drug was injected into the blood stream and data relating to the levels of the drug were collected in the blood and the rumen over a 10-minute period. The subtlety to this problem was that because of the time required for the bile to pass through a catheter, a delay was needed in the model. Data are shown in Table 5.2 and the model in Fig 5.16.

Table 5.2 Data

Time	Blood	Bile
0	100	0
3	23.3	21.8
4	15.1	52.1
5	7.9	71.8
6	4.3	84.3
7	2.5	90.9
8	2.2	96.0
9	2.1	97.3
10	0.6	98.4

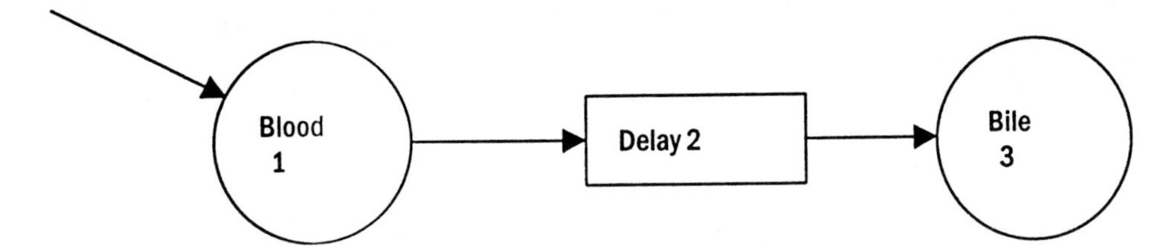

Fig 5.16 WinSAAM graphic model illustrating delay element.

The WinSAAM Text Input model

```
4:  L(2,1)   5.015604E-01  0.000000E+00   1.000000E+02
5:  DT(2)    2.424603E+00  1.000000E+00   1.000000E+01
6:  DN(2)    2
7:  L(3,2)   1
8:  IC(1)    100
```

The final values for the iteratively adjusted parameters and their uncertainties:

```
> fsd(i)
* VALUES MAY NOT RELATE TO CURRENT PARAMETERS
* L ( 2, 1)   5.016E-01     FSD( 1)   4.323E-03
  * DT( 2)       2.425E+00     FSD( 2)   8.651E-03
```

The observed (QO) and predicted (QC) values:

```
> prin q(1,3)
-------------------------------------------------------------------
*** NAME :   1
CURRENT KOMN
 # COMP TC  CATEGORY      T          QC         QO         QC/QO
  1   1   0  F ( 1)    3.000E+00  2.221E+01  2.330E+01    .9532
  3   1   0  F ( 1)    4.000E+00  1.345E+01  1.510E+01    .8907
  5   1   0  F ( 1)    5.000E+00  8.145E+00  7.900E+00   1.0310
  7   1   0  F ( 1)    6.000E+00  4.932E+00  4.300E+00   1.1470
  9   1   0  F ( 1)    7.000E+00  2.987E+00  2.500E+00   1.1948
 11   1   0  F ( 1)    8.000E+00  1.809E+00  2.200E+00    .8222
 13   1   0  F ( 1)    9.000E+00  1.095E+00  2.100E+00    .5216
 15   1   0  F ( 1)    1.000E+01  6.634E-01  6.000E-01   1.1056
-------------------------------------------------------------------
*** NAME :   3
CURRENT KOMN
 # COMP TC  CATEGORY      T          QC         QO         QC/QO
  2   3   0  F ( 3)    3.000E+00  2.162E+01  2.180E+01    .9918
  4   3   0  F ( 3)    4.000E+00  5.328E+01  5.210E+01   1.0227
  6   3   0  F ( 3)    5.000E+00  7.308E+01  7.180E+01   1.0179
  8   3   0  F ( 3)    6.000E+00  8.450E+01  8.430E+01   1.0023
 10   3   0  F ( 3)    7.000E+00  9.111E+01  9.090E+01   1.0023
 12   3   0  F ( 3)    8.000E+00  9.525E+01  9.600E+01    .9922
 14   3   0  F ( 3)    9.000E+00  9.788E+01  9.730E+01   1.0060
 16   3   0  F ( 3)    1.000E+01  9.956E+01  9.840E+01   1.0118
```

Results are plotted in Fig 5.17.

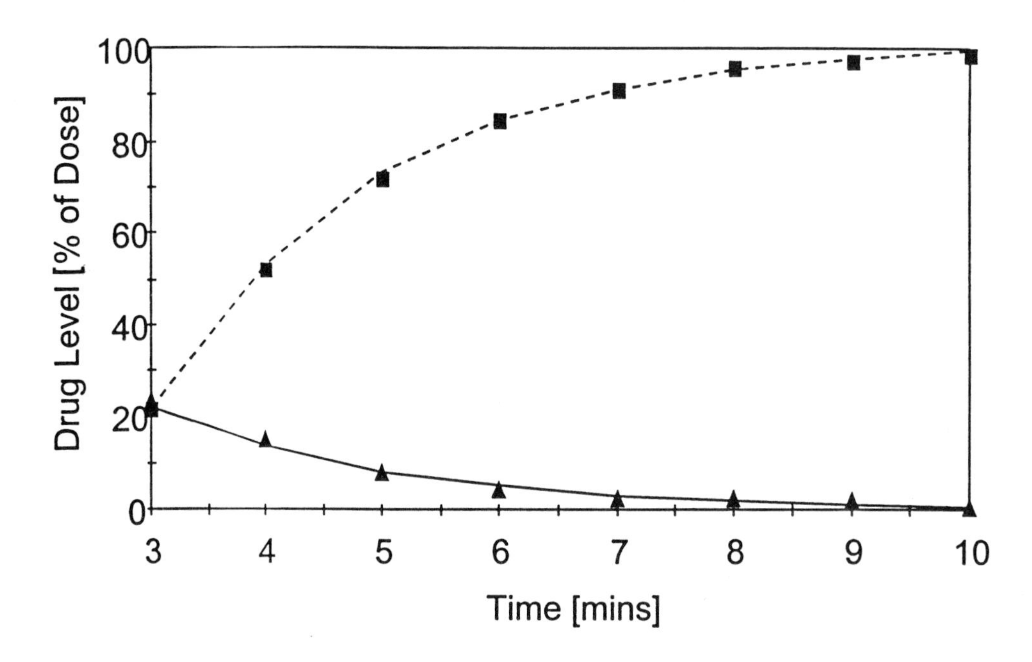

Fig 5.17 WinSAAM plotted results.

Other strengths of WinSAAM include:

1. the absence of a need to explicitly specify differential equations for a differential equation model
2. the support for a powerful set of highly refined modeling constructs, such as delays, to meet a wide range of modeling needs. Indeed none of the other software reviewed was able to solve the above problem
3. the smoothness of its integration into the Windows environment

2. MLAB

Perhaps the greatest strength of MLAB is its tremendous scope of operation. Its capacity to fit data, graph responses, and explore models *per se* simply places it in a league of its own. Possibly the best way to illustrate the power of MLAB in this regard would be to use it to run Bergman's IVGTTminimal-model for glucose kinetics (Bergman and Bowden, 1981). In MLAB syntax we can create this model (from the MLAB Application Manual) as shown with the following control file (also called a 'do' file in MLAB) and results are plotted in Fig 5.18.

```
1.  filedir="c:\meryl\book\chap_5\"
2.  dofiledir=filedir
3.  m=read("diabetes2.txt",24,3)
4.  m
5.  gludat=m col (1,2)
6.  insdat=m col (1,3)
7.  fct g't(t)=-g(t)*(p1+x(t))+p1*gb
8.  fct x't(t)=-p2*x(t)+p3*(if (i(t)-ib)<=0 then 0 else (i(t)-ib))
9.  fct i(t)=lookup(insdat, t)
10. init g(0)=g0
11. init x(0)=0
12. gb=90
13. ib=7
14. p1=0.03; p2=0.05; p3=0.0000725; g0=282
15. constraints c={p1>0, p2>0, p3>0}
16. for loop=1:21 do {ws[loop]=1/(0.015*gludat[loop,2])}
17. fit (p1,p2,p3), g to gludat with weight ws with constraints c
18. gluout= g on m col 1
19. gluout=m col 1 &' gluout &' m col 2
20. gluout
21. draw gludat, pt 2, lt 0
22. draw insdat, pt 4, lt 0
23. draw gluout col (1,2), pt 0, lt 2
24. view
```

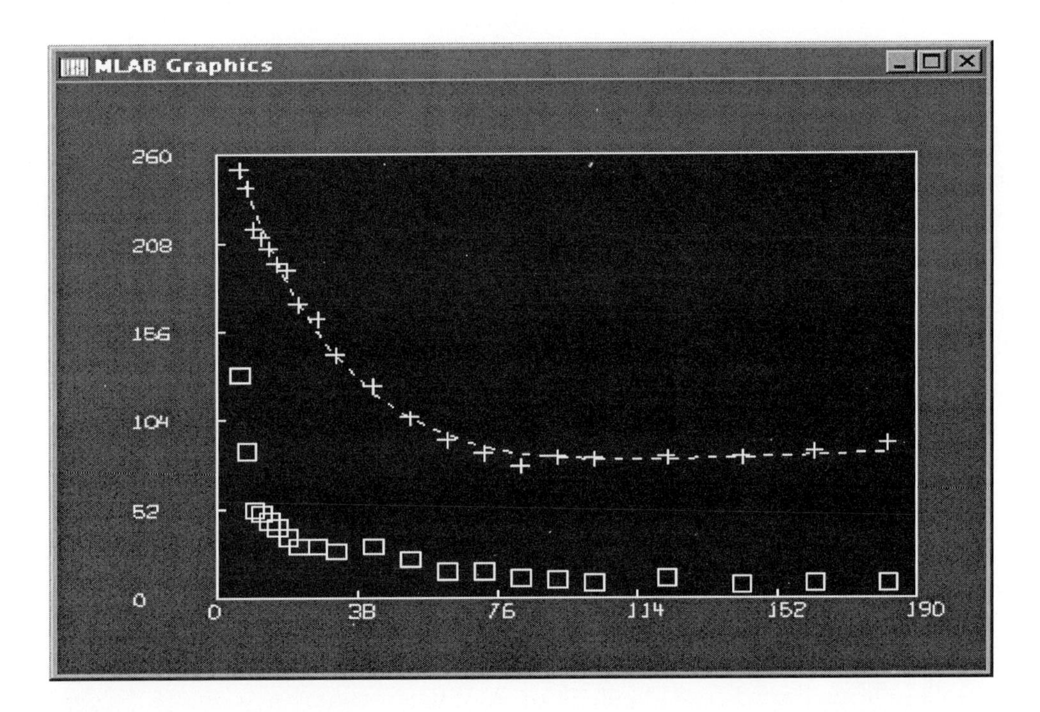

Fig 5.18 MLAB plotted results.

and the output generated by this control file (using the default MLAB echo status) is as follows:

```
M: a  21 by 3 matrix

 1: 6    251  130
 2: 8    240  85
 3: 10   216  51
 4: 12   211  49
 5: 14   205  45
 6: 16   196  41
 7: 19   192  35
 8: 22   172  30
 9: 27   163  30
10: 32   142  27
11: 42   124  30
12: 52   105  22
13: 62   92   15
14: 72   84   15
15: 82   77   11
16: 92   82   10
17: 102  81   8
18: 122  82   11
19: 142  82   7
20: 162  85   8
21: 182  90   8

Redefining G DIFF T
Redefining X DIFF T
Redefining I
Redefining G
Redefining X
final parameter values
     value            error            dependency   parameter
  0.02529126375    0.004829141013      0.968934914   P1
  0.02585655196    0.01221890692       0.9765266812  P2
 9.430235213e-06   4.29461862e-06      0.9905379551  P3
13 iterations
CONVERGED
best weighted sum of squares = 1.915963e+02
weighted root mean square error = 3.262551e+00
weighted deviation fraction = 2.288963e-02
R squared = 9.944136e-01
no active constraints

 GLUOUT: a  21 by 3 matrix

 1: 6    250.043707  251
 2: 8    238.682877  240
 3: 10   227.543961  216
```

```
4: 12   216.916563  211
5: 14   206.860355  205
6: 16   197.384479  196
7: 19   184.25299   192
8: 22   172.405202  172
9: 27   155.301429  163
10: 32  141.034766  142
11: 42  119.297768  124
12: 52  104.176087  105
13: 62  94.0454279  92
14: 72  87.6046471  84
15: 82  83.6530203  77
16: 92  81.4639081  82
17: 102 80.4751153  81
18: 122 80.5169695  82
19: 142 81.5158024  82
20: 162 83.0718544  85
21: 182 84.5524097  90
```

For the reader interested in insulin-glucose kinetics the same problem setup in WinSAAM follows, and the data and solution plotted by the command, >plot q(6,8), are shown in Fig 5.19.

```
1: A SAAM31 BERGMANS INSULIN SENSITIVITY MODEL USING SAAM
2: 2    25
3: H PAR
4: C
5: C P(1)=SG
6: C P(3)/P(2)=SI
7: C P(4)=GB
8: C P(5)=IB
9: C P(6)=G0
10: C P(7)=SI
11: C
12:  P(1)   2.748238E-02 0.000000E+00  1.000000E+02
13:  P(2)   4.864673E-02 0.000000E+00  1.000000E+02
14:  P(3)   2.670302E-05 0.000000E+00  1.000000E+02
15:  P(4)   9.000000E+01
16:  P(5)   3.500000E+00
17:  P(6)   2.820000E+02
18:  L(0,7)=P(2)
19:  UF(7)  1                       8G 7
20:  P(7)=P(3)/P(2)
21: C
22: C COMPT. 6 = GLUCOSE
23: C COMPT. 7 = REMOTE INSULIN
24: C COMPT. 8 = PLASMA INSULIN
25: C
```

```
 26: IC(6)=P(6)
 27: IC(7)=0
 28: C
 29: H DAT
 30: X UF(6)=-(P(1)+F(7))*F(6)+P(1)*P(4)
 31: X G(7) = P(3)*(FF(8)-P(5))
 32: X FF(8)= F(8)
 33:   P(7)
 34: 106                    WT=0.0
 35:        2       178
 36:        3       262
 37: 106                    FSD=.04

glucose data is entered here ...
 60: 108QL
 61:        2       2.46
 62:        3       43.33
 63:        4       68.43
```

The WinSAAM data fitting results are accessed as follows:

```
> fsd(i)
* VALUES MAY NOT RELATE TO CURRENT PARAMETERS
* P ( 1)   2.748E-02   FSD( 1)  7.347E-02
* P ( 3)   2.667E-05   FSD( 2)  2.028E-01
* P ( 2)   4.869E-02   FSD( 3)  2.094E-01
```

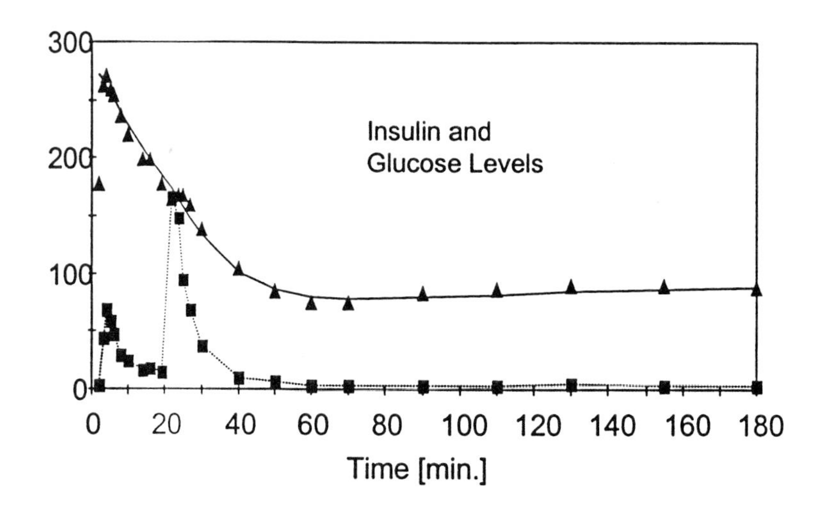

Fig 5.19 WinSAAM plotted results.

3. Scientist

A great advantage of Scientist is the provision of access to the total functionality of the software via the interface. Indeed, 'commands' are distributed, as menu buttons, and selectable items from pull down lists in conjunction with the model window, the parameters window, the spreadsheet window, and the graphics window. Furthermore, the menu bar is appropriately reconfigured to match the sensible options needed in conjunction with the 'uppermost' window. For example it is not until the 'data and solutions' spreadsheet window is brought to the foreground that we gain access to the data transformation and data weighting services. This, together with the sensible assignment of 'hot keys', means that after loading the model, the data, and the parameters, we can advance from a job compilation to a display of the fit just by pressing three keys (f8, f7, and f9). This assumes that the grunt work of model entry has been completed.

A very powerful feature of Scientist is evident from the following analysis. We wish to model the response of a linear system (e.g. a person) to 4 'drug' boluses, one at t=0, and one at each of 3 equal intervals (of 4 units) after. Setting up of the model to accomplish this using Scientist is quite straight forward, especially by incorporating the 'Laplace inverse' function in conjunction with the summation function.

Firstly the response to a single bolus: We seek to simulate the response of a two-compartment system (expressed here as the sum of two exponentials) to a bolus injection at t=6. We assume (as in the Scientist Manual) that the response of the system to a single injection has the form

$$C(t) = A.e^{(-alpha*t)} + B.e^{(-beta*t)}$$

Then the Laplace transform of the response of such a 'system' to the bolus injection would be given by:

c = (A / (s + alpha) + B / (s + beta))/InjectionDose

and numerical values describing the actual response can be generated in Scientist using,

LaplaceInverse(T, c*Dose, s)

Where: Dose may be equal to the variable InjectionDose or may simply bear some relation to it. Now, if the injection takes place some time Tau after the start of the 'experiment' then the LaplaceInverse function argument 'T' is replaced by 'T-Tau':

LaplaceInverse(T-Tau, c*Dose, s)

The Scientist model below will simulate the desired response and a plot of the response is shown in Fig 5.20:

```
// Pharmacokinetic Simulations
// Single bolus dose administered at t=6
IndVars: T
LaplaceVar: S
DepVars: C
Params: TAU, DOSE, A, B, ALPHA, BETA
INPUT=DOSE/4
DIST=(A/(S+ALPHA)+B/(S+BETA))/DOSE
C=LAPLACEINVERSE(T-TAU, INPUT*DIST, S)
// Parameter values
TAU=6.0
DOSE=1.0
A=8.0
B=1.0
ALPHA=1.0
BETA=0.2
0<t<30
***
```

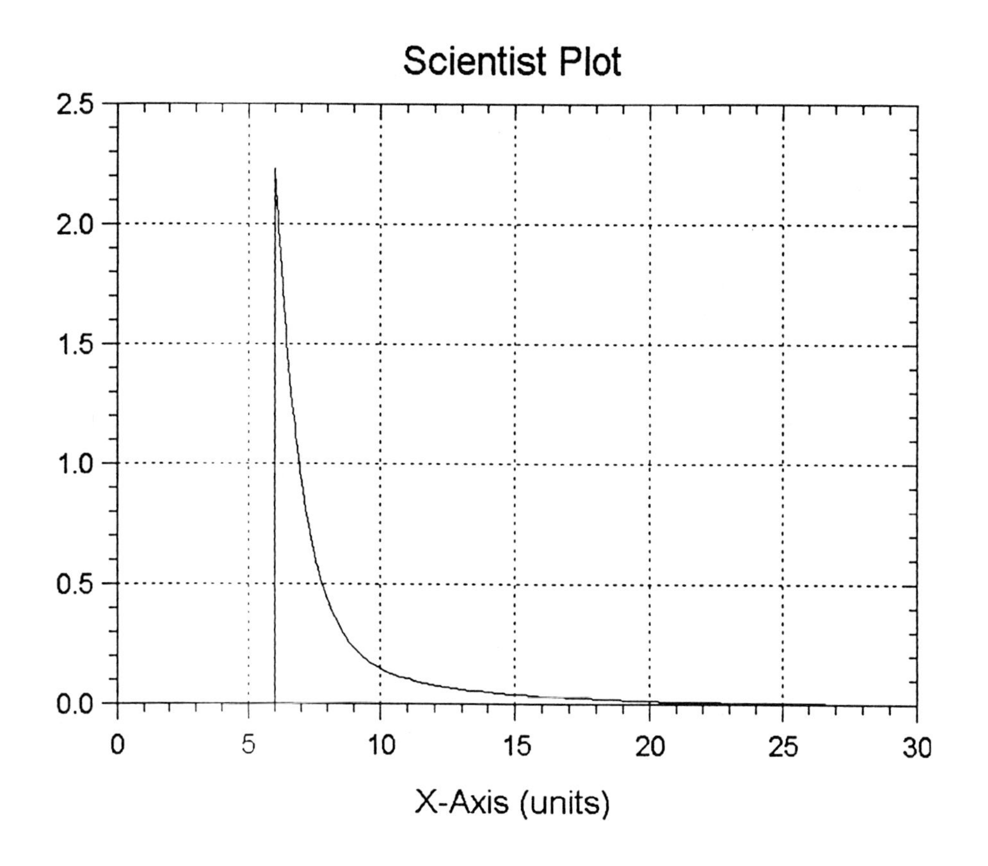

Fig 5.20. Scientist actual response function plot.

Using the principle of superposition, the Laplace transform of the response of the system to a series of injections such as the above is the sum of the individual Laplace transforms. Furthermore using Scientist's 'summation' function in conjunction with the above demonstration we can generate the response of the system to the original 4 injections as follows:

SUMMATION(0, 3, LAPLACEINVERSE(T-I*TAU, INPUT*DIST, S), I)

where this statement implies 4 injections, t=0, TAU, 2*TAU, and 3*TAU, of magnitude INPUT into a system with the Laplace transform of the response given by DIST. Note the I, and S are respectively summing, and integrating variables for the summer and the integrator. The Scientist model needed to generate response;

```
// Pharmacokinetic Simulations
// Multiple bolus doses
IndVars: T
LaplaceVar: S
DepVars: C
Params: TAU, DOSE, A, B, ALPHA, BETA
INPUT=DOSE/4
DIST=(A/(S+ALPHA)+B/(S+BETA))/DOSE
C=SUMMATION(0, 3, LAPLACEINVERSE(T-I*TAU, INPUT*DIST, S), I)
// Parameter values
TAU=6.0
DOSE=1.0
A=8.0
B=1.0
ALPHA=1.0
BETA=0.2
0<t<30
***
```

The plot of the responses is shown in Fig 5.21.

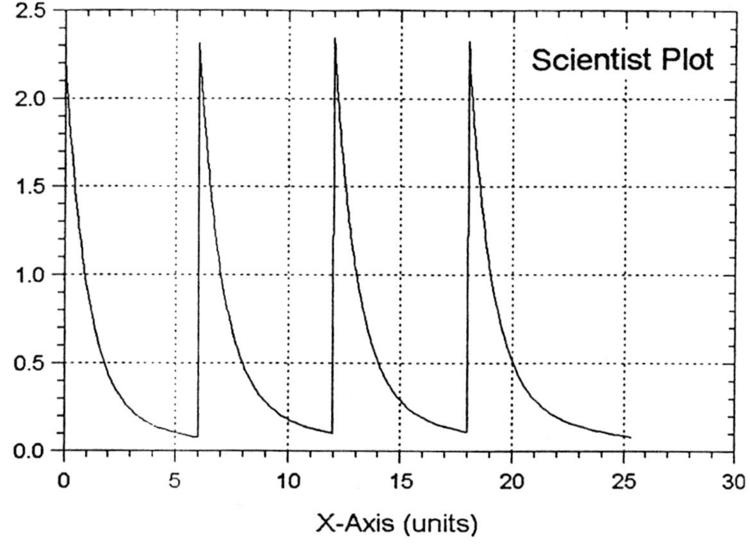

Fig 5.21 Scientist response function plot.

4. ACSL

There are two great features of ACSL which separate it from the other three programs reviewed here; 1) its capacity to manage very large models, and 2) its array of tools with which a comprehensive analysis of model responses can be undertaken. The array of extremely large and complex models developed and examined by defense investigators and by agricultural researchers has distinctly established it as the modeling tool of preference here. Baldwin (1995), for example has spent many years developing and refining a comprehensive model of the dairy cow including digestive, metabolic, and production-related sub-systems and this system, involving more than fifty-five state equations, is exclusively maintained and developed in the ACSL environment. It is unlikely that versions of the model could be developed for use with any of the other simulators. One of the earliest agricultural models to be expressed in the ACSL syntax was France's sheep digestion model (France and Thornley, 1984). We present this model and some solutions relating to it here and discuss additional features of it elsewhere. There are nine state variables in this model (including 4 carbon, 3 nitrogen, a fluid, and a microbial pool) and the purpose is to examine the influence of dietary composition on levels of carbon-related and nitrogen-related rumen pools including bacterial contributions to- and removals from the system. Solutions can be derived and displayed though we haven't used a control file for this purpose. Here we present plots of some of the nitrogen (N, Pr) pools and carbon pools (A, Br, and C) and the microbial (M) pool (Fig 5.22).

```
PROGRAM FranceSheep Model
! Units: kg/m**3, day, m**3, /day
! Definitions:
!     C:  Rumen conc. of Water Soluble CHOH: kg/m**3
!     M:  Rumen conc. of Microbial Dry Matter: kg/m**3
!     N:  Rumen conc. of NPN: kg/m**3
!     V:  Metabolic vol. of Rumen: m**3
!     A:  Rumen conc. of alpha hexose: kg/m**3
!     Bn: Rumen conc. of non-rumen degradable b-hexose: kg/m**3
!     Br: Rumen conc. of rumen degradable b-hexose: kg/m**3
!     Pn: Rumen conc. of non-rumen degradable protein: kg/m**3
!     Pr: Rumen conc. of rumen degradable protein: kg/m**3
! Driving Variables:
!     Di: Dietary input of i: kg or m**3 of i /day
!     Si: Salivary input of i: kg or m**3 of i /day
! Parameters:
!     fC:  Metabolic composition parameter: kg CHOH /kg microbial dm
!     fN:  Metabolic composition parameter: kg P /kg microbial dm
!     kC, kN, kCN: parameters of the microbial growth equation
!     kAM, kBrM, kPrM: Rate constants for microbial degradation of
!                 a-hexose, b-hexose, and p m**3/kg microbial dm /day
!     kMu: Constant determining dependence of microbial catabolism of
!         microbial growth rate: days
!     YMs:  Conversion efficiency of substrate CHOH into microbial CHOH
!         CHOH / kg CHOH
```

```
!      lM:  Maximum specific rate of microbial catabolism: / day
!      mM:  Maximum microbial specific grwoth rate: / day

INITIAL

!      ACSL variables:
       IALG = 5                    ! R-K 4th order
       CINTERVAL CINT = 0.1        ! Output interval
       NSTEPS   NSTP = 1           ! Calculation interval
       MAXTERVAL MAXT = .0025   ! Maximum integration step
       CONSTANT  TSTP = 6.0        ! Termintaion time

! Model constants
       kAM  = 1.5
       kBrM = 2.0
       kPrM = 2.0
       mM   = 5.0
       kC   = 0.2
       kN   = 0.2
       kCN  = 0.0
       lM   = 0.2
       kMu  = 5.0
       fC   = 0.3
       fN   = 0.5
       YMs  = 0.1

! Dietary Inputs
       DV  = 4.2e-03
       DBn = 67.3e-03
       DPn = 24.3e-03
       DA  = 26.6e-03
       DBr = 493.7e-03
       DPr = 72.9e-03
       DC  = 215.0e-03
       DN  = 42.2e-03

! Salivary Inputs and Rumen Outout
       vo   = 16.3e-03
       SV   = 12.1e-03
       SN   = 6.1e-03
       SPr  = 12.1e-03

! Initial Values for State variables
       Vi  = 5.0e-03
       Pni = 4
       Pri = 1.5
       Ai  = 5
       Bni = 36
       Bri = 30
       Ci  = 7
       Ni  = 2
       Mi  = 26.5

END !INITIAL

DYNAMIC

DERIVATIVE
```

```
! Rumen fluid pool: V

     dVdt = DV + SV - vo
     w = vo / V     ! w = washout or dilution rate (/ day)

! Alpha-Hexose: A

     dAdt = (DA-A*(DV+SV))/V - kAM*A*M
     UA = V*kAM*A*M  ! UA = alpha-hexose utilization rate

! Beta-Hexose: Br, and Bn

     dBrdt = (DBr-Br*(DV+SV))/V - kBrM*Br*M
     UBr = V*kBrM*Br*M

     dBndt = (DBn-Bn*(DV+SV))/V

! Protein: Pn, and Pr

     dPndt = (DPn-Pn*(DV+SV))/V

     dPrdt = (DPr+SPr-Pr*(DV+SV))/V - kPrM*Pr*M
     UPr = V*kPrM*Pr*M       ! UPr = degradable protein utilization rate

! Water-soluble Carbohydrates: C

     dCdt = (DC-C*(DV+SV))/V + kAM*A*M + kBrM*Br*M ...
     + l*fC*M - mu*fC*M/YMs

! Non-protein Nitrogen: N

     dNdt = (DN+SN-N*(DV+SV))/V + l*fN*M - mu*fN*M ...
     + kPrM*Pr*M

! Microbial pool: M

     mu = mM/(1+(kC/C)+(kN/N)+(kCN/C/N))
     l  = lM/(1+kMu*mu)
     dMdt = mu*M-l*M-M*(DV+SV)/V

! Integration Section

     V  = INTEG(dVdt, Vi)
     A  = INTEG(dAdt, Ai)
     Bn = INTEG(dBndt,Bni)
     Br = INTEG(dBrdt,Bri)
     C  = INTEG(dCdt, Ci)
     Pn = INTEG(dPndt,Pni)
     Pr = INTEG(dPrdt,Pri)
     N  = INTEG(dNdt, Ni)
     M  = INTEG(dMdt, Mi)

END    ! Derivative section

     TERMT (t .GE. TSTP)

END    ! Dynamic section
END    ! Program
```

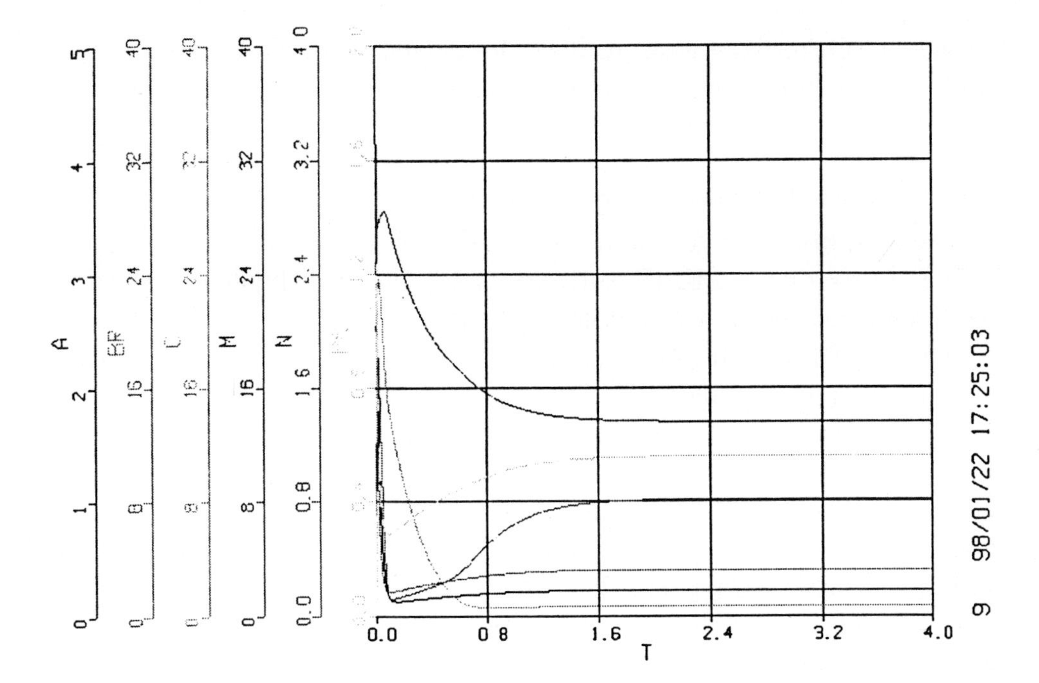

Fig 5.22 ACSL solution plots for nitrogen, carbon, and microbial pools

XIII. Weaknesses of Each Package

To some extent the weaknesses of a modeling package are a consequence of its strengths. Packages, for example focussing on flexibility will need to provide a fairly modeless style of operation for the user and this in turn implies that the novice user may become buried in a maze of windows and dialog boxes with no evident way of backtracking. Packages offering a more protected mode of operation will need to anticipate all the modeling maneuvers an investigator needs and this clearly would be a daunting if not impossible task.

1. ACSL

The most troublesome features of ACSL are as follows:

1. There are no prefabricated modeling constructs, i.e. all aspects of the model are conjured from the ground up using variables and constants
2. The need for efficiency in the mechanics of software compilation and linking would be of considerable concern to a naïve investigator with no background in computer science
3. The need for a security key to operate ACSL can be quite troublesome, particularly if your printer port needs to be set to a variety of different configurations to meet communication and printing needs
4. Until recently there was no data fitting capability available from MGA accompanying ACSL. A third party data fitter had to be acquired. This has now been remedied.

5. Many of the analytic tools available for detailed investigation of the model will have very little meaning to biomedical investigators

2. MLAB

The most troublesome features of MLAB are:
1. The syntax of the MLAB software is extremely difficult to master and seems not to be based on a consistent paradigm
2. The output from MLAB is somewhat disorganized and untidy ... with a little attention to alignment of columns and other output (see for example the statistical software STATA) dramatic improvements to the output appearance could be achieved
3. As with ACSL, MLAB requires a security key for its operation and remembering to have this connected and the parallel port appropriately configured is frustrating to say the least
4. The documentation for MLAB, while copious and accurate, is not user-friendly and neither is the onboard help system. Civilized Software would be well advised to invest efforts developing helpful, instructional, friendly tutorials
5. Again as with ACSL there are no pre-fabricated modeling constructs, i.e. no existing objects from which models can be developed. Never the less, there are a vast array of tools to access once the user-orchestrated structures are in place
6. Almost all of the common services of MLAB expect matrices as their input, and produce matrices as their output. Biomedical investigators who are not trained in mathematics may have difficulty mastering this aspect of MLAB.

3. Scientist

The greatest shortcomings of Scientist are as follows:
1. The software fails ungracefully under many situations. While it may be attributed to user inexperience, after discussing the program with other instructors, it appears that there are some operational problems with the software (although these may have been corrected in later versions later than the one tested).
2. The number of open windows becomes unwieldy rapidly as the modeling session advances. Each iteration may generate a new set of plot, and parameters windows and after 10 such sessions there may be more than 20 opened windows. The problem is that you aren't sure which plots relate to which parameter sets and accidentally deleting the wrong windows causes considerable time loss.
3. There are no pre-fabricated modeling constructs. As with ACSL and MLAB models are built with parameters and variables of the users own

creation. There is some help here though because some of the inappropriate use of modeling blocks can be detected in conjunction with the assignment of variables and parameters in the declaration section of the model.

4. WinSAAM

The greatest disadvantages of WinSAAM are as follows:

1. Just as the non-existence of pre-fabricated modeling blocks is disconcerting in some software, the constraint to use pre-fabricated tools in the proper way subject to their unique set of rules can also be disconcerting to the novice investigator. On balance, an array of both pre-fabricated tools and non-assigned tools probably best serves the users' needs. WinSAAM is a little awkward in the latter department
2. There are relatively few statistical and other functions available in WinSAAM
3. WinSAAM supports only one optimizer, and gives relatively little output in regard to statistics associated with data fitting

XIV. Conclusion

The modeling software we have reviewed, ACSL, MLAB, Scientist, and WinSAAM, are all excellent programs maintained and developed by groups dedicated to advancing the application of modeling methodology to biology, medicine, pharmacology, and other areas. It is impossible for a single individual to be an expert in each of these systems and neither is it possible in a single chapter in a book to detail all the functionality, subtleties and nuances of software as sophisticated as we have attempted to present. We have tried to show the way problems are tackled with each of the packages, to highlight the respective strengths of the packages, and to draw attention to the apparent shortcomings of the software. But all of this has been from a specific perspective, namely one where a novice biomedical investigator is attempting to investigate a problem using the respective software.

The reader is reminded that learning to use software properly isn't a matter of just becoming familiar with it but also making a commitment (albeit possibly a lifetime commitment) to understand the entire structure, organization, and use of the software. We start by going through tutorials, then we move on to solving problems for which we know solutions (e.g., from a textbook) and trying to solve these problems in increasingly smart and efficient ways. Finally when we are absolutely sure we know what's going on, we gingerly apply the software to our real problems, or real data, but never settling for the first solution, nor the second, but only solutions which stand up to rigorous appraisal from many directions. Mastering software is similar to mastering any new skill or accomplishing a new technique. However, it is more than that, in that you will rapidly find that conquering software, such as the four programs we just discussed, will dramatically enhance not just your scientific productivity but your entire way of thinking about science *per se*.

REFERENCES

1. Baldwin, RL 1995. Modeling Ruminant Digestion and Metabolism. Chapman & Hall, London. 578 pp.
2. Bergman, R.N., and Bowden, C. R. 1981. The minimal model approach to quantification of factors controlling glucose disposal in man. In: Carbohydrate Metabolism. C. Cobelli and R.N. Bergman, Editors, Wiley, 269-296.
3. Boston et al., 1998 WinSAAM (Windows version of Simulation, Analysis, And Modeling).
4. Civilized Software,1996 Windows MLAB, Bethesda, MD.
5. Coburn, S., and D. Townsend, 1996 Advances in food and nutrition research. Vol. 40. Mathematical modeling in experimental nutrition (Publ. Academic Press).
6. Collins, J.C. Resources for getting started in modeling. J. Nutr. 122:695-700. 1992.
7. Dijkstra, J.: 1993 Mathematical modeling and integration of rumen fermentation processes (Publ. private).
8. Foster, D.M., Barrett, H.R., Beltz, W.F., Cobelli, C., Golde, H, Jacquez, J.A., Phair, R.D. 1994 SAAM II : Simulation, analysis, and modeling software. BMES Bulletin 18: 19-21.
9. France, J., and J. H. M. Thornley: 1984 Mathematical models in agriculture (Publ. Butterworth).
10. Forbes, J. M., and J. France: 1993Quantitative aspects of ruminant digestion (Publ. CAB International).
11. MicroMath, Inc, 1995 Scientist, Salt Lake City , Utah.
12. Mitchell & Gauthier, 1993 ACSL/PC: Automated Continuous Simulation Language for Windows, Concord, MA.
13. NONMEM Project Group, 1994 NONMEM, University of California.
14. Subramanian, K. N., and M. E. Wastney: 1995. Kinetic models of trace element and mineral metabolism during development (Publ. CRC Press).
15. Thornley, J. M. H., and I. R. Johnson: 1990 Plant and crop modeling: a mathematical approach to plant and crop physiology (Publ. Oxford Scientific Publications).

6

WinSAAM

This chapter describes the commands and conventions for using WinSAAM, the Windows version of SAAM (Simulation, Analysis And Modeling), a software package developed and supported by NIH over the last 30 years. It is intended as a reference on how the software works, i.e., how to run the software, the conventions used for entering models, and the commands for solving models, fitting data and analyzing the fits. Other chapters in this book describe how to actually apply the software for modeling. Specifically, Chapter 8 provides examples on modeling systems, and Chapter 10 discusses how to fit models to data.

The chapter begins with a primer that gives an overview of the program for new users, notes for users already familiar with WinSAAM, descriptions of commands for using the program, definitions of the 'building blocks' for describing models, and finally, a demonstration of a session using WinSAAM. Within this chapter the name SAAM and WinSAAM may be used interchangeably. Terms in CAPITAL LETTERS refer to items on a Window menu bar, italic terms enclosed in angled brackets (<*Text*>) or preceded by a right angled bracket (>*Text*) refer to text the user needs to type. WinSAAM is included on a CD distributed with this book and upgrades can be downloaded over the Internet (see CD for details on installation).

I. WinSAAM Primer

1.1 Introduction

WinSAAM (WS) is a general equation solving package that is available free over the Internet. It is based on SAAM which was originally written as a batch-style executable mainframe program. The input for SAAM was interpreted as a set of commands for the program to perform. CONSAM (CONversational version of SAAM) was ported to DOS and it allowed the user to interact with the program using commands. WinSAAM (the Windows version of SAAM), allows interaction through commands but also has great flexibility for exchanging information with other Windows programs. As an analogy, batch processing is like giving work to a typist, material is submitted and the results are obtained while the DOS version was like giving work to the typist but making suggestions for changes while the job is being performed. Windows is like typing the material oneself. There is complete freedom to change the tasks, and sequence of tasks, while the job is being performed.

1. 2 What does WinSAAM do?

Firstly, WinSAAM simulates, or solves equations (i.e., models). The equations can be entered directly or, for compartmental models which are represented by differential equations, the equations will be set up automatically by the program based on the parameters that are specified by the user. Secondly, WinSAAM fits a model to data by adjusting parameter values. This can be done manually by the user by setting values for the parameters, or by iteration by the program using a least squares procedure. Thirdly, the program enables the model-calculated solution to be compared to observed data by printing or plotting the results. Fourthly, the program provides statistical measures on how well the model fits the data. These include sums of squares, parameter errors, correlations etc.

To use WinSAAM, a user needs to know;

i) how the program is structured.

ii) how to enter a problem.

iii) how to solve the model and view the solutions.

1.3 Program structure and use

As shown in Fig 6.1, instructions in WinSAAM are communicated through a control center called the Terminal Window. Some instructions and commands are accessed from the menu bar at the top of the window. Other instructions need to be typed in. A problem can be started by using the FILE|NEW command on the menu, or the FILE|OPEN command to open a WinSAAM Text Input file (named *.saam). The Text Input file then becomes the Working Text Input file which can be viewed on the Terminal Window screen.

On starting the program (by opening the program icon, or by opening an existing file named with a *.saam extension), the user will enter an area called the Terminal Window (control center). A problem, consisting of the model parameters, equations and data, can be created by choosing 'EDIT|TEXT INPUT FILE' from the menu bar (See below). Once the problem has been entered, it needs to be saved using the 'FILE|SAVE INPUT FILE' command. A permanent copy can be saved to disk using the 'FILE|SAVE COPY AS' command.

Once the Text Input file is created, the user will then select COMMANDS|DECK from the command menu (or type *>DECK*). This translates the problem into a format that can be solved by WinSAAM. The model is solved by *>SOLVe*. Output can be viewed by using *>PLOT* or by viewing a spreadsheet via OUTPUT|SPREADSHEET. Parameters can be listed and the values changed, as described below, without editing the Text Input file. This allows the model to be re-solved and a new solution to be printed (*>PRIN*t) or plotted (*>PLOT*). To save any changes in the parameter values to the Text Input file, the command *>UPDATE* needs to be typed. Alternatively, upon exiting WinSAAM, a dialog box appears which asks the user "Do you wish to save Text Input File?"

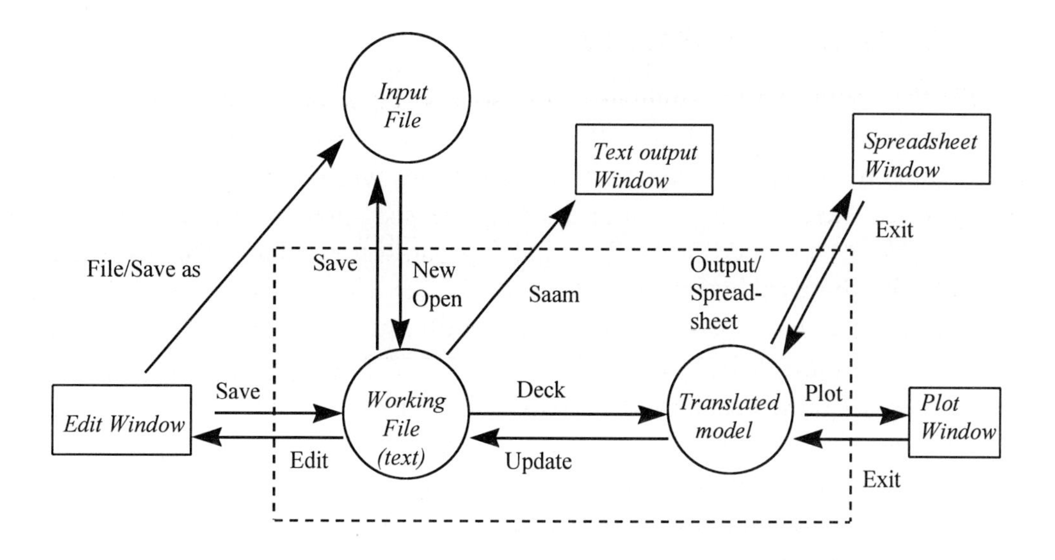

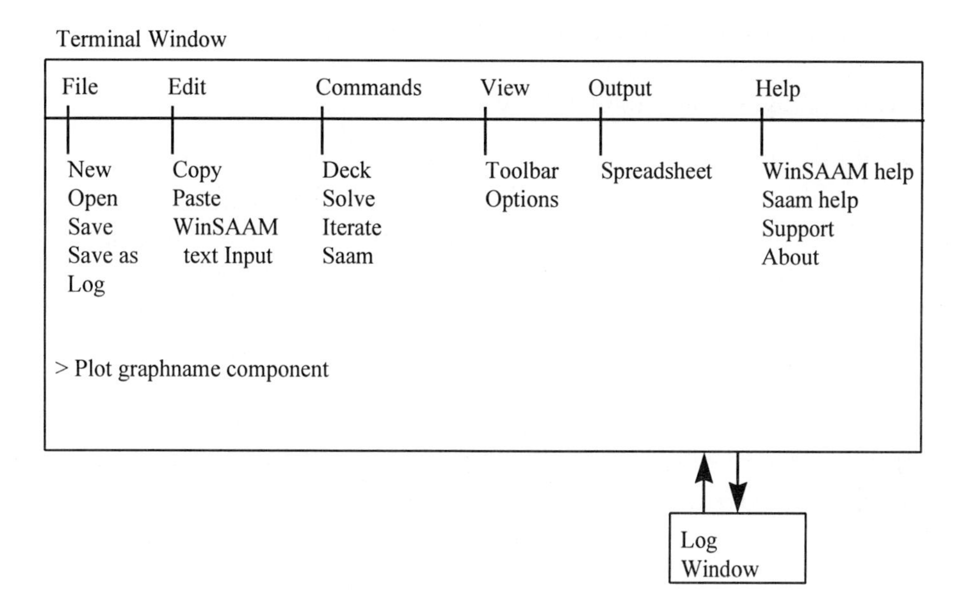

Fig 6.1 Structure of WinSAAM. The squares enclose windows. The Terminal window, or control center includes all the processes enclosed by the dotted square. Some commands can be accessed within the Terminal window via the drop-down menus. Other commands, such as PLOT are typed in, as this command requires additional information on what to plot (component number) and template (graphname).

1.4 How to enter a problem

Problems are entered in a Text Input file ('text' because it is not formatted but is ASCII or generic format). The Edit window is opened by choosing EDIT|WinSAAM TEXT INPUT from the menu bar (Fig 6.2).

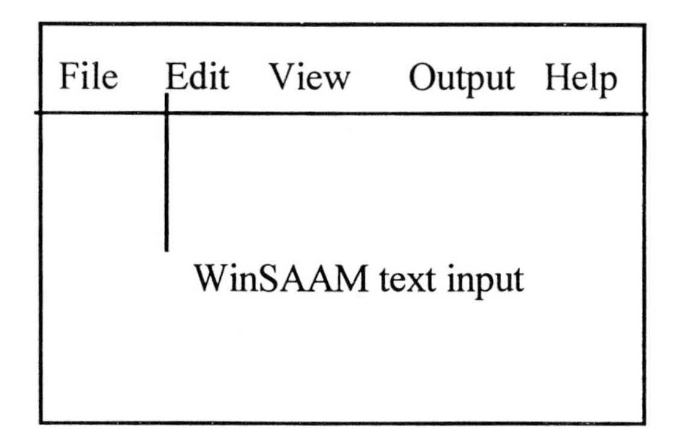

Fig 6.2 Terminal window showing how to access the Edit window.

The input file consists of the model (parameters and/or equations) and data (Fig 6.3). Information is entered under headings: H PAR for parameters and H DAT for data. The first line begins with 'A SAAM31'. This indicates to the program that it is a new problem. It is suggested that the name of the file and the date be recorded in the first line.

Information is placed within fields on each line. A line is considered to be 80 spaces (or columns) wide. Some information is entered in column 1. For example, the first line contains A in column 1, and comments are defined by a 'C' in column 1. Fields for other information begin in columns 4, 13, 27, 42, 56, 60, and 62. The TABS are set at these values by default to assist with entering information. For example, parameter names begin in column 4 (1 TAB), and the value in column 13 (+1 tab). A parameter can be made adjustable by entering MIN and MAX values for the parameter in columns 27 (+1 tab) and 42 (+1 tab), respectively.

A Text Input file illustrating the use of headings and fields is shown for a one-compartment model (Fig 6.4). The parameters used in the example are: IC(I), initial condition of compartment I and L(I,J), fractional transfers (or rate constants) into compartment I from compartment J (0 refers to the outside).

```
A SAAM31              <name.saam>          <date>
H PAR
     <model parameters>
H DAT
    <link between model and data, data>
```

Fig 6.3. A skeleton Text Input File showing the headings for parameters (H PAR) and data (H DAT).

```
A SAAM31         Ch6Eg01.saam              1/13/98
H PAR
   IC(1)      100.
   L(0,1)     0.01
H DAT
C          time (min)       Data (% of dose)
101F(1)
                0.
                1.             98
                5.             95
                10.            90
                15.            85
                30             75
                60             55
                90             40
                120            30
C ****** end of data
```

Fig 6.4 Example of a lexical one-compartment model in WinSAAM.

Equations are entered under the H DAT heading as G-functions (Fig 6.5). In this problem, the equation is specified under H DAT. Two general parameters P(1) and P(2) are used in the equation and are listed under H PAR. No observed data are entered in this problem, but values can be calculated (generated) at specified times (i.e., in the example, at time 0, 5, 10, 15, and 20). Calculated values can also be generated automatically as shown in the example by stating a starting time (here, time zero in column 13) and then specifying on another line a '2' in column 1, the time interval in column 13 (5 units in the example) and the number of solutions to be generated in column 42 (4 in the example), (Fig 6.5).

```
A SAAM31         CH6EG02.saam                  1/13/98
H PAR
   P(1)       1.
   P(2)       2.
H DAT
XG(1)=P(1)+P(2)*EXP(-T)
C Values will be calculated for G(1) every 5 time units
101G(1)
              0.
              5.
              10.
              15.
              20.
C Values can be generated for G(1) by placing a '2' in column 1
C and generating values every 5 time units (column 13),
C for 4 times (column 42), I.e., up to 20 time units
102G(1)
              0.
2             5.                            4
C ****** end ******************************************
```

Fig 6.5 Example of Text Input file showing a model that specifies an equation model.

It is necessary to link observed data to the solution of a compartment or equation in the model. To understand how this is done in WinSAAM, we need to define the '3 C's of SAAM'. The first 'C' is for compartment. These are the circles on the model graphic and in SAAM terminology are represented by F(I). The second 'C' refers to component. A component is the address of a solution, and is referred to as Q(I), where Q(I) is the address for the calculated and observed values. Note that QC(I) refers to calculated values and QO(I) to observed values. The third 'C' is for category. Category refers to what is stored at a solution address (e.g., it could be F(I), the solution to a compartment, or G(I), the solution to a general function or equation). A line under the H DAT heading with a '1' in column 1 indicates to the program that this line contains information linking a solution, Q(I), to a compartment, F(I) or equation, G(I);

H DAT
1<component no.><category>

e.g.,

101F(1) - solution of compartment 1 is stored at solution address 1, or in Q(1) .
102F(1) - solution of compartment 1 is stored in Q(2).
103G(1) - solution of equation G(1) is stored in Q(3).

For experimental data the independent variable (usually time) is entered in column 13 (2 TABS), and the observed values 1 TAB over (Fig 6.5). A second independent variable (TH, THETA) can be entered in Field 6. Data can also be entered by copying from a spreadsheet and, in the Edit Window, choosing EDIT|PASTE SPECIAL|DATA FROM SPREADSHEET. Once the model parameters and data have been entered, the Text Input file is saved (FILE|SAVE). A copy can be saved to disk by choosing FILE|SAVE COPY AS.

1.5 How to solve a problem and view the solution

Once a working file has been opened (or created in the editor) it is translated into a form for solving using the >DECK command, and then solved using the >SOLVE Command. Solutions can be viewed with the command >*PLOT*, e.g., <*PLOT Q*(*1*) will plot the solution of data stored in component 1 (Fig 6.6). The plot can be customized and saved by specifying a template name after the plot command, e.g.,

>*PLOT FIG1 Q*(*1*)

The template (in this case called FIG1) will be saved in the same directory as the Text Input file. To access options for customizing the plot, select the OPTIONS menu from the plot window, or right click the mouse while placed over

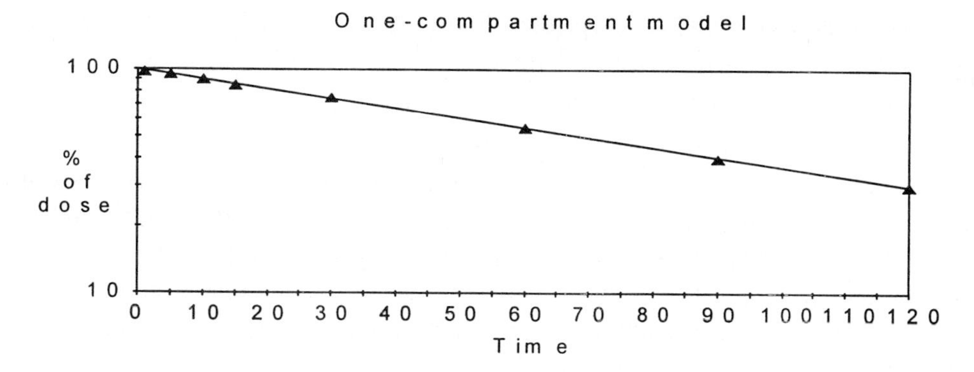

Fig 6.6 Visualization showing a plot of the observed data (symbols) and model calculated values (line) for a one compartment model. The Text input file for generating this plot is shown in Fig 6.4

the plot window. Options include labeling axes, choosing fonts and colors, as well as making the graphs log or linear (see release notes below for more details). Plots can be EDIT|COPY and EDIT|PASTE into other applications.

From the Terminal Window, the OUTPUT|SPREADSHEET menu item, parameter values, and data (observed(QO) and calculated(QC)) can be accessed by clicking the appropriate tab at the bottom of the spreadsheet window (Fig 6.7)

DATA SPREADSHEET

Component	Category	TC	Time	QC	QO
1	F(1)	0	0	100	0
1	F(1)	0	1	99	98
1	F(1)	0	5	95	95
1	F(1)	0	10	90	90
1	F(1)	0	15	86	85
1	F(1)	0	30	74	75
1	F(1)	0	60	55	55
1	F(1)	0	90	41	40
1	F(1)	0	120	30	30

PARAMETER SPREADSHEET

Category	Form	CurrentValue
L(0,1)	F	0.01

Fig 6.7 Data spreadsheet and parameter spreadsheet with output from fitting a one-compartment model. The component number is listed in the first column, the category in the second column, the Time Change (TC) in the third, the time values, the calculated values (QC) at each time point, and the observed values (QO) at each time point. On the Parameter spreadsheet, the parameter is listed followed by the form (F is fixed, adjustable parameters would be tagged as A, and dependent parameters as D). The current value is then listed. For adjustable parameters, the standard deviation (SD) and fractional standard deviation (FSD, as value divided by the SD) would also be listed.

Batch-style solutions can be printed to another window by typing the >*SAAM* command. The Text Input file will be solved and the solution will be printed in the Text Output window. The Text Output window will show the Text Input file and the solution (Fig 6.8). The Text Output window shows the name of the problem , the date it was run, a copy of the text input file (lines 1-20) and, under Final Values, the value for the fixed parameter L(0,1), the initial condition (initial component value for compartment 1 is 100). The solution then lists in sequential columns; data numbers (1-9), component number (C), category number, time (T), values for a second independent variable (THETA), calculated values (QC), observed values (QO), the difference between the observed and calculated values (QO-QC), the ratio of the observed and calculated values (QC/QO), and the weight assigned to each datum (uniform weighting is assumed if no weight is specified). The solution then lists the sum of squares of the differences between the observed and calculated values.

1.6 Guidelines for Upgraders from CONSAM/SAAM

WinSAAM is a Windows version of SAAM31 that supports almost all the features of this version of the program and at the same time, capitalizes on features of Windows to bring added functionality and efficiency to a modeling session. For example, WinSAAM incorporates a new graphics subsystem that increases flexibility in configuring plots and allows plots to be exported directly from the plot window to a word processor or graphics editor. Similarly, a new spreadsheet output facility allows results to be exported from WinSAAM directly to EXCEL or other spreadsheet systems, or to a statistical tool that supports spreadsheet-style input for further data processing and analysis of results.

The most exciting change is the addition of a new tool for entry of data and models into WinSAAM (the WS Text Input Editor). Previously, models were entered into SAAM31 via the system 'editor' (e.g., Notepad) or the CONSAM line `editor’. Now there are four additional methods for data and model entry; WinSAAM’s Text Input Editor now accepts tab-delimited text from either the clipboard (CUT|PASTE) or files and inserts it into the appropriate fields of the Text Input File. Tabulation has also been implemented for routine model and data entry.

1.7.1 Using WinSAAM

Although WinSAAM functions in the same fashion as CONSAM, some aspects of WinSAAM are new and require further description. Table 6.1 lists definitions of some terms:

```
A SAAM31        CH6EG01.SAAM          1/13/98
                                            PROBLEM DECK
                                            ------------
A SAAM31        CH6EG01.SAAM          1/13/98
                                   SAAM31.0 / 11-MAY-95 - 5000 DATA POINTS 75 COMPARTMENTS/ - 386
***ALL WEIGHTS=1.*2*
 INITIAL VALUES
   (IF A COMP. IS NOT LISTED COMP.TYPE=REG., ALL OTHER VALUES ARE ZERO)
I     F(I,0)     COMP.TYPE    M(I)          U(I)          UF(I)   DF(I)   FF(I)
 1  1.00000E+02    REG      0.00000E+00 0.00000E+00
CARD NO.               CARD IMAGES
   1             A SAAM31         CH6EG01.SAAM          1/13/98
   2    ADDED    2
   3    ADDED    3
   4    ADDED    4  2        1
   5             H PAR
   6                IC(1)     100.
   7                L(0,1)     1.000000E-02
   8             H DAT
   9             C           TIME (MIN)      DATA (% OF DOSE)
  10             101F(1)
  11                          0.
  12                          1.            98
  13                          5.            95
  14                          10.           90
  15                          15.           85
  16                          30            75
  17                          60            55
  18                          90            40
  19                          120           30
  20             C ****** END OF DATA



                                            FINAL VALUES
                                            ------------
 MODEL CODE=  2  ESTIMATE OF SIG FROM READ-IN DATA=    .0000000E+00
   TIME=22:08:26
SOLUTION TIME=      .720 SEC.


PARAMETER VALUES
ADJUSTABLE PARAMETERS                 DEPENDENT PARAMETERS                  FIXED PARAMETERS
                                                                       L ( 0, 1)=  1.0000000E-02
INITIAL COMPONENT VALUES
  1=  1.0000E+02
  D  C ---CAT--        T          THETA        QC              QO          QO-QC      QC/QO         WT
  1  1       0    0.0000E+00  0.0000E+00  1.00000E+02    0.00000E+00  -1.000E+02        *      1.00E+00
  2  1       0    1.0000E+00  0.0000E+00  9.90050E+01    9.80000E+01  -1.005E+00    1.010     1.00E+00
  3  1       0    5.0000E+00  0.0000E+00  9.51229E+01    9.50000E+01  -1.229E-01    1.001     1.00E+00
  4  1       0    1.0000E+01  0.0000E+00  9.04837E+01    9.00000E+01  -4.837E-01    1.005     1.00E+00
  5  1       0    1.5000E+01  0.0000E+00  8.60708E+01    8.50000E+01  -1.071E+00    1.013     1.00E+00
  6  1       0    3.0000E+01  0.0000E+00  7.40818E+01    7.50000E+01   9.182E-01     .988     1.00E+00
  7  1       0    6.0000E+01  0.0000E+00  5.48812E+01    5.50000E+01   1.188E-01     .998     1.00E+00
  8  1       0    9.0000E+01  0.0000E+00  4.06570E+01    4.00000E+01  -6.570E-01    1.016     1.00E+00
  9  1       0    1.2000E+02  0.0000E+00  3.01194E+01    3.00000E+01  -1.194E-01    1.004     1.00E+00
 C    DATA     SS          NORM.SS      MEAN SD       MEAN FSD
 1     9   1.0004E+04   1.1115E+03  3.3340E+01   1.1821E-01
SUM SQUARES AFTER     0 ITERATIONS IS   1.00037100E+04
      COMP  1=    1.0003710E+04
SIG AFTER     0 ITERATIONS IS   1.11152300E+03
                                            0.00000E+00    0.00000E+00    0.000E+00        *      0.00E+00
***NO ADJUSTABLE PARAMETERS**67*
                                   SAAM31.0 / 11-MAY-95 - 5000 DATA POINTS 75 COMPARTMENTS/ - 386
```

Fig 6.8 The Text output window generated by typing the >SAAM command for the one-compartment model shown in Fig 6.4. See text for a description of this output.

Table 6.1 Table of WinSAAM definitions.

Term	Definition
Associated Files	Files which travel with the Input Text File. Currently these include the Plot Files.
Text Output Window	Window displaying output produced by the 'SAAM' command.
Home Directory	Directory where WinSAAM has been installed
Log File Window	Window displaying the current contents of the open Log File.
Options Window	Window presenting the options that can be changed in WinSAAM (VIEW|OPTIONS).
Output Spreadsheet	Spreadsheet containing the output as of WinSAAM's last solution.
Plot Options Window	Window displaying the options available for the presentation of a plot.
Plot Template File	A file containing the template for a WinSAAM plot. Default file extension is "schrt"
Plot Window	Window displaying a named WinSAAM plot. Each named plot has a corresponding Plot Template File recalling the settings for that name.
Terminal Window	Main window for the WinSAAM program.
Text Input File	Text Input File for WinSAAM (formerly known as a SAAM deck). Default file extension is ".saam"
Work Directory	Directory where WinSAAM runs
Working File	A temporary file called "fort.3" located in the Work Directory on which all WinSAAM commands operate.

WinSAAM works exclusively from a user-defined Work Directory, which by default is called SaamWork under your Home Directory (ie. C:\ProgramFiles\Winsaam\SaamWork). All of WinSAAM's actions, excluding file manipulations, take place in this Work Directory. WinSAAM operates on the Working File (fort.3) and its Associated Files located here. The WinSAAM Work Directory can be changed through the Options Window. You can select it from the Terminal Window under VIEW|OPTIONS|DIRECTORIES. A change in this parameter will not take effect until WinSAAM has been restarted.

The FILE|NEW menu command opens a new Working File and causes a time-stamp notification to be displayed immediately in the Terminal Window. When an existing Input Text File is opened, its contents are copied into the Working File and the associated files located in the Input Text File's directory are copied into the Work Directory. There are three mechanisms for opening an existing Input

Text File with WinSAAM; I) by using the Windows95 or Windows NT Explorer and opening any file with the ".saam" extension by double clicking, ii) navigate to an Input Text File from within WinSAAM and use the FILE|OPEN menu service, iii) select from a list of the most recently accessed Input Text Files in the FILE menu.

When the Working File is saved, it is copied back to its original Text Input File. If the Working File is created *de novo* in a current WinSAAM session, a permanent copy of this Working File can be retained by using the SAVE AS... facility and designating both the name and the directory of the Input Text File. The associated files will be copied to the Input Text File's directory.

The graphics of WinSAAM are somewhat more sophisticated than those of past versions of SAAM. Although the *PLOT* command has not changed, plot renditions can be greatly altered by capitalizing on the following features of the graphics subsystem:

The chart control can be accessed when the chart window is in the foreground (i.e. a plot is currently displayed) in either of two ways; select VIEW|PLOT|OPTIONS from the chart menu bar, or simply click the right mouse button. In a WinSAAM session previous commands can be retrieved by pressing the 'alt' and '-' keys. Note that WinSAAM allows you to set the number of stored commands using the OPTIONS Window.

All text windows and text boxes support selection and copying in the usual Windows fashion. Select by dragging across text of interest, and copy that text to the clipboard using the 'cntrl-c' key combination or EDIT|COPY. The Batch Output Window (WS Batch Output) and the Text Input Window (WS Text Input Window) additionally, support an array of explicit text management tools listed in the EDIT menu.

The SAAM command works slightly differently than in Consam. Now, at the completion of the command, a text box (WS Batch Output) is filled with SAAM's batch output. This text can be copied from the text box for further editing.

A sequence of spreadsheets are being developed to support result display as well as conveyance of WinSAAM results to other environments. The spreadsheets are accessed via the OUTPUT|SPREADSHEET menu sequence and currently a 'data', and a 'parameter' spreadsheet are provided. The 'data' spreadsheet displays all the common computed results from WinSAAM and the 'parameters' spreadsheet displays all aspects of the model parameters currently available. Other spreadsheets under development include a 'partials' spreadsheet and a 'covariance' spreadsheet. There are two options for transporting computed results from the WinSAAM data spreadsheet (for example) to another graphic environment:

i) in the spreadsheet 'file' menu use the 'save as' service to select a portable spreadsheet output format (e.g. Excel 5.0),

ii) use the EDIT|COPY menu sequence of the designer spreadsheet (available via the EDIT menu) to place selected cells on the clipboard and deposit these directly into the appropriate cells of the target software.

TABLE 6.2 Plotting Features of WinSAAM

Special Feature	Tab an Item selection for access to special feature
Accessing Plot Options Window	Click right button with cursor over plot
Scaling a plot	Hold 'Shift' key and drag
Zooming a plot	Hold 'Ctrl' key and drag
Restoring a plot	Press 'r' key
Toggle between log and linear display	Axes-Tab | Select "X or Y" | General-Tab | IsLogarithmic = "Check"
Joining data points	ChartStyles-Tab | Select "ChartGroup1-Style n" | LineStyles-InnerTab | Pattern = "Dotted..."
Labeling axes	Axes-Tab | Select "X or Y" | Title-InnerTab | Text = "Your Axis Label"
Labeling plots	Titles-Tab | Select "Header or Footer" | Label-InnerTab | Text = "Your Plot Label"
Adding grids	Axes-Tab | Select "X or Y" | Grid-InnerTab | Spacing = " 1 " (See GridStyle-InnerTab for more control)
Adding borders to plot	PlotArea-Tab | General-InnerTab | IsBoxed = "Check"

| Denotes followed by Tab, " Denotes tab section, InnerTab Denotes tab section under another tab section, = " " Denotes user supplied value

The test scroll space in the Terminal Window, can be increased using the entry in the 'Characters Displayed' box via the VIEW|OPTIONS|TERMINAL menu sequence.

1.7.2 Fonts

Because of the diversity of purposes amongst the four text windows, independent font control is available for each. Accordingly we have provided a font management service which enables you to set the fonts for the Terminal Window, the Text Input Window, the Batch Output Window and the Log File Window. The font management service for any of these windows can be accessed from the Options Window. To display this window, you can use the menu sequence VIEW|OPTIONS, or ctrl-L. You should note that all the option values persist until explicitly changed in the Options Window. We recommend that you experiment with settings here.

1.7.3 Logging WinSAAM sessions

Once a strategy for achieving a fit to your data is found, you are immediately confronted by two serious problems; how to re-invoke that strategy for subsequent data sets, and how to recall the modeling approaches that did not lead you to a satisfactory representation of the data. The former is obviously a serious problem, but the need to re-access the futile strategies may not be so obvious. The answer to this rests on the fact that everything that we do that does not work tells us in an incremental way about the system we are exploring, certainly with regard to the data we have on this system anyway. In essence we learn least about a system if our efforts to model it are immediately successful ... we learn most about a system through systematically exploring an array of feasible and scientifically justifiable hypotheses regarding our system until we reach one that cannot be rejected.

To assist the user we have created a logging facility. From the File menu in the Terminal Window you can trigger logging by selecting the 'LOG|OPEN' service. This will then give you the chance to open any existing log file for appending or overwriting, or to create a new log file. Once logging gets under way you can 'VIEW' the log file by selecting FILE|LOG|VIEW from the terminal window. You will notice that if a session logging is successfully underway the 'RECORDING' item is checked in the log sub menu. If you are viewing the log file you will see that terminal window entries are more or less duplicated in the Log File Window. To stop viewing the logging you can either simply close the Log File Window. In order to discontinue logging select close from the log sub menu of the terminal window file menu.

1.7.4 Plot Templates

In WinSAAM the appearance of plots is controlled using a 'template' type of procedure. However, unlike CONSAM, the mechanism whereby these are created is now located exclusively in the chart control tool and is available to you from a Plot Window. We remind you that you can access the chart control tool using either the VIEW|PLOT Options menu sequence or by simply pressing the right button once the plot is visible. By default all non-null plot templates are saved under the name given to them in the ‘plot’ command. For example if you enter a ‘*PLOT*’ command

PLOT CHRT Q(SET1, 1, SET2)

then the Plot Template File chrt.schrt will be saved for you in the directory to which you save the permanent copy of your WinSAAM Input Text File.

To recover a plot template simply navigate (or otherwise find your way) to the directory where it (and the allied WinSAAM Input Text File) is located and select it from the drop down list box in the Terminal Window. Note that selecting a template actually builds the command and template section of the full *PLOT* command and all you need to do to complete the plot command is enter the plot data path (e.g. Q, W, P etc.), the data segment (1, set1, etc.) and, if necessary, the record list. We have not implemented a method for deleting template files from within WinSAAM so in the experimental stages you are reminded to explore this feature systematically. Note, if the caps lock is on, you cannot restore a plot with the r key.

1.7.5 Communicating with the program developers

Under the HELP menu in the terminal window you will see the item 'feedback'. This facility has been specifically established in WinSAAM for users to e-mail messages to WinSAAM developers.

1.7.6 Summary of WinSAAM for upgraders:

There are six objects you'll be dealing with when you use WinSAAM; four text windows, a spreadsheet window and a chart (or plot) window. The four text windows are the terminal window, the text input window, the batch output window and the log file window. The terminal window is, more or less what was formerly the Consam window. The text input window is an input tool facilitating the input of tab-delimited clipboard based, or tab-delimited file based data and models into WinSAAM. Tabulation of text items in the text input window has been preset for you to ensure appropriate alignment with the SAAM data and parameter fields. The batch output window will display the ordinary SAAM output achieved formerly from the Consam saam command. The log file window will display all your interaction with WinSAAM for the duration of the session logging.

The spreadsheet window contains two workbooks, the data workbook and the parameters workbook. Information accessible in either of these workbooks represents all the information that WinSAAM has access to in these domains in regard to your state of processing. Advanced spreadsheet manipulation of either the parameter information or the data information is available via the 'spreadsheet designer window'. Here you will be able to perform all the spreadsheet-like calculations and manipulations on the respective information set. In particular, selecting and copying all or subsets of the cells from the workbooks will enable you to export WinSAAM results to other processing media. To access the 'designer' window simply press the edit menu item on the output spreadsheet window The chart window comes up whenever you plot a WinSAAM data set (or subset). If you issue your plot command using the plot name then the plot template you subsequently configure for your plot is subsequently saved for re-use. To access the chart tool enabling you to modify your plot you either click the right button or select thc VIEW|OPTIONS menu service. The control emerges and you are then able to select or toggle features as needed. Once satisfactory plots are obtained they can be printed (FILE|PRINT) or copied to the clipboard (EDIT|COPY) as needed.

II. WinSAAM Commands

Table 6.3 lists the commands, their basic features and an example of and their use in WinSAAM. Other features can be invoked by using the on-line Help facility. A summary of the use of basic commands to solve a model in WinSAAM is;

- To start program: Open the WinSAAM icon
- To create or view an input file: Choose 'Edit' on the menu bar
    - Enter parameters under H PAR: Rate constants, L(I,J), general parameters, P(I), linear parameters, K(I) and initial conditions, IC(I)
    - Enter data under H DAT: Independent variable (time), observed data
    - Save Text Input file
- To transfer the problem for solving: >DECK
- To solve: >SOLV
- To view solutions:
    - To list parameters: >L(I,J)
    - To list solutions e.g., of component 1 >Prin Q(1)
    - To plot solution of e.g., component 1: >Plot Q(1)
        - To plot using a template: >Plot <name> Q(1)
- To change a parameter value: >L(0,1)=.5
- To update latest parameter values: >UPDA
- To stop: Choose FILE|SAVE from the command menu, Then FILE|EXIT

A more detailed illustration of the use of the commands is given later in this chapter, in Section VI.

TABLE 6.3 Summary of WinSAAM commands

CLASS	COMMAND	ILLUSTRATIONS	PURPOSE
Printing	Print	Print QC(1,3,7) Print S(4)L(1,2),P(3)	Prints values pertaining to specific data segments
Plotting	Plot	Plot Q(1,2,3,4,5) Plot Data QC(3,4)1,2,3	Plots values pertaining to specific data segments
Solution Storage Commands:	New	New	Open a new solution storage device
	Keep	Keep	Save a solution on the solution storage device
	Reco	Reco details	Display details of saved solutions
	Coun	Coun 2	Resets the number of saved solutions
	Old	Old	Open an existing solution storage device
	Reli	Reli 2	Relist aspects of saved solution 2
	Rest	Rest 4	Restore saved solution 4
	Move	Move 2 to 1	Move saved solution 2 to become saved solution 1
Parameter Manipulation Commands:	Par(I,J)	P(4) L(2,1) K(3)	Display the current value of a fixed or adjustable parameter
	Par(I,J)=x	P(4)=3 L(2,1)=7 K(8)=10.2	Reset the current value of a fixed or adjustable parameter
	Min/Max	Min L(3,4)=14 Max P(3)=27	Reset the lower/upper bound for an adjustable parameter
	Adju	Adju Adju P	Display details of the adjustable parameters, or a subset of adjustable parameters
	Excl/Incl	Excl L(5,9) K(27) DT(3) Incl All	Exclude/Include listed groups of parameters from/in the adjustable set. Parameters included in the adjustable set will have their values automatically altered in conjunction with iterations
History Commands:	Hist	Hist	List the 20 most recent commands
	Redo	Redo 4	Redo command 4 from the past command list
	Step	Step	Following a 'Redo' will cause the execution of next commands to be sequentially executed following each 'Step'

TABLE 6.3 Summary of WinSAAM commands cont.

CLASS	COMMAND	ILLUSTRATIONS	PURPOSE
Matrix Commands	Calc LI	Calc Li	Calculates the inverse of the L matrix (residence time) and stores it in LI
	Calc Eig	Calc Eig Li	Calculates the eigenvalues of the indicated matrix, here Li
	Egv	Egv(I)	Display the current eigenvalues
	Eig	Eig(I,J)	Display the current eigenvectors
	FCR	FCR(I,J) FCR(I,I)	Display the FCR values of the current L matrix
	TR	TR(I,J)	Display a transfer matrix for the current L matrix
Local editing commands	Upda	Upda	Copy memory attributes of parameters into the appropriate fields in the edit buffer*
	Resc	Resc	Retrieve the working file and overwrite recent edits
	Reta	Reta	Copy the edit buffer text to the working file destroying its original contents
	Fix	Fix P Fix L(1,2)	Remove limit field entries from classes of parameters in their specification lines in the edit buffer
	Free	Free all Free L Free K(37) .3 10	Include limit field entries for classes of parameters in their specification lines in the edit buffer
	P	P P end P 4 10	Display blocks of lines of the input file in the edit buffer
	Dele	Dele Dele 5 19	Delete blocks of lines of the input file in the edit buffer
	I	I I 3	Insert lines into the input file in the edit buffer
	Dupl/Relo	Dupl 4 7 12 Relo 8 10 17	Duplicate/Relocate blocks of lines in input file to the target line
	A I/R/D/C	A I A R A D A C	Inline edit modes allowing insertion, replacement, deletion, and alteration C) of text in lines in the input file

TABLE 6.3 Summary of WinSAAM commands cont.

CLASS	COMMAND	ILLUSTRATIONS	PURPOSE
Other Commands:	Deck	Deck	Compile the input problem
	Solve	Solve	Solve the system equations
	Depe	Depe	Solve the parameter dependencies
	Part	Part	Evaluate the partials
	Iter	Iter	Iteratively adjust the parameters
	Unwe	Unwe 6 17 27	Unweight a block of data
	Rewe	Rewe 6 17 27	Reweight a block of data
	Swit	Swit stor on Swit part off	Swit on, or switch off a WinSAAM switchable feature
	Proj	Proj	Use a set of approximate partials to project new values for adjustable parameters
	Stat	Stat	Display the status of the WinSAAM switches
	Eval	Eval	Access the WinSAAM scalar (HP style) calculator and stay in this mode until an Exit is entered
	Saam	Saam	Invoke a full batch style run of the current input file
	Rand	Rand Q(3) Rand/FSD=.1 Q(5)	Generate a set of observations with defined error structure
	Info	Info	Display the relative information associated with each calculated response
	Cons	Cons	Display extent to which current problem is using available program spaces
	File	File	Access WinSAAM's file switcher to assert control over inputs and outputs
	Mode	Mode=23 Mode=1	Set the solution mode to one other than selected by WinSAAM

*The Editor buffer is a temporary area where text resides after editing. It is copied to the Working File when the DECK command is used.

III. WinSAAM Terminology

Models are built using building blocks or modeling constructs. These can be divided into parameters (L, IC, K, S, P, DT, DN, M, and U), operational units (QO-, QL- and QF-operations, UF, F, FF, QO-function, TC, SA, INF, T, and TH), data elements (parameters, G, and UF), statistical weights, and function dependency (Table 6.4). Each construct serves a specific purpose in defining the model. In addition to the building blocks, the terms compartments, components, and categories are often used. Compartments have responses F(J) and are derived from the system of differential equations defining the model. Components are domains that store data. Categories are the model solutions which are to be related to the actual data. To associate categories with a specific data sequence we encapsulate them into the same component as the data sequence.

Each item will be described below in terms of its name, purpose, special features, including limitations, and the fields used to describe them. Each name has an index, or subscript, and this serves two purposes; to establish an association of modeling blocks with one another, and to distinguish between modeling blocks of the same type. Special features relate to the number of blocks of a particular type which may be used and any limitations of the building block such as the way its values are assigned and estimated are defined in the Text input file. In general these properties cannot be changed without editing this definition line. However, the values of the parameters, and their adjustable boundaries can be changed interactively

Dependent parameters can be specified using equations. When a parameter is dependent on adjustable parameters the value of the dependent is updated with each solution to ensure that it reflects the current value of the adjustable parameter. When a parameter is dependent on a constant, its value is fixed at that constant value, e.g.,

K(3)=29 (See below for details).

Information is entered into fields, where a field is a range of character spaces, and corresponds to the number of TABS (when set to 4, 13, 27,42,56, 60,and 62 character spaces across a line) (Fig 6.5). The first line of a WinSAAM problem is;

A SAAM31

This line can contain other information, such as the file name and date. The second, third and fourth lines are optional. The second line can contain the number of iterations, if the problem is to be run using the >Saam command. The third is optional and can contain a setting for the internally allotted time for solution of differential equations, and the fourth line is also optional for output information (see WinSAAM help for more information).

3.1 Fractional turnover rate, L(I,J)

Purpose: defines model connectivity, existence of compartments, transport mechanism between any two compartments. If a model contains a parameter, L(1,2), where L is the fractional transport rate (also called rate constant, or

Table 6.4 WinSAAM Terminology: Parameters, Operational Units, Data Elements and Statistical Weighting Schemes

	NAME	USE	EXAMPLE	DESCRIPTION
PARAMETERS				
L	Fractional transfer rate	L(I,J)	L(1,2)	Transfer into compartment I from compartment J
IC	Initial condition	IC(I)	IC(10)	Initial condition in compartment I
K	Linear parameter	K(I)	K(6)	Parameter associated with component I. If there is no component I, it is treated as a general use parameter.
S	Summer	S(I,J)	S(3,2)	Sums compartment J into compartment I
P	General parameter	P(I)	P(10)	Parameter for use in equation
DT	Delay time	DT(J)	DT(8)	Delay time of delay J
DN	Delay number	DN(J)	DN(4)	Number of cells in delay J
M	Mass	M(J)	M(4)	Mass of compartment J
U	Input	U(J)	U(5)	Tracee input into compartment J
OPERATIONAL UNITS				
QO-operation	QO-operation	F(J)QO	110QO	Resets F(J) to specified value
QL-operation	QL-operation	F(J)QL	102QL	Resets FF(I) to QO(I)
QF-operation	QF-operation	F(J)QF	111QF	Resets F(J) to FF(J)
UF	Input function	UF(I)=f(x)	UF(8)=P(1)	Input into a compartment
F	Compartment response	F(I)	F(9)	Solution of a differential equation for a compartment
FF	Forcing function	FF(I)	FF(8)	Forces compartment in a system to see analytic solution of compartment I.

Table 6.4 cont.

QO-function	Data generation function	QO(I) =f(x)	QO(8) =F(25)	Generates 'observed' values in component 8
TC	Time-block	TC(I)	TC(3)	Separates data and parameters into blocks that correspond to different experimental conditions
SA	Specific activity	Category field entry	105SA	Divides compartment response by mass
INF	Infinity	INF	INF	Enter in time field to calculate value at T=1E35
T	Time variable	T	P(1)*T	Use in equations, reset time blocks
TH	Theta	TH	P(1)/TH	Second independent variable
DATA ELEMENTS				
(any parameter, G(I), UF(I))	(as above)			Values and their error can be entered, and fitted as data
STATISTICAL WEIGHTS				
FSD	Fractional standard deviation	FSD=x	FSD=0.2	
SD	Standard deviation	SD	SD=x (x)	SD value x can be assigned to data in field 4.
RQO		RQO		
WT	Assigned weight	WT=x	WT=1	
FUNCTION-DEPENDENCY			x=f(y)	Modifies a parameter of function x, by a second function, y

Table 6.5 Entry of Information on a Model into a WinSAAM Text Input File

Field	*Field 0*	*Field 1*	*Field 2*	*Field 3*	*Field 4*	*Field 5*	*Field 6*	*Field 7*
		Name	Value*	Min. Value	Max Value	Fn-Dep		Error
TABS	0	1	2	3	4	5	6	7
Character spaces	0	4	13	27	42	56	60	62
First Line (xx)	A SAAM31							
Second (x)	2							
Third (x)	3							
Fourth (x)	4							
Comments	C							
Parameters	H PAR							
L		xx	xx	x	x	x	x	x
IC		xx	xx					
K		xx	xx	x	x			x
S		xx	xx	x	x			x
P		xx	xx	x	x			x
DT		xx	xx	x	x	x		
DN		xx	xx					
M		xx	xx					
U		xx	xx					
UF		xx	xx				x	
FF		xx	xx					
G		xx	xx					
Data	H DAT							
Control line	1nn	x						
Data Generation	2		xx		xx			
Experimental data			xx	x				
Steady state	H STE							
U or M		xx	xx	x	x			x
M or U		xx	xx	x	x			x
R(I,J)								
Time Bloc m								
Parameter	H PCC			TC(m)				
Data	H DAT			TC(m)				
		T	xx					

Name: name of parameter, including index or subscript(s), Value: initial value of parameter, Min: lower limit, and Max: upper limit, of adjustable parameters, Fn. Dep: function dependency of parameter, Error: error of parameter. xx - required entry if this parameter is used; x - optional entry; On the second line, number of iterations can be specified in character space 9 and 10. *Value can be entered directly in Field 2, or expressed via an equation. (See description of individual elements for examples).nn is component number, m is time block number; After H DAT, need to specify time that block starts, by T in character space 4 and the starting value in field 2. M, U, and IC can be indirectly adjustable (e,g., by equating to a P(I)).

fractional transfer coefficient) into compartment 1 from compartment 2, then this implies the generation/existence of the following parts of differential equations;

$$F(1)' = L(1,2).F(2)+\ldots, \text{ and } F(2)' = -L(1,2).F(2)+\ldots$$

Fields available for L(I,J): all fields, and L(I,J) can be dependent.

Special features: L(I,J) are compartment associated; can be fixed, dependent, and adjustable; can be function dependent; are nonlinearly estimated. Have an index range 75>=I>=0, and 75>=J>=1. L(I,J) can be used to define zeroth order kinetics, and equilibrium kinetics by entering 'z', or 'k' in field 5.

Examples:

Field 1	*2*	*3*	*4*	*5*	*6*	*7*
L(3,4)	14.2	10	21			1.7
L(21,22)	1	.1	1.4	*F23		.2
L(0,7)	12		100			
L(0,2)	0.03	.001	.1	Z		0.002
L(3,4)=P(7)+G(23)/UF(2)						

3.2 Initial condition, IC(J)

Purpose: IC(J) is the initial condition of compartment J (i.e., the amount of tracer, or material of interest in a compartment at the beginning of the study). IC initialize compartments prior to solving, viz: F(J,0)=IC(J), where F(J,0) is the value of the solution to the Jth component of the system of differential equations at T=0.

Fields available for IC(J): fields 1 and 2 only. IC can be dependent.

Special features: IC(J) are compartment associated; can be fixed and dependent. Have index range 75>=J>=1. IC(J) can be indirectly adjustable by assigning them to a nonlinearly adjustable parameter, e.g. IC(1)=P(47)

Examples:

Field 1	*2*	*3*	*4*	*5*	*6*	*7*
IC(75)	15					
IC(1)	100					
IC(2)=P(47)						

3.3 Linear parameter, K(J) (for J<76)

Purpose: scales the solutions of the differential equations, viz: QC(J,T)=K(J).F(J,T), where F(J,T) is the solution to the Jth component of the system of differential equations, K(J) is the linear scale parameter, and QC(J,T) is the scaled solution of the differential equation component.

Fields available for K(J): all fields except field 6.

Special features: K(J) are compartment- and component-associated; can be fixed, dependent, or adjustable; are linearly estimated; can only be used in equations in a linear context. Have index range 75>=J>=1. K(J) are closely associated with IC(J), and also, with M(J) when specific activity data are being modeled. Care needs to be exercised when estimating of any combination of these three. It is important to confirm that these values are identifiable for the model being fitted and the data at hand. The specific role of K(J) is to allow a 'unit of measurement' association of the differential equation solutions and the data being modeled. By default, K(J) =1 if it is not present in the model. Examples:

Field 1	*2*	*3*	*4*	*5*	*6*	*7*
K(75)	1					
K(23)	125.2	100	150			10.7
K(16)=25						

3.4 Summers, S(I,J)

Purpose: sums and scales the solutions of the differential equations;

QC(I)=sum[J=1,N]{S(I,J).F(J,T)}

where sum[ ] denotes sum over J; S(I,J) denotes the value of the scaling summing parameter, and F(J,T) is the solution of the Jth compartment of the system of differential equations.

Fields available for S(I,J): all fields except field 6.

Special features: S(I,J) are compartment-, and component-associated; can be fixed, dependent, and adjustable; are linearly estimated; can only be used in equations in a linear context. Have index ranges 75>=I>=1, and 75>=J>=1, $J \neq I$. S(I,J) are useful in assembling a response for some organ, or space, which may not be directly accessible for measurement. Here, based on a linearity assumption (which will be valid for systems perturbed to a small degree from their steady state), and using the principle of superposition we assemble a computed response for the inaccessible space from the computed responses of the accessible spaces. An example would be calculating a liver response using a combination of responses from accessible tissues and plasma.

Example:

Field 1	*2*	*3*	*4*	*5*	*6*	*7*
S(1,2)	10					
S(4,63)	12.2	10	15			1.7
S(3,5)=1-S(4,5)						

3.5 General parameter, P(J)

Purpose: an unassociated, nonlinearly estimated parameter to which the user assigns model significance, viz: G(J)=P(J1)/(P(J2), where G(J) denotes a dependent parameter, and P(J1) and P(J2) denote user defined parameters.

Fields available: all fields except '6'.

Special features: P(J) are neither component nor compartment associated; can be fixed, dependent, and adjustable; can be linear or nonlinear. Have index ranges 99>=J>=1. The user must attach her own significance to P(J). These parameters are especially useful in defining nonlinear aspects of systems, and for estimating those aspects. For example to build a function dependency for an L to represent a Michaelis Menten mechanism we might use P(4)/(P(5)+F(3)), where P(4) represent Vmax, P(5) represent Km, and F(3) represents a substrate level. Care needs to be exercised in using P's to ensure that they are used consistently, and that the units that they 'accept' in conjunction with simulating data, are as would be expected.

Examples:

Field 1	*2*	*3*	*4*	*5*	*6*	*7*
P(26)	15.2		29.7			
P(99)	1.2	1	1.5	.7		
P(1)=P(2)/(L(2,1)+P(49))						
P(3)=27						

3.6 Delay time, DT(J)

Purpose: establishes a delay chain associated with 'transit' of a particle between two compartments. If a model contains a delay chain DT(J) between two compartments, F(J-1) and F(J+1), then, in essence, F(J+1,T+DT(J))=F(J-1,T). This means that the delay holds up the appearance of a particle between its departure from a source (F(J-1)) and its movement to the source (F(J+1)). Here we are assuming that all transfer from compartment J is to the delay, and all transfer out of the delay is to compartment J+1. More complicated relations can easily be built though where multiple transfers exist. If there are indeed multiple transfers out of the delay then the relative transfer along any path is simply in proportion to the 'L' value for that path.

Fields available for DT(J): all except 6.

Special features: DT(J) are compartment (and can be component) related, can be fixed, dependent, and adjustable. Have index ranges, 75>=J>=1. DT(J) is a very convenient way of representing transit processes. Adjustment values for delays are often quite well estimated from data when the need for such a process is evident. Care should be taken to ensure that in conjunction with either fitting a model containing a delay, or even solving such model data, solution points are called for beyond the 'effect' of the delay. DT(J) is a relative delay, in that its effect is to hold up transfer from the 'mean' entry time, for the duration of the delay. WinSAAM model codes 1, 21, 22, and 23, have a slightly different implementation of the delay machinery that do model codes 8 and 10.

Examples:

Field 1	*2*	*3*	*4*	*5*	*6*	*7*
DT(17)	7		29.7			
DT(75)	12.5	1	15.5			
DT(14)=10.7						
DT(13)=4/L(3,1)						

3.7 Delay number, DN(J)

Purpose: The specification of a delay, DT(J), is only complete after specification of DN(J) as well. Specifies the resolution (number of pseudo compartments) associated with a delay.

Fields available for DN(J): only fields '1' and '2'.

Special Features: Strictly associated with a delay. Fixed and dependent forms permitted. The same restrictions of the DN(J) index (J) apply as do for the DT(J) index. The magnitude of DN(J) can be thought of as influencing the abruptness of DT(J) or, on a per compartment basis, the delay chain may be perceived as a series of compartments each feeding the next with an L(K+1,K)=DN(J)/DT(J). The resolution time of the delay is actually DT(J)/DN(J), which equals 1/L(K+1,K). DN(J) should never be given a value less than 2, and for all intents and purposes should be set somewhere in the range 2 <=DN(J)<=10. Odd values for the delay seem to cause it to behave less predictably than even values.

Examples:

Field 1	*2*	*3*	*4*	*5*	*6*	*7*
DN(4)	15.2					
DN(8)	1.2					
DN(6)=2						

3.8 Input, UF(J) (parametric form)

Purpose: sets the input to a differential equation or differential system, also allows discrete switching of differential inputs. Inclusion of UF(J) in the model leads to the establishment of F(J)′=UF(J)+.... (see also operation units).

Fields available for UF(J): all fields are available. However, in field '6', does not support either zeroeth order (Z) or equilibrium kinetic form (K) assignment.

Special features: Compartment associated; can be fixed, dependent, adjustable, and function dependent. Index of UF(J) is 75>=J>=1. With the capacity to set the right hand side of a differential system, and with the capacity to be discontinuously altered, UF(J) allows the user to emulate highly complex system inputs very flexibly. Particular variations of UF(J) refined in association with either its equation form, or its 'data' related assignment (see QO below) further enhance the flexibility this construct offers for simulation purposes.

Examples:

Field 1	*2*	*3*	*4*	*5*	*6*	*7*
UF(4)	14.2	10	21			1.7
UF(22)	1	.1	1.4	AG23		.2
UF(7)	12		100			
UF(7)=P(1)/(L(0,3)+L(2,4))						

3.9 General parameter, G(J) (parametric form)

Purpose: general parameter; Intended as a vehicle to permit the incorporation of 'static' determinations into the data fitting process. For example, if there is evidence that a particular area under the curve (AUC) value should be noted in regard to a particular study then, G(1)=K(78)/P(1)+K(79)/P(2) calculated from the model could be forced to accommodate the known value where this is provided as a 'datum' value for G(1).

Fields available for G(J): only fields '1' and '2'.

Special features: G(J) are unassigned and their context is established by the user. Index range is 99>=J>=1. The parametric version of G(J) is used most often in assigning some constant which is to be used in conjunction with data fitting. Examples:

Field 1	*2*	*3*	*4*	*5*	*6*	*7*
G(44)	14.2					
G(7)	122					
G(93)=P(7)/(UF(10)+K(21)+L(3,1))						

3.10 Compartment mass, M(J)

Purpose: request computation of steady state pool size, force a known value for the pool size to influence the fitting process, assign a steady state pool size value.

Fields available for M(J): all except '5' and '6'. Assignments of uncertainty are achieved as a datum entry.

Special features: Extreme care needs to be exercised with the use of M(J). If M(J) is to be taken as a 'fixed' value then, for each fixed M(J), an associated input needs to be freed to enable solution of the equation, U = L.M. If M(J) is taken as a datum then equations defining M(J) are statistically constrained to yield the set value subject to its uncertainty and the pressure of other data forces relating to the fitting process. Index range of M(J), 75>=J>=1. M(J) is most often used when fitting of specific activity data. Since M(J) appears in the denominator of a 'fractional' response calculation, neither M(J) nor IC(J) can be adjusted directly during the fitting process.

Examples:

Field 1	*2*	*3*	*4*	*5*	*6*	*7*
M(14)	14.2					
M(21)	1	.1	1.4			
M(4)=21						

3.11 Steady state input, U(J)

Purpose: request computation of steady state input, force a known value for the steady state input to influence the data fitting process, assign a fixed steady input value. For example, if the known daily intake of a nutrient is 90 mg/d, units of data are hours, and nutrient enters compartment 3, then U(3) is 3.75 mg/h.

Fields available for U(J): as for M(J).

Special features: see M(J). Limitations: see M(J)

Examples: U(3)=10, or

Field 1	*2*	*3*	*4*	*5*	*6*	*7*
U(34)	1					
U(12)	1	.1	1.4			
U(4)=L(0,4)/M(4)						

IV. WinSAAM Operational Units

To assist with runtime, or solution time, model adjustments and definitions, WinSAAM supports an array of operational units. They are: the resetting facilities (QO, QL, QF) the time dependent system equations, (G(J), UF(J), F(J), FF(J), and QO(J)), the time block descriptor (TC); the specific activity specifier (SA); time domain management (INF and T), and second independent variable, TH.

4.1 QO-operation

Purpose: resets the next F(J) value (time T+) to the QO(J) (time T) divided by K(J), where F(J) is the compartment response value at time T+.

Form: QO is added to the category field of component J.

Application: Generation of responses from data. Often used to reset integrated compartment responses to permit modeling of removed collections, e.g. as with daily collections of stool or urine. (See Chapter 8, Example 6).

Special features: Available for index values 75>=J>=1. Exercise care in regard to the delayed action of the resetting process.

4.2 QL-operation

Purpose: resets the FF(J,Ti) to the data based linearly interpolated value associated with component J viz: FF(J,Ti)= (QO(J,Tn+1)-QO(J,Tn).(Ti-Tn)/(Tn+1-Tn) /K(J). Here FF(J) is the forcing function value at time Ti. (See Chapter 8, Example 19).

Form: QL is added to the category field of component J.

Application: generation of responses from data. Often used to generate values for components of a system of differential equations where model details yielding the response are not in the current focus of an investigation. For example Cobelli uses this approach in driving the glucose response in both his minimal model and his two-pool minimal model (1, 2).

Special features: available for index values 75>=J>=1. Ensure that the initialization of the forcing function achieves what the data would suggest.

4.3 QF-operation

Purpose: resets the F(J) to FF(J).
Form: QF is added to the category field of component J.
Application: generation of responses from a mathematical form of the model. Often used to reset integrated compartment responses to permit modeling of removed collections, e.g. as with daily collections of stool or urine.
Special features: available for index values 75>=J>=1.

4.4 G(J) (general function)

Purpose: general simulation and fitting of functions, and functions of system equations.
It is through this feature of WinSAAM that its highest modeling utility is achieved viz: QC(J,T)=K(J).H(F, T, Par) may be solved and fitted to data where, QC(J,T) is the component J solution value for time T, under whatever category H forces it to fall. F is a compartment solution, and Par is the set of parameters relevant to QC(J,T) in this part of the model. For example if H is G(14), then the particular category for QC(J) would be G(14) and we would be attempting to simulate some data in this component using the functional or mechanistic form represented by G(14). Iterative adjustment here means progressively altering adjustable set within Par until the residual sum of squares can be reduced no further.
Form: G equations are entered as system equations in the model section of the input file. Where they are essentially terminal, i.e. they can be no further resolved, they are usually component assigned in order to have a data association. The objective with such functions is to create a mechanistic representation of the data features in the structure of the G(J). Typical forms of G(J) are;

```
G(14)=P(77)*F(19)+P(18)*F(1)/(P(16)+F(1)),
G(3)=K(16)*F(82)+F(17),
G(23)=AMAX1(G(7),G(27))/(1+AMIN(F(2),F(3))+ERF(1+P(2)*G(7))
```

Special features: available for index values 99>=J>=1. However, index values may not clash with indices for the parametric form of G(J).

4.5 UF(J) (input function)

Purpose: generates input into a compartment, e.g., all or part of the right hand side of the differential equations, to facilitate easy and flexible creation of differential systems. Avoids the complexity of factoring in the implied F(J) when function dependent L's are used to create a system of differential equations.

Form: UF(J) is entered as a system equation in the model section of the input. To illustrate the use of UF(J) we show how it may be used as an alternate means to L(I,J) of generating differential systems. Consider the model with the parameter L(1,2), L(2,1), and L(0,1) viz:

Field 1	*2*	*3*	*4*	*5*	*6*	*7*
L(1,2)	14.2	10	21			1.7
L(2,1)	1	.1	1.4	*F02		.2
L(0,1)	12		100			

This same model could be set up using:

UF(1)=P(1)*F(2)-(P(2)*F(2)+P(3))*F(1)

UF(2)=-P(1)*F(2)+P(2)*F(1)*F(2)

Where P(1) is given an initial value of 14.2, P(2) is given an initial value of 1, and P(3) is given an initial value of 12. Thus we see that we have broken down the very powerful, albeit somewhat confusing L, into its component parts, the value, the mechanism, and the direction of movement.

Special features: compartment associated, and hence the index is restricted as follows: 75>=J>=1.

4.7 F(J) (compartment response)

Purpose: renders the solutions of the differential system available for manipulation in conjunction with other constructs and operational units. For example, G(41)=K(79)*F(2).

Form: these operational units can only appear on the right hand side of expressions or function definitions and they only exist if the structures implying their presence exist in the model. F(J) only exists if IC(J), L(I,J), L(J,I), UF(J), QO(J), QF(J), or QL(J) exist in the model. It is legal to simply include an IC(J) and no other compartment J reference to simply generate a fixed function, for whatever reason.

Special features: compartment associated and hence index is restricted to 75>=J>=1, can only be set indirectly using QO, QL, and QF.

4.8 FF(J) (forcing function)

Purpose: sets the exposure of a compartment to the rest of the model to an analytic or 'fixed' form. If we define FF(J) then all compartments connected to compartment J respond to FF(J), rather than F(J). Compartment J itself responds to the other compartments connected to it though. Thus if FF(J) is defined then for all compartments I connected to compartment J we have, F(I)′=L(I,J)*FF(J)+... The purpose of the forcing function is to decouple the J subsystem from the remainder of the model. For example in the case where an array of compartments are exchanging with plasma, the plasma driving function causing the various response profiles in the connected compartments (subsystems) may be simulated and, where data permits, fitted using a plasma forcing function. The plasma forcing function is generated separately using the plasma data first. Models for zinc by Foster et al.(3) and for retinol by Lewis et al.(4) used plasma forcing functions.

Form: the forcing function is defined as either an explicit, or an implicit (F(I) dependent) function in the system equations definition area of the model. To see the forcing function it needs to be attached to a component and solutions requested in the domain of interest. Some examples of forcing functions follow:

FF(14)=K(77)*EXP(-P(77)*T)+K(78)*EXP(-P(78)*T)
FF(43)=P(1)*G(3)/(1+F(17))
FF(59)=F(59)

Special features: forcing functions are compartment associated and hence have indices constrained to 75>=J>=1.

4.9 QO(J) (data generation function)

Purpose: generates data for a component reflecting a particular explicit, or implicit form, allows the fitting of one function to another and hence the replacement of a complex function with a simpler one.

Form: QO(J) is defined using an equation in the system equations section of the model and this leads to the generation of data points for component J. For example QO(4)=P(1)*EXP(-P(2)*T) would lead to the generation of observations, at nominated time points given by the exponential expression. If this were accompanied by an association of G(47) with component '4' then we could fit the model defined by G(47) to the data of component '4', and hence the expression given by QO(4). Other examples are:

QO(43)=F(25)+F(22), and

QO(28)=P(27)*EXP(-P(23)*T)+K(79)*FF(2)/F(2).

Special features: QO functions are component related and hence have the following index constraint; 75>=J>=1.

4.10 TC (time blocks)

Purpose: separates data associated with different phases of an experiment into 'sub units', allows the fitting of a model accommodating aspects which may change in conjunction with the protocol. For example if in an alcohol digestion experiment alcohol is consumed in three bursts each of 15 minutes duration over a 2 hour period, then we may want to, on the one hand, establish a strategy delivering the alcohol into the system in this quasi-regular fashion (probably using UF constructs), and, on the other hand, admit the possibility that alcohol accumulation may itself block its emptying and hence we would want to explore the possibility that elimination may change in conjunction with the successive intakes. Each process involves abrupt changes at particular points in time and it is via WinSAAM's TC block control scheme that allows us to introduce these changes.

Form: time blocks are identified as data segments, or data input units and have their specific time encapsulation are marked by specifying a TC(J). Time blocks can be progressively ordered, in which case the start for block TC(J+1) is given by the last time point for block TC(J), or they can be independently set in which case time is 'reset' at the start of each block. In either circumstance the index (here J or J+1) bears no relationship to the actual temporal offset of the time block. The index is just a mechanism for associating all actions linked to a particular time block, i.e. data collection, parameter re-assignment, re-initialization of system equations and resetting of steady state parameters. It should be pointed out that re-assignment of adjustable parameters across a time block allows the parameter to independently adjust data associated with each time block, for example, if gastric emptying, L(0,1) changes in association with alcohol intakes, then L(0,1) might take on adjustable form P(2) in time block '2' (TC(2)), P(3) in time block '3', and so on. Then the values P(J) adjusts to in time block 'J' will be strictly in regard to the original context of L(0,1), i.e. P(J) each take on the modeling role of L(0,1).

Limitations: TC(J) can be created for 99>=J>=0. There is by default always one time block implied in the data unit, and this is usually TC(0). TC(0) need not be specified.

4.11 SA (specific activity)

Purpose: calculation of the specific activity for a compartment viz:
SA(T)=F(J,T)/(M(J)*F(J,0).

Form: associated with specific activity data in a particular component via the component's category entry simply as 'SA'.

Special features: only achieves its purpose by explicit specification as a component's category. Takes neither arguments nor indices.

4.12 INF (equilibrium solution request)

Purpose: requests the solution for a category at a very large T value (e.g. 1.0E35) leading potentially to equilibrium solution for the category.

Form: the text entry 'INF' is entered in the Time field for the category of interest.

Special features: if the particular category has no equilibrium solution then no solution will be found for T=INF.

4.13 T (the Time variable)

Purpose: allows access to the time (independent variable) for expressions, permits resetting time in conjunction with time blocks.

Form: to use time in expressions refer to it as 'T' in the expression, e.g. G(7)=2*T, or G(47)=P(26)/(P(28)+T). To reset time for a TC block simply specify 'Time Reset' in conjunction with time block definition, or in the former data input organization insert a 'T' in the category field immediately after the TC definition line.

Limitations: any time block can have its time flow re-initialized.

4.14 TH (second independent variable)

Purpose: to enhance the flexibility of explicit functions, to admit a second monitored variable to influence the profile of a response. Referred to as TH, this second independent variable may appear in expressions in the following ways:

G(31)=K(77)*F(2)+EXP(-P(29)*TH), or

G(49)=K(78)*T+K(79)*TH+K(80)*T*TH.

Form: the second independent variable is accessed via the 'TH' reference.

Limitations: differential equations cannot implicitly depend on TH, only T. Values for TH must be provided in the appropriate field of the relevant data segment. Usually values of T will be accompanied by values of TH.

V. Other WinSAAM Features

5.1 Statistical Weights

To allow data precision to reflect the influence of data domains on the data fitting process WinSAAM uses a generalized least squares (this description of fitting has changed in recent times to reflect generalized linear modeling, WinSAAM does not currently directly support generalized linear modeling) data fitting scheme. This means that data weighting is incorporated into the fitting process. The weights applied in WinSAAM are of the form W(K)=1/SD(K)**2, where SD(K) is the standard deviation of datum K. There are 4 strategies in WinSAAM for generating weights, FSD, SD, RQO, and WT. (See Chapter 8 for examples).

5.1.1 FSD: assigns data standard deviations on a C*QO(K) basis. This means that the weights are of the form W(K)=1/(C*QO(K))**2, and this weighting scheme smaller data points influence the fit more heavily than do larger points.
5.1.2 SD: assigns data standard deviations on a C basis. This means that the weights are of the form W(K)=1/C**2, and hence using this weighting scheme larger data points influence the fit more heavily than smaller points.
5.1.3 RQO: assigns data standard deviations on a C*QO(K)**0.5 basis. This means that the weights are of the form W(K)=1/(QO(K)*C**2), and hence using this weighting scheme smaller data points influence the fit more heavily that larger points.
5.1.4 WT: assigns weights as specified directly to data, and thus W(K)=WT(K). The user of this weighting scheme can, on a datum by datum basis, directly establish the observation influence on the data fitting process.

Grouped versus Individual Weighting: WinSAAM supports weighting of data by groups or individually. Group weighting amounts to defining the error structure, i.e. weight structure, of the observation of a group basis and here FSD, SD, RQO, and WT weighting options are supported. On an individual basis SD weighting only is supported. Whereas we mentioned (see SD) that group SD weighting favors larger observations, in that with this scheme these observations influence the fit more heavily, on an individual SD application basis this may not be so. Indeed, with individual observation SD weights the actual errors associated with observations themselves define the weights in conjunction with the intrinsic application of this process.

5.2 Data Fitting

WinSAAM supports a number of features to help the user with data fitting. Adjustable parameters can be constrained to a region by specifying upper and lower bounds. This prevents wild erratic parameter behavior during fitting. The linear and nonlinear adjustable parameters are fitted separately. In each iteration sequence, first the nonlinear adjustable parameters are iteratively adjusted and then the linear parameters are updated in a single step. Adjustments for the nonlinear parameters found in association with each iterative step are in fact linearly refined prior to their boundary analysis and application to the actual parameters. This increases the sensitivity of the fitting process. Each estimation of the matrix of partial derivatives is used twice in estimating changes of various parameters. This saves considerable processing time. For large data sets and large models, partials estimation can be quite time consuming. Parameter convergence criteria are managed automatically by WinSAAM, and values for these criteria have been determined in conjunction with considerable exposure of the software to real problems of considerable magnitude. The fitting scheme involves a variation on the Gauss-Newton algorithm which forces the procedure to follow 'steepest decent' when the fit seems to be at considerable distance for convergence, and then moves to a more pure Gauss Newton procedure as convergence looms. Steady state, and other static observations can be incorporated into the fitting process, and this means that all relevant data can influence the final fit.

A helpful approach to data fitting is Berman's reference fit scheme; first, an unidentified model is fit, not to determine values for its parameters but rather to give a guide to the minimal sum of squares possibly achievable with a model of the type considered. Then a somewhat contracted model, with fewer parameters, and more than likely to have a good chance of being identified, is fit to the data. Particularly significant here are both the residual sum of squares and the parameter estimates. The residual sum of squares ought to be compared with the minimal sum of squares to get some reference concerning the degree to which this identified model explains the data versus what is potentially (nearly) achievable.

With data fitting one is always concerned with the question of whether the residual sum of squares is as small as is possibly achievable, or alternately framed, whether the parameter values of our iteration scheme has settled to represent the best possible set given the model and the data. While there is no way of absolutely answering this question there are guidelines to help us evaluate the likelihood of such a situation, e.g., how rapidly did the iteration sequence reach its 'resting' point? How correlated are the parameter estimates? How large are the relative uncertainties of the parameters? What sort of structure exists in the residuals, what is the balance of the sum of squares across the various components?

5.3 Function Dependency (Fn-dep)

The following tables shows function dependencies supported in WinSAAM

Field 1	Field 2	Field 3	Field 4	Field 5	Field 6	Field 7
Name	Value	Lo-Lim	Hi-Lim	Fn-Dep		Error

Fn-Dep consists of the operator and the operand. The operators supported are as shown below and the operand is of the form GXX, or FXX, where XX denotes the index of the appropriate function.

Operator	Interpretation
+	x + z
-	x - z
*	x * z
/	x / z
5	x ^ z
6	exp(x * z)
7	if z>=0 then x else 0
8	if z>0 then x * z else 0
9	if z> 0 then x / z elseif z = 0 then 1.0e12 else 0
A	if z > 0 then z else 0
B	if z > 0 then 1 else if z < 0 then -1 else 0
C	if z < 0 then 0 else z
D	time of last t-crossing of z
E	time of first t-crossing of z

Some examples of function dependent parameters are;

Field 1	Field 2	Field 3	Field 4	Field 5	Field 6	Field 7
L(0,1)	0.1	.01	.4	*F01		.02
L(2,1)	.3			*G02		
UF(4)	1			+G06		

In terms of a UF(J) these achieve the following:

UF(1)=-P(1)*F(1)**2, {UF(2)=P(2)*F(1)*G(2) … UF(1)=-P(2)*F(1)*G(2) …},
and
UF(4)=1+G(6)

VI. An introductory walk through some WinSAAM instructions

After a WinSAAM Text Input file has been created using the WinSAAM editor, the following listing shows an annotated WinSAAM session for the solution of a two-compartment model. The bold lettering indicates a command typed in by the user (or selected from the menu bar) and the letters in italics describe the action that results from the command.

p deck *display the WinSAAM input file*

```
 1: A SAAM31
 2: H PAR
 3:     L(0,1)    .1          3.333334E-02   3.000000E-01
 4:     L(2,1)    .05         1.666667E-02   1.500000E-01
 5:     L(1,2)    .01         3.333333E-03   3.000000E-02
 6:     IC(1)     1
 7: H STE
 8:     U(1)=100
 9:     M(1)                                10000
10: H DAT
11: 101                                     FSD=.4
12:               1        .7605
13:               2        .6753
14:               4        .5704
15:               8        .2847
16:               16       .09946
17:               24       .04400
18:               32       .02807
19:               48       .01711
20:               96       .01403
21:               144      .008711
22:               288      .004021
```

\> **deck** *compile the input*

```
UPDATE?
* DECK BEING PROCESSED
PRE-PROCESSING TIME :        .160 SECS
```

\> **solv** *solve the system equations*

```
*** MODEL CODE 10 SOLUTION
SOLUTION TIME :        .110 SECS
```

\> **adju** *display the adjustable parameters*

```
PARAMETER    VALUE     LOW-LIMIT   HI-LIMIT
L ( 0, 1) 1.0000E-01 3.3333E-02 3.0000E-01
L ( 2, 1) 5.0000E-02 1.6667E-02 1.5000E-01
L ( 1, 2) 1.0000E-02 3.3333E-03 3.0000E-02
```

\> **ss(i)** *display the component sums of squares*

```
SS( 1)    8.6989E-06
```

\> **plot q(1)** *plot the observations and prediction for component 1*

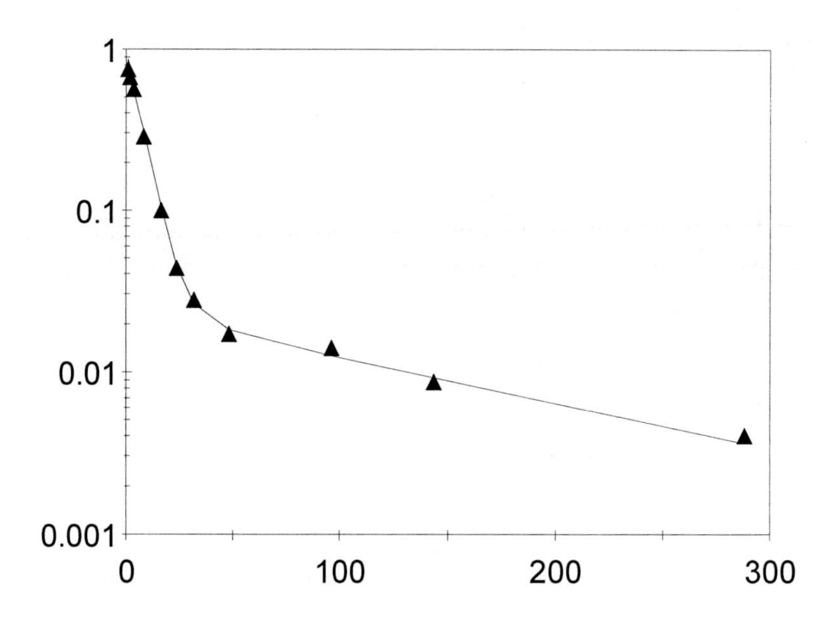


```
> iter  iteratively adjust the parameters
* PARTIALS ESTIMATED
* CORRECTION VECTOR ESTIMATED

CONVERGENCE MEASURES
 IMPROVEMENT IN SUM OF SQUARES =    14.67(%)
 FINAL VALUE OF CONAB =    1.000E+00
 LARGEST CHANGE (     5.58 %) WAS IN PAR( 2, 1)

* CORRECTION VECTOR ESTIMATED

CONVERGENCE MEASURES
 IMPROVEMENT IN SUM OF SQUARES =      .62(%)
 FINAL VALUE OF CONAB =    5.683E+00
 LARGEST CHANGE (     1.30 %) WAS IN PAR( 1, 2)

ITERATION TIME :        .550 SECS

DISTRIBUTION OF SQUARES
 COMP   SUM OF SQUARES
   1     7.3771E-06
> fsd(i)    display the parameter fsd's
* VALUES MAY NOT RELATE TO CURRENT PARAMETERS
* L ( 0, 1)    1.000E-01      FSD( 1)    2.295E-02
* L ( 2, 1)    5.310E-02      FSD( 2)    6.222E-02
* L ( 1, 2)    9.822E-03      FSD( 3)    8.556E-02
> plot  overlay a new plot showing improvements

> m(i)  display the pool sizes
M( 1)     9.9980E+02
M( 2)     5.4051E+03
> r(i,j)    display the fluxes
R(  0, 1)  1.0000E+02 A
R(  2, 1)  5.3091E+01 A
R(  1, 2)  5.3091E+01 A
> calc li   calculate the L-inverse matrix
```

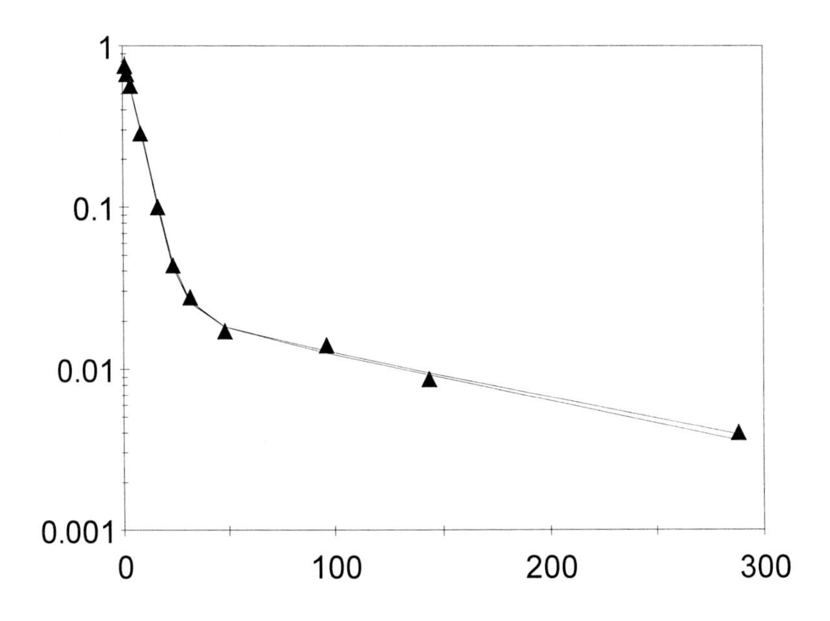


```
   *** MODEL CODE 10 SOLUTION
   SOLUTION TIME :        .000 SECS

> li(i,j)   display the L-inverse matrix

         1             2
 1  9.9980E+00    9.9980E+00
 2  5.4051E+01    1.5586E+02

> calc eig li  calculate the eigenvalues & eigenvectors for the system
STOPPED AFTER     3 ITERATIONS

> egv(i)        display the eigenvalues … note that these are the reciprocals of
                the actual system eigenvalues
EGV( 1) =    6.3827E+00
EGV( 2) =    1.5947E+02

> eig(i,j) display the eigenvectors for the system

         1             2
 1  9.7638E-01    2.3615E-02

 2 -3.5306E-01    3.5306E-01
> p 1 20        display the first 20 lines of the input in preparation for some
                input revision.  From here much of the effort will be directed at
                modifying the description of the data to have the form of an
                exponential model

   1: A SAAM31
   2: H PAR
   3:     L(0,1)    .1              3.333334E-02    3.000000E-01
   4:     L(2,1)    .05             1.666667E-02    1.500000E-01
   5:     L(1,2)    .01             3.333333E-03    3.000000E-02
   6:     IC(1)     1
   7: H STE
   8:     U(1)=100
   9:     M(1)                                  10000
  10: H DAT
  11: 101                                       FSD=.4
```

```
  12:              1              .7605
  13:              2              .6753
  14:              4              .5704
  15:              8              .2847
  16:              16             .09946
  17:              24             .04400
  18:              32             .02807
  19:              48             .01711
  20:              96             .01403

> i 11  insert new text beginning at line 11


INSERTING AT LINE   11
  10: H DAT
      12345678 1 2345678 2 2345678 3 2345678 4 2345678 5 2345678 6...

  11: x g(1)=k(77)*exp(-p(1)*t)+k(78)*exp(-p(2)*t)
  12:



  10: H DAT
  11: X G(1)=K(77)*EXP(-P(1)*T)+K(78)*EXP(-P(2)*T)
  12: 101                                         FSD=.4
Edit the deck to add K(77), k(78), P(1) and P(2)
> p 1 20

   1: A SAAM31
   2: H PAR
   3: Y
   4:     L(0,1)    .1              3.333334E-02   3.000000E-01
   5:     L(2,1)    .05             1.666667E-02   1.500000E-01
   6:     L(1,2)    .01             3.333333E-03   3.000000E-02
   7:     IC(1)     1
   8: Y-
   9:     P(1)      .1                             10
  10:     P(2)      .01                            10
  11:     K(77)                                    10000
  12:     K(78)                                    10000
  13: Y
  14: H STE
  15:     U(1)=100
  16:     M(1)                                     10000
  17: Y-
  18: H DAT
  19: X G(1)=K(77)*EXP(-P(1)*T)+K(78)*EXP(-P(2)*T)
  20: 101 G(1)                                     FSD=.4

> p deck

   1: A SAAM31
   2: H PAR
   9:     P(1)      .1                             10
  10:     P(2)      .01                            10
  11:     K(77)                                    10000
  12:     K(78)                                    10000
  18: H DAT
  19: X G(1)=K(77)*EXP(-P(1)*T)+K(78)*EXP(-P(2)*T)
  20: 101 G(1)                                     FSD=.4
  21:              1              .7605
  22:              2              .6753
  23:              4              .5704
  24:              8              .2847
  25:              16             .09946
  26:              24             .04400
  27:              32             .02807
  28:              48             .01711
```

```
  29:             96           .01403
  30:            144           .008711
  31:            288           .004021

> deck
UPDATE?
* DECK BEING PROCESSED
*** L(0,1) ADDED AS A DUMMY.*3*
PRE-PROCESSING TIME :        .160 SECS
> solv
*** MODEL CODE  2 SOLUTION
SOLUTION TIME :        .000 SECS
> prin q(1)
-------------------------------------------------------------------------
*** NAME :    1
CURRENT KOMN
 #  COMP TC  CATEGORY        T          QC          QO          QC/QO
  1   1   0  G ( 1)     1.000E+00  4.381E-01  7.605E-01         .5761
  2   1   0  G ( 1)     2.000E+00  3.987E-01  6.753E-01         .5905
  3   1   0  G ( 1)     4.000E+00  3.307E-01  5.704E-01         .5798
  4   1   0  G ( 1)     8.000E+00  2.293E-01  2.847E-01         .8053
  5   1   0  G ( 1)     1.600E+01  1.149E-01  9.946E-02        1.1549
  6   1   0  G ( 1)     2.400E+01  6.255E-02  4.400E-02        1.4216
  7   1   0  G ( 1)     3.200E+01  3.820E-02  2.807E-02        1.3610
  8   1   0  G ( 1)     4.800E+01  2.050E-02  1.711E-02        1.1984
  9   1   0  G ( 1)     9.600E+01  1.040E-02  1.403E-02         .7415
 10   1   0  G ( 1)     1.440E+02  6.419E-03  8.711E-03         .7369
 11   1   0  G ( 1)     2.880E+02  1.521E-03  4.021E-03         .3782
> iter
* PARTIALS ESTIMATED
* CORRECTION VECTOR ESTIMATED

CONVERGENCE MEASURES
 IMPROVEMENT IN SUM OF SQUARES =     94.09(%)
 FINAL VALUE OF CONAB =    9.915E-01
 LARGEST CHANGE (    95.66 %) WAS IN PAR( 2, 0)

* CORRECTION VECTOR ESTIMATED

CONVERGENCE MEASURES
 IMPROVEMENT IN SUM OF SQUARES =     41.61(%)
 FINAL VALUE OF CONAB =    1.867E+00
 LARGEST CHANGE (    17.41 %) WAS IN PAR( 2, 0)

ITERATION TIME :        .060 SECS

DISTRIBUTION OF SQUARES
 COMP  SUM OF SQUARES
   1    6.4671E-06
> iter
* PARTIALS ESTIMATED
* CORRECTION VECTOR ESTIMATED

CONVERGENCE MEASURES
 IMPROVEMENT IN SUM OF SQUARES =    28.29(%)
 FINAL VALUE OF CONAB =    1.084E+00
 LARGEST CHANGE (     5.05 %) WAS IN PAR( 1, 0)

* CORRECTION VECTOR ESTIMATED

CONVERGENCE MEASURES
 IMPROVEMENT IN SUM OF SQUARES =      .17(%)
 FINAL VALUE OF CONAB =    5.507E+00
 LARGEST CHANGE (      .66 %) WAS IN PAR( 2, 0)
```

```
ITERATION TIME :          .050 SECS

DISTRIBUTION OF SQUARES
 COMP   SUM OF SQUARES
   1     4.6296E-06
> fsd(i)

* VALUES MAY NOT RELATE TO CURRENT PARAMETERS
* P ( 1)          1.495E-01       FSD( 1)    3.356E-02
* P ( 2)          6.142E-03       FSD( 2)    7.081E-02
* K (77)          8.909E-01       FSD( 3)    3.507E-02
* K (78)          2.299E-02       FSD( 4)    6.146E-02
> prin
------------------------------------------------------------------------------------
*** NAME :      1
CURRENT KOMN
 #   COMP  TC   CATEGORY         T           QC           QO            QC/QO
  1    1    0   G ( 1)      1.000E+00   7.900E-01   7.605E-01       1.0388
  2    1    0   G ( 1)      2.000E+00   6.834E-01   6.753E-01       1.0119
  3    1    0   G ( 1)      4.000E+00   5.123E-01   5.704E-01        .8982
  4    1    0   G ( 1)      8.000E+00   2.913E-01   2.847E-01       1.0231
  5    1    0   G ( 1)      1.600E+01   1.023E-01   9.946E-02       1.0286
  6    1    0   G ( 1)      2.400E+01   4.447E-02   4.400E-02       1.0108
  7    1    0   G ( 1)      3.200E+01   2.634E-02   2.807E-02        .9383
  8    1    0   G ( 1)      4.800E+01   1.780E-02   1.711E-02       1.0405
  9    1    0   G ( 1)      9.600E+01   1.275E-02   1.403E-02        .9088
 10    1    0   G ( 1)      1.440E+02   9.494E-03   8.711E-03       1.0899
 11    1    0   G ( 1)      2.880E+02   3.921E-03   4.021E-03        .9750
>
```

Note that the results in terms of the data and parameter values can also be accessed using the OUTPUT|SPREADSHEET item on the menu of the Terminal Window. Contents of the spreadsheet can be edited, and EDIT|CUT and EDIT|PASTE into other programs, such as a statistical package supporting spreadsheet data entry or word processing program.

Future Developments of WinSAAM

WinSAAM will have Project Management and Linker tools to link models with different data sets, to facilitate multiple studies analysis, a graphical interface to assist new users, and sample optimization for study design as a drop-down menu tool.

REFERENCES

1. Caumo, A., and C. Cobelli. 1993. Hepatic glucose production during the labeled IVGTT: estimation by deconvolution with a new minimal model. *Am. J. Physiol.* 264:E829-E841.
2. Caumo, A., G. M. Morgese, G. Pozza, and C. Cobelli. 1991. Minimal models of glucose disappearance: Lessons from the labeled IVGTT. *Diabetic Medicine*. 8:822-832.
3. Foster, D. M., R. L. Aamodt, R. I. Henkin, and M. Berman. 1979. Zinc metabolism in humans: a kinetic model. *Am J Physiol*. 237:R340-R349.
4. Lewis, K. C., M. H. Green, J. Balmer Green, and L. A. Zech. 1990. Retinol metabolism in rats with low vitamin A status: a compartmental model. *J Lipid Res*. 31:1535-1548.

SECTION III

Concepts and Tools of Modeling

7
BUILDING MODELS IN SECTIONS

This chapter introduces the concept of building models of systems in sections. First, model decomposition, or the breaking-down of a system into smaller parts is discussed. Secondly the mathematical tools needed to represent the inputs into a system and the processes within a system are explained. Thirdly by using an example for alcohol metabolism, the steps are presented for setting up a model and checking it for accuracy. Fourthly, the steps for coupling model subsystems into a larger model are discussed using glucose and insulin metabolism as an example.

I. Model Decomposition

The notion of problem decomposition has been used in science and engineering for many years. Problem decomposition is based on the idea that collections of small problems are more easily solved, and their solutions may be more useful than those of large complex problems. For example, large scale computer software can only be produced, with any guarantee of reliability or efficiency, by using the constructional tools in computing languages of subroutines, procedures, functions, modules, classes, and objects to facilitate program decomposition. Under the title of 'divide and conquer', problem decomposition consists of the following steps:

1. Decomposing the problem to be solved into small, more manageable, sub-problems.
2. Programming solutions to each sub-problem, e.g., abstract data-typing or creating objects.
3. Interfacing these programmed solutions in such a fashion as to provide a solution path for the original complex problem.

Of course, in conjunction with these steps a number of allied and fairly sophisticated concepts will need to be negotiated. For example, in regards to modeling, it is not unreasonable to suggest that a mathematical proof of the correctness of the decomposition step be undertaken and possibly for the interfacing step as well. Furthermore, white-box testing at the unit creation stage (Step 2) and integration testing at the interfacing stage will definitely add to our confidence in the final product. The idea of problem decomposition lends itself to modeling in a number of ways; firstly in confirming that the modeling constructs that we are using are indeed appropriate for the job at hand and secondly to reduce the estimation process complexity by using decoupling.

II. Confirming the Modeling Constructs

Models need to be specified, created, and tested in sections and response to model inputs, or assaults need to be independently synthesized and tested. There is no point to manipulating the structure of a model to generate a desired response when the inputs to that model are inappropriate, just as it is inappropriate to manipulate aspects of a model to generate a described response when the model itself is inaccurately assembled. The finer we can decompose our synthesis and testing exercise the more confidence we will have in the final outcome.

Modeling constructs are the tools offered by the simulation language for creating and analyzing models. The relevance and use of the modeling constructs depends upon the type of problem. When a simulation language embraces the problem domain with ease and flexibility, it becomes natural to conjure the modeling constructs as physical processes themselves. It is critical to the modeling processes that the ramifications of the selected modeling constructs, in regard to modeling the system under investigation, are understood fully.

Let us consider an array of operational and functional features of a modeling language that we may need access to in order to realistically portray a system and some assaults to it. In terms of processes, those listed in Table 7.1 are required for modeling aspects of biological responses.

Viable systems in steady state don't appear to respond in any way unless they are perturbed. Indeed, in tracer and stable isotope experiments, one goal is to investigate the manner in which a system returns to its steady state when perturbed. By monitoring the 'perturbation' as it both damps away at its site of application and ripples through the rest of the system, we gain insight into the system in regard to the number of exchanging pools and the rate of exchange of material amongst those pools.

Table 7.1 Modeling constructs.

PROCESS	FORM	PARAMETER UNITS[a]
Zeroeth order kinetics	$\acute{y} = -k$	$[k] = MT^{-1}$
First order kinetics	$\acute{y} = -ky$	$[k] = T^{-1}$
Second order kinetics	$\acute{y}_1 = -ky_1.y_2$	$[k] = M^{-1}.T^{-1}$
Multi-reactant kinetics	$\acute{y}_1 = -ky_1^{a}.y_2^{b}$	$[k] = M^{-(a+b)}.T^{-1}$
Michaelis-Menten kinetics	$\acute{y}_1 = -v.y.(k+y)$	$[v] = M.T^{-1}$, $[k] = M$
Inhibited elimination	$\acute{y}_1 = -ky/(1+a.y^2)$	$[k] = T^{-1}$, $[a] = M^{-2}$
Bi-substrate Ping Pong inhibition	$\acute{y}_1 = -v/(1+k_1/y_1 + k_2/y_2)$	$[v] = M.T^{-1}$, $[k_j] = M$
Inhibition	$\acute{y}_1 = -v/(1+k_1/y_1)(1 + y_2/J_2)$	$[v] = M.T^{-1}$, $[k_1] = M$, $[J_1] = M$
Allosteric binding	$\acute{y}_1 = v/(1+ (k_1/y_1)^a)$	$[v] = mt^{-1}$, $[k_1] = M$

[a]M denotes the units of y, e.g. mass, concentration, or volume.

Different perturbations to the system lead to different responses. On the one hand, smaller perturbations which don't materially change the size of the pools of a system can be characterized in terms of first order kinetics. Larger perturbations, on the other hand, which do materially change the size of the pools of a system are characterized in terms of nonlinear kinetics. While there is an infinite number of ways to perturb a system, some perturbations used to explore system responses using models are listed in Table 7.2.

The perturbations described above can be used to portray either a scientific protocol to explore the system e.g., a tracer injection of ^{45}Ca to study calcium kinetics, or, an actual digestion or metabolic assault to a system, e.g., a person whose blood alcohol is monitored as they imbibe wine.

III. Example of Model Construction and Testing

Although Chapter 8 is dedicated to the use of modeling constructs, ideas underpinning model construction and testing are described here.

Example: A male drinks three 100 ml glasses of wine of 10% alcohol concentration each over 15 minutes at times t=0, 35, and 70 minutes and it is desired to determine the blood alcohol level reached and the persistence of blood alcohol over 0.01%.

Wagner et al. (8) have shown that alcohol is metabolized at a rate of;

$$C` = -V_m C/(K_m + C)$$

where C is the blood alcohol concentration (g/L), V_m is the maximum catabolism rate (0.18 g/L/hr), and K_m is the Michaelis-Menten parameter (0.05 g/L).

Table 7.2. System perturbation tools

Perturbation	Form
Bolus	$u = \delta(t)$
Pulsed infusion	$u(t) = c, \; 0<t<T$
Pulsed ramp	$u(t) = c.t, \; 0<t<T$
Exponential Decay	$u(t) = c.e^{-d.t}, \; 0<t<\infty$
Random bolus	$u = \delta(t_1), t_1 = T_1$
Gamma function	$u(t) = c.t^n.e^{-d.t}, \; 0<t$
Constant infusion	$u(t) = c$

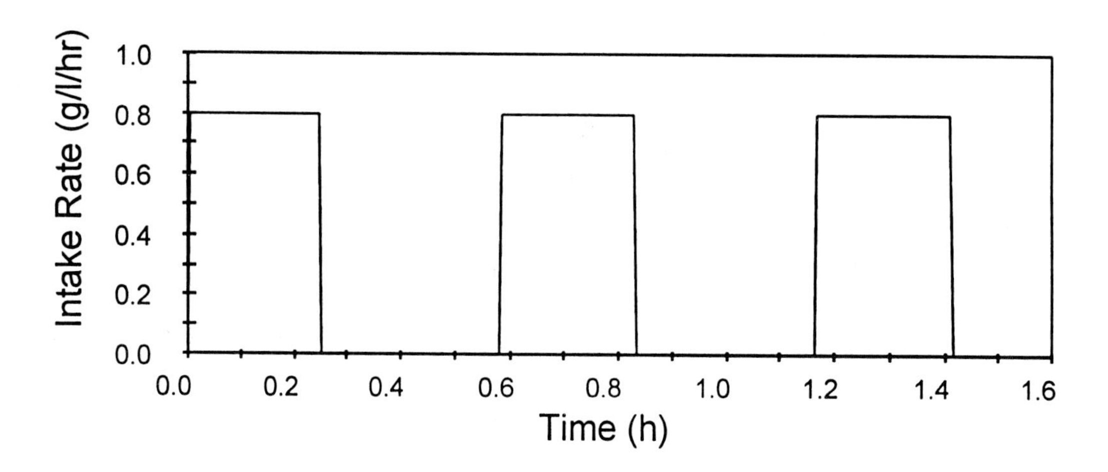

Fig 7.1. Alcohol delivery schedule.

Furthermore with a 50 L distribution space (8) the input rate (g/L/hr) is (assuming 100% absorption);

$$U(t) = 0.8 \text{ g/L/h} \qquad 0<t<0.25,\ 0.58<t<0.83,\ 1.17<t<1.42$$
$$= 0 \text{ g/L/h} \qquad 0.25<t<0.58,\ 0.83<t<1.17,\ 1.42<t$$

The alcohol delivery schedule is illustrated in Figure 7.1.

Testing the Model

We have several good guides with this model to assist us in validating it:

$$C' = -V_m$$
$$= -0.18 \qquad \text{if } C>>km$$
$$C' = -V_m \cdot C / Km$$
$$= -3.6\ C \qquad \text{if } C<<Km$$

Finally, it is suggested (Scientist User Manual, MicroMath 1995) (6) that a point of clear change of kinetics for C is at C = Km. Using the WinSAAM model depicted in Figure 7.2A (see also Chapter 8), we monitored the profile of C subject to an initial bolus of 1 g/L (i.e., a value 20 times that of Km). We see in Fig 7.2A that the initial fate of C is independent of C and has the value of -0.18 g/L/hr. Then in Fig 7.2B from our plot of C and exp(-Vm.t/km), we see that the pattern suggested by our approximation above emerges.

In Fig 7.2C, C is plotted semi-logarithmically against time and we see that the point Km (0.05 g/L) does indeed represent some point of change in the pattern of C against t. Thus it would appear that the model is displaying the features we need and our next step is to create the output pattern.

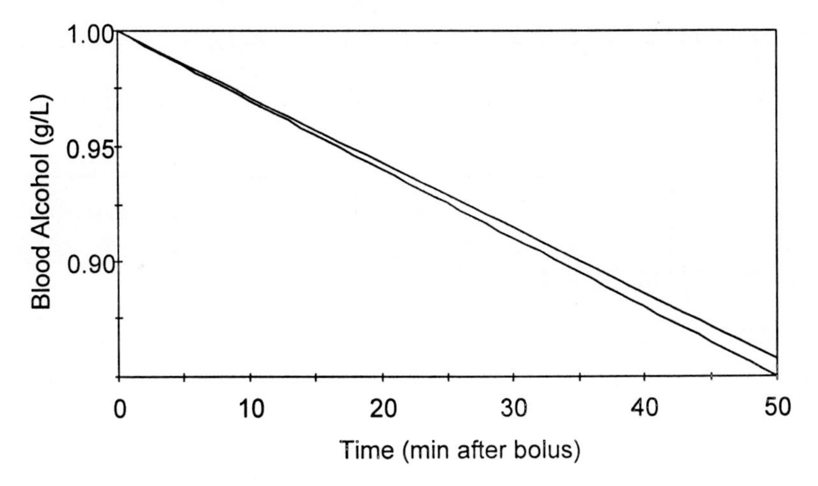

Fig 7.2a Early stages of alcohol intake.

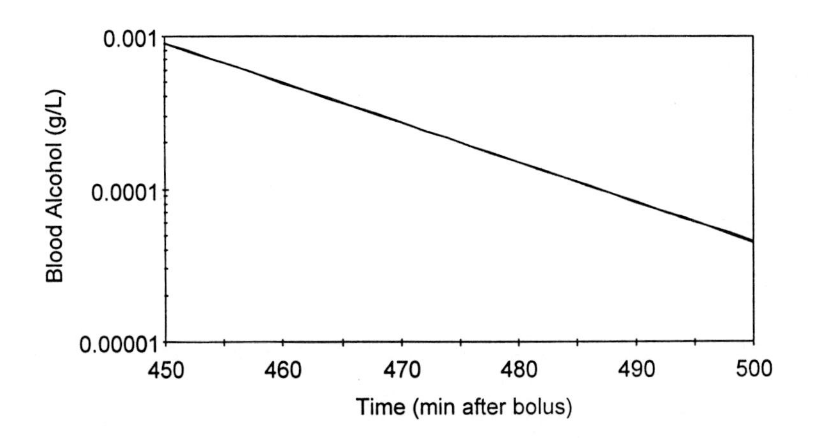

Fig 7.2b Later stage of alcohol metabolism.

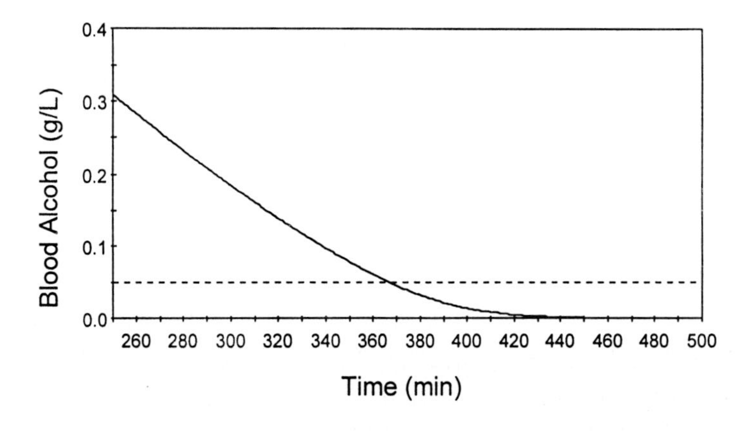

Fig 7.2c C plotted against time, with Km (dotted line).

Testing the input

The Text Input file used to model the input function representing the alcohol delivery scheme is shown in Fig 7.3.

```
A SAAM31                    CH07Eg1.saam
H PAR
C
C VMAX = 18 MG/DL/HR = 0.18 GM/L/HR
C KM   =  5 MG/DL    = 0.05 GM/L
C P(1) = VMAX
C P(2) = KM
C P(4) = ALCOHOL INPUT RATE [GM/HR]
C P(5) = DISTRIBUTION SPACE [/L]
C P(6) = ALCOHOL ABSORPTION
C
 P(1)=.18
 P(2)=0.05
 P(4)=40
 L(0,1)=P(1)                                            *G 1
   P(5)      50
   P(6)      1
H DAT
X G(1)=1/(P(2)+F(1))
X UF(1)=P(4)*P(6)*F(2)/P(5)
X UF(3)=F(1)/(3*P(4)*P(6)/(4*P(5)))
102 QO       /60
             0             1
             15            0
             35            1
             50            0
             70            1
             85            0
101
             0
2            .02                        200
101 UF(1)
             0
2            .01                        200
101 F(3)
             0
2            .02                        200
103 F(3)
             100
104 F(1)
             0
             .25
             .58
             .83
             1.16
             1.41
```

Fig 7.3 WinSAAM Text Input file

We see from Fig. 7.1 that we achieved the needed input, and in principle, all we need to do now is to use this input to drive the model to reflect the fate of alcohol. Before doing so, there are a couple of points we should address. Firstly, whereas the intake of alcohol has been via oral consumption into the gastrointestinal tract, we have set up a model of metabolism known to apply in the distribution space of the subject. This implies on the one hand that delivery of the alcohol to the circulation is immediate, and on the other hand, that all alcohol consumed actually enters the circulation metabolically intact. The former point (intravenous delivery) is simply a shortcoming of our model and to deal with this deficiency Wilkinson et al. (9) and Pieters et al. (7) have proposed a digestive side enhancement to this model. We will return to this point when we have discussed more about the modeling tools needed to negotiate the more advanced work.

Absorption, for now, is represented by parameter P(6) and we have set this parameter to 1. However, passage and alcohol metabolism is perhaps somewhat less, about 90% of the imbibed alcohol arriving in the circulation intact, ie., a P(6) value of 0.85 (see Holford (4)) may be appropriate.

Finally, connecting our metabolic model and our input model to form the above input file we obtain the following plot upon solving the system (Fig 7.4). The input, the blood alcohol level, and the area under the curve (AUC) are displayed in one plot. Furthermore though not shown on the plot we determine the asymptotic value of the AUC (0.86 g) and the factor by which this exceeds the intake (1.433). The three plots each demonstrate the characteristics we would expect (Figs 7.4). The input plot switches on to the value 0.8 g/L/hr at the appropriate times and in each case stays turned on for the 15 minute duration. The blood alcohol concentration plot moves with the inputs and attains levels consistent with the 0.2 g/L pulsed infusions (which the input represent). The peak blood alcohol concentration is around 0.4 g/L or 0.04% and remains above 0.01% or 0.01 g/L until around the 3.5 hr.

In Fig 7.5 the blood alcohol concentration curves are plotted for presumed absorption levels of 0.8, 0.9 and 1.0 and this and Table 7.3, reveal the approximate pattern of blood alcohol consumption response with absorption level. The disproportionate peak response has been discussed by Wagner et al. (8) and is a consequence of the nonlinear and non-zeroeth order reaction of alcohol metabolism. If the alcohol elimination was either linear or zeroeth order the peak response would scale precisely in proportion to the absorption (Figure 7.5).

Table 7.3 Effect on fraction of alcohol absorbed on peak height (see Fig 7.5)

Absorption level	Peak 1	Peak 2	Peak 3	Peak 1/ Peak 1	Peak 2/ Peak 2	Peak 3/ Peak 3
1.0	0.173	0.292	0.403	1	1	1
0.9	0.154	0.259	0.349	0.89	0.87	0.87
0.8	0.135	0.220	0.295	0.78	0.75	0.73

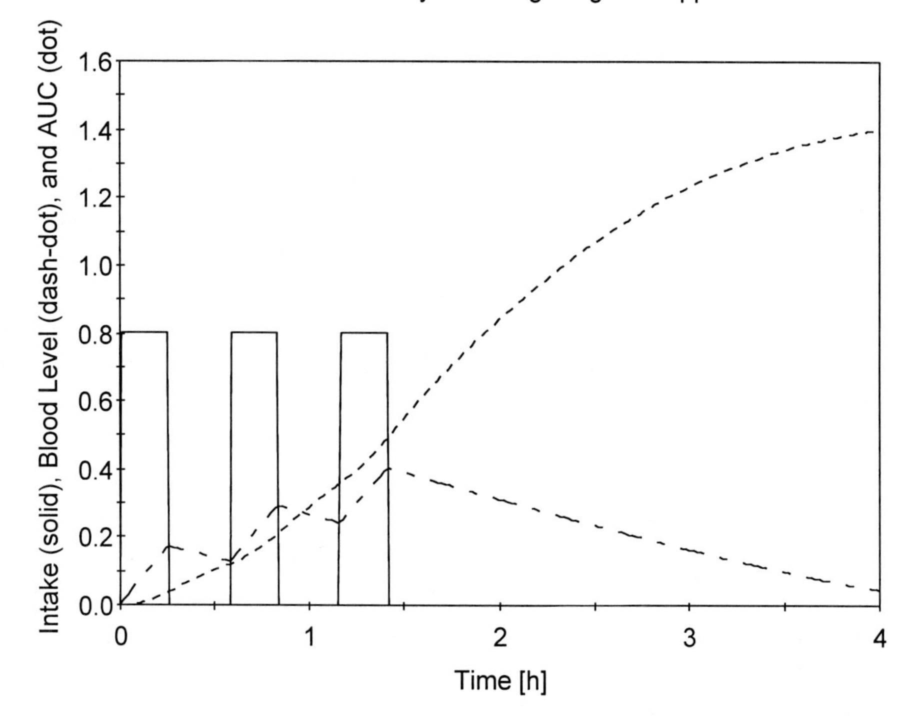

Fig 7.4 Alcohol metabolism analyzed using Wagner's approach

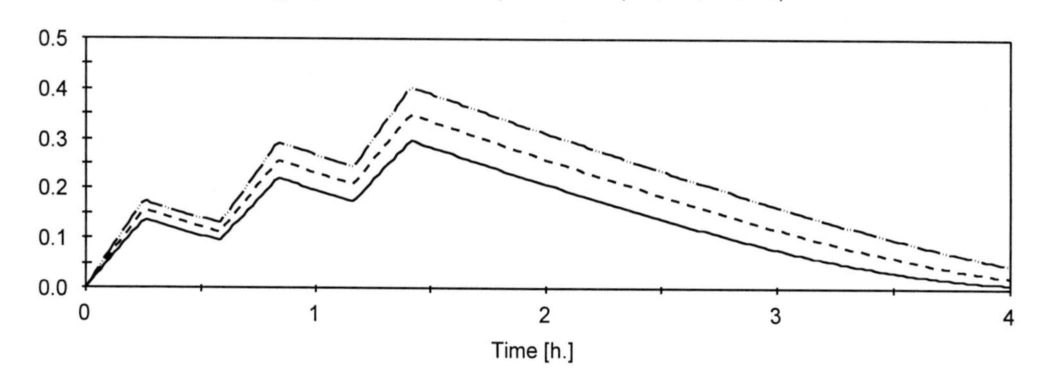

Fig 7.5 Blood alcohol concentration

IV. Decoupling examples

Assembling models in sections thus leads to i) a better understanding of the behavior of the model components, ii) more confidence than the algebraic model due to assembly of the key functions properly, and iii) a set of potentially reusable building blocks from which other complex models can be assembled. But the advantages of developing models in sections actually extend beyond this. Berman et al. (2) and later Boston et al. (3) demonstrated that uncoupling models using the notion of forcing functions greatly stabilizes the model fitting process. When a 'large' model involving some tenuously (unidirectional) connected sub-models is to be fitted to data from various sources, the fitting process is drastically eased if the sub-models can be independently fit. As each sub-model is fit it is replaced by a forcing function which essentially uncouples the fit of the sub-model, but at the same time, drives the connected sub-model allowing its parameters to be fit virtually independent of the sub-model underlying the forcing function. Of course, where the sub-models are highly interconnected, the application of the approach is more complicated (see Chapter 8 for examples).

We illustrate the ideas here as follows. Consider the model of Fig 7.6 in which the two sub-models are unidirectionally coupled (recycling from sub-model II to sub-model I exists but is quite small relative to the forward cycling fraction (5)). A strategy for fitting this complex model would be as follows.

1. Fit the response data from compartment 2.
2. Map the kinetics model as shown to an exponential model.
3. Define the exponential model for compartment 2 to be a forcing function.
4. Fit the data from compartment 5 using sub-model *2*.

The advantage in using this approach is that whereas fitting the single model at once involves estimating 9 parameters, one of which may be only marginally resolved (numerically identifiable), when we fit the model in two stages, we fit the parameters in groups of 5 and 4 to the respective data sets pertaining to compartments 2 and 5.

Analytically-based approaches are not the only ones available to decouple sections of a model. Where data are available for two connected sub-models, it may be appropriate to use the data associated with each sub-model, in turn to drive the other sub-model, enabling sequential fitting of the sub-models. The justification of this approach is that if each sub-model describes the associated data acceptably well, then biases in conjunction with sequential fitting will be small.

The advantage offered by this approach is that sub-model estimation is usually simpler than estimation of complex models. Novel applications of this idea have been discussed by Bergman and Cobelli in a series of papers (1). In investigating

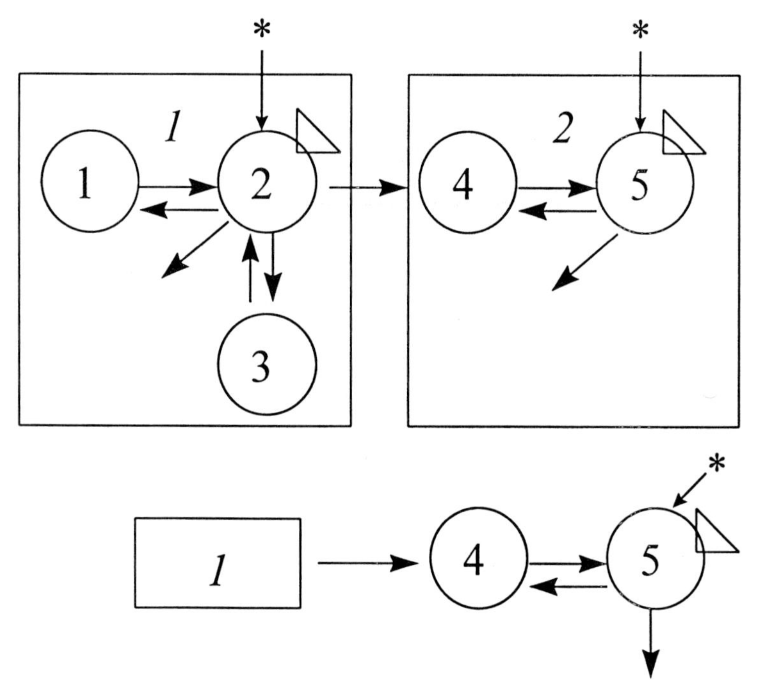

Fig 7.6 Coupled compartmental models.

the glucose-insulin system, they explored data-based uncoupling of the two sub-models for each glucose and insulin (Fig. 7.7).

To illustrate their approach, consider the glucose disposal sub-model of Fig 7.7. Here the insulin supply term in the equation for remote insulin (also known as insulin action is X) involves the circulating insulin observations as opposed to values calculated from the effect of glucose pancreatic beta-cells.
The differential equations for this model become:

$$G' = -(P_1 + X)\,G + P_1.G_b$$
$$X' = -\,P_2.X + P_3\,.(I\text{-}I_b)$$

where G is the computed glucose time course, X is the computed insulin action time course, and I is observed circulating insulin.
The reverse is deployed by Bergman and Cobelli (1) when pancreatic response is to be explored. Here insulin response is driven by observed circulating glucose and the model in Fig 7.8 is used.

The equation describing the insulin response is;

$$I' = -nI + \gamma(G\text{-}h)t$$

where G is now a functional form created from the circulating glucose observations, and n, h, and γ are insulin sub-model parameters.

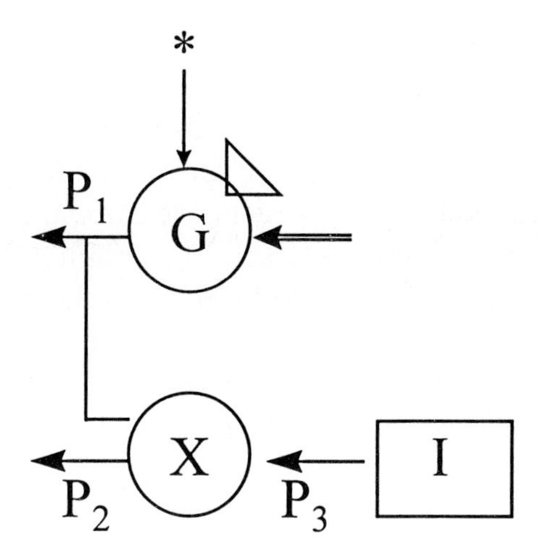

Fig. 7.7 Bergman minimal-model of glucose/insulin kinetics; G-glucose, I-insulin, X-remote insulin (1).

A remaining issue is the use of data to form forcing functions, i.e., what is the degree to which the nature of the formed function influences the fitting process and hence the parameter values emerging. Bergman and Cobelli (1) appear to use step functions in which the function only changes at observation points in keeping with the new values obtained there. Clearly a more realistic approach would be to either linearly interpolate between observation points or to create a running smoothing function and these and other alternative approaches are being explored by several groups. Evidence is mounting that the form of the data/forcing function indeed influences parameter estimates.

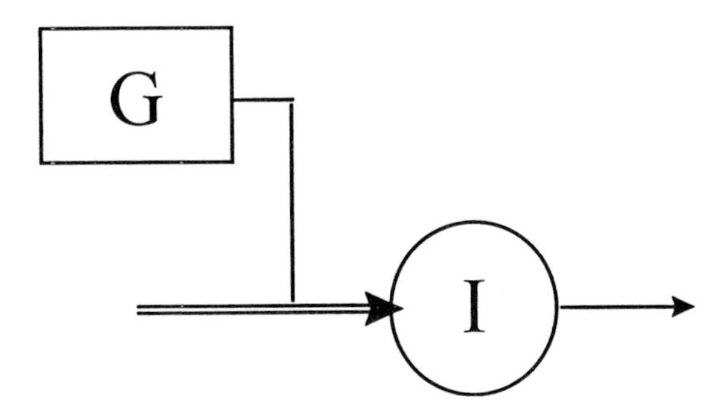

Fig 7.8 Bergman insulin response model (1).

V. Summary

In this chapter we have discussed the ideas underpinning building models in sections or by system decoupling. We have shown that for large models involving an array of complicated modeling constructs, we can only have confidence in the model predictions, if we are sure that each of the building blocks of the model function as required. Furthermore, the links between the models must also reflect what we desire to accomplish. We also stressed the importance of separating the model from its inputs, or experimental perturbations, as we attempt to assemble the model. By doing this we can confirm that each, individually, reflects our needs and thus we can have more confidence that the two combined will function appropriately.

Finally we reviewed the value of manipulating models in sections as regards to model fitting. We used the notion of decoupling and model section replacement with forcing functions and saw that the complexity of the fitting process is substantially lessened. Two types of decoupling were described: model decoupling in which the model section was replaced by a sub-model-based forcing function and data decoupling in which the model section was replaced by a sub-system data-based forcing function. Instances of the application of each were described. Details on how to apply these techniques are described in the following chapter.

REFERENCES

1. Bergman, R. N., and C. R. Bowden. 1981. The minimal model approach to quantification of factors controlling glucose disposal in man. *In* Carbohydrate Metabolism. C. Cobelli and R. N. Bergman, editors. Wiley. 269-296.
2. Berman, M., M. F. Weiss, and E. Shahn. 1962. Some formal approaches to the analysis of kinetic data in terms of linear compartmental models. *Biophys J.* 2:289-316.
3. Boston, R. C., D. D. Leaver, and J. J. Quilkey. 1976. A DEC System 10 version of SAAM: its implementation and application. *In* DEC User's Society. Vol. 2. 1097-1102.
4. Holford, N. H. G. 1987. Clinical pharmacokinetics of ethanol. *Clin Pharm.* 13:273-292.
5. McGuire, R. A., and M. Berman. 1978. Maternal, fetal, and amniotic fluid transport of thyroxine, triidothyronine, and iodide in sheep. *Endocrinol.* 103:567-576.

6. MicroMath Scientific Software. 1995. Scientist, Mathematical Modeling, Differential and Nonlinear equations., Salt Lake City, UT.
7. Pieters, J. E., M. Wedel, and G. Schaafsma. 1990. Parameter estimation in a three compartment model for blood alcohol curves. *Alcohol & Alcoholism*. 25:17-24.
8. Wagner, J. G., P. K. Wilkinson, and D. A. Ganes. 1989. Parameters Vm and Km for elimination of alcohol in young male subjects following low doses of alcohol. *Alcohol & Alcoholism*. 24:555-564.
9. Wilkinson, P. J., A. J. Sedman, and J. G. Wagner. 1977. Fasting and non-fasting blood ethanol concentrations following repeated oral administration to one adult male subject. *J. Pharm Biopharm*. 5:41-52.

8
TECHNIQUES AND TOOLS TO FACILITATE MODEL DEVELOPMENT

The purpose of this chapter is to show how to use models to represent systems using WinSAAM. The examples begin by introducing one technique or tool at a time and then show how the tools are applied to model biological systems (Table 8.1). It is suggested that the reader setup and solve each model because this is the best way to understand the tools. Each model is discussed in terms of the problem being addressed, followed by the strategy used to solve the problem, and a demonstration of the result.

Table 8.1 WinSAAM Examples

	NAME	DEMONSTRATE
1	One-compartment model	IC, L, F
2	Two-compartment model	IC, L, F
3	Steady state solution	U, R, M
4	Equations	P,G
5	Continuous infusion	UF
6	QO-operation	QO
7	Delays	DT, DN
8	Nonlinear system	Function-dependent L
9	Tracer and Tracee model	Model non-steady state
10	Equivalence of tracer and tracee supply	Effect of site of tracee input on mass
11	Multiple dosing	QO
12	Infusion	Time-blocks
13	a. Infusion	QO
	b. Infusion	Function dependence
14	Nonlinear Equations	Parameter sensitivity, integrals, area under curve
15	Michaelis-Menten	Function dependency
16	Dosing Regimes	Modeling a system using UF's
17	Fit data from 2 sites	UF-equation vs. L model
18	Area under curve	FF
19	Forcing function	FF, QL
20	Function dependencies	Function operators (+-/*, 5-9, A-E)

Example 1. One-Compartment Model
Problem: To model data represented by a monoexponential loss.
Solution: Set up a one-compartment model.
Demonstration: See Fig 8.1a for the model graphic and Text input file and Fig 8.1b for a plot of the solution.

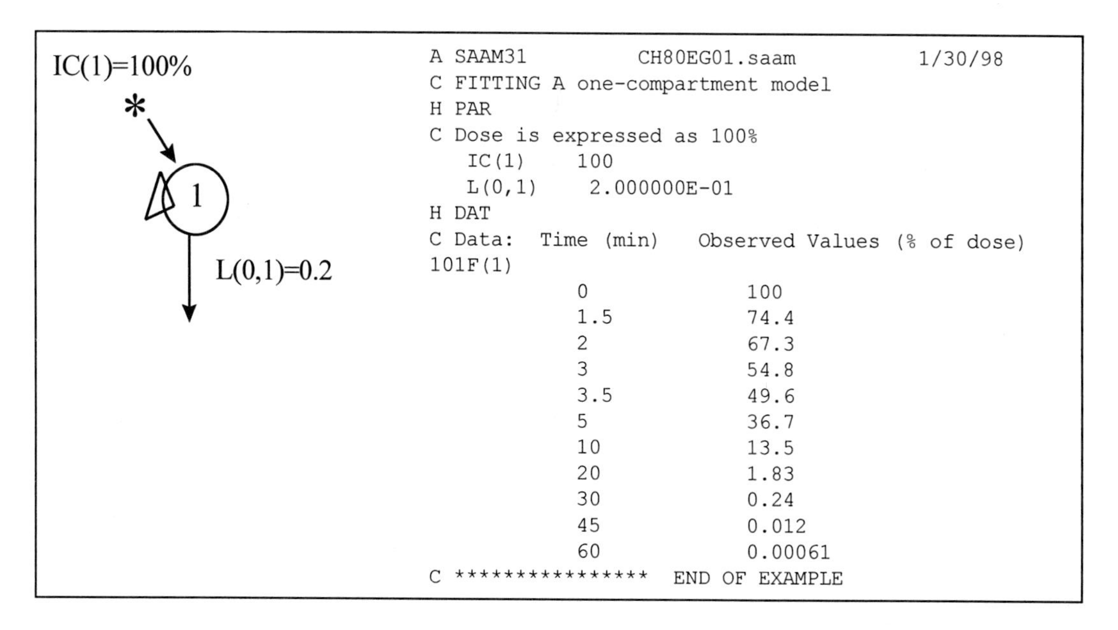


```
A SAAM31              CH80EG01.saam             1/30/98
C FITTING A one-compartment model
H PAR
C Dose is expressed as 100%
   IC(1)     100
   L(0,1)     2.000000E-01
H DAT
C Data:  Time (min)    Observed Values (% of dose)
101F(1)
            0              100
            1.5            74.4
            2              67.3
            3              54.8
            3.5            49.6
            5              36.7
            10             13.5
            20             1.83
            30             0.24
            45             0.012
            60             0.00061
C ****************  END OF EXAMPLE
```

Fig 8.1a One-compartment model in graphic at left is represented as a WinSAAM Text input file on right. In the graphic the asterisk indicates the entry site of the compound of interest, or the initial condition, IC(1), the circle is the compartment, F(1), the triangle represents the site of data sampling, the arrow out of the circle represents the monoexponential loss from the compartment, designated as L(0,1), i.e., into the outside from compartment 1.

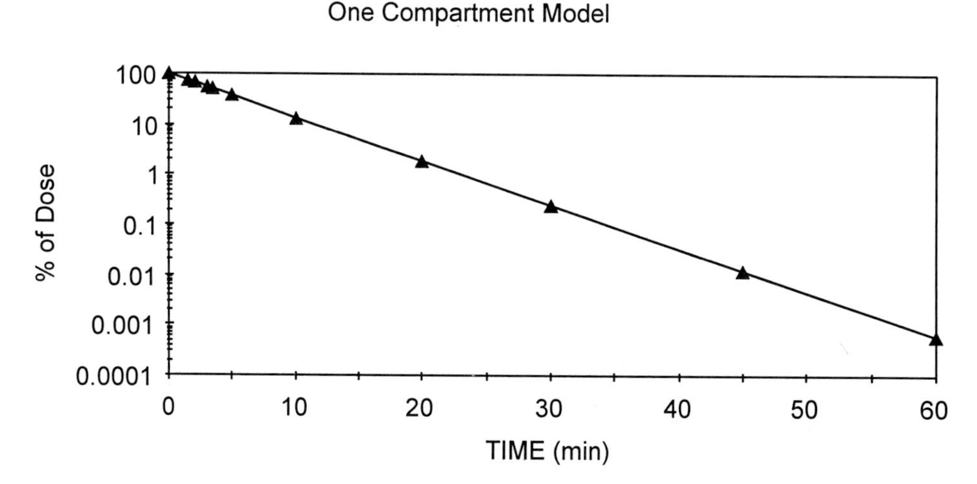

Fig. 8.1b. Plot of a monoexponential loss, solved using the model in Fig 8.1

Example 2. Two-Compartment model

Problem: A tracer is injected into blood. The amount of tracer in blood is measured for 30 min. Determine the fractional transfer rates of exchange with blood.

Strategy: Use a 2-compartment model

Demonstration: See Fig 8.2.

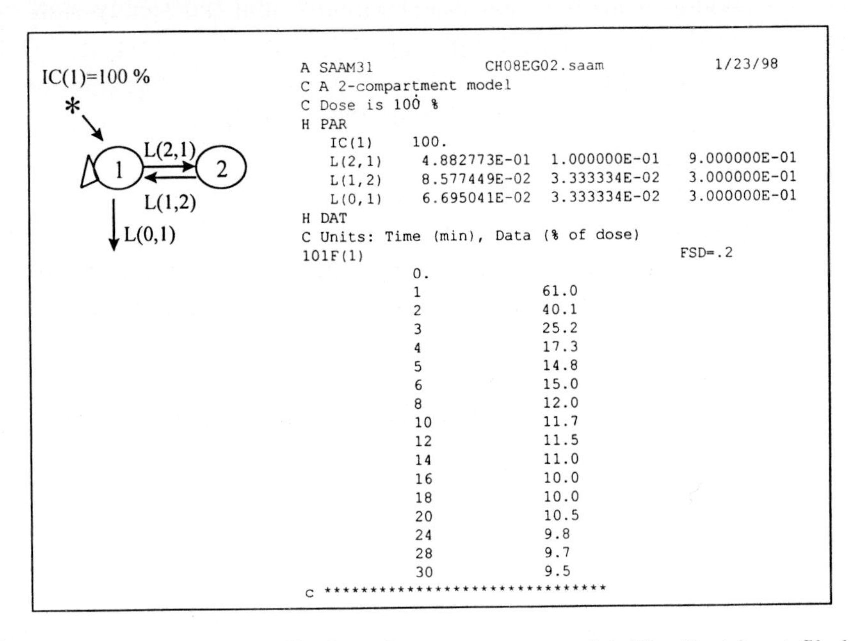


```
A SAAM31                 CH08EG02.saam                 1/23/98
C A 2-compartment model
C Dose is 100 %
H PAR
   IC(1)     100.
   L(2,1)     4.882773E-01  1.000000E-01   9.000000E-01
   L(1,2)     8.577449E-02  3.333334E-02   3.000000E-01
   L(0,1)     6.695041E-02  3.333334E-02   3.000000E-01
H DAT
C Units: Time (min), Data (% of dose)
101F(1)                                          FSD=.2
           0.
           1               61.0
           2               40.1
           3               25.2
           4               17.3
           5               14.8
           6               15.0
           8               12.0
          10               11.7
          12               11.5
          14               11.0
          16               10.0
          18               10.0
          20               10.5
          24               9.8
          28               9.7
          30               9.5
C ********************************
```

Fig 8.2a Graphic and text input file for a 2-compartment model. The Text input file has upper and lower limits on the parameters, and a weight representing a fractional standard deviation of 20% assigned to the data.

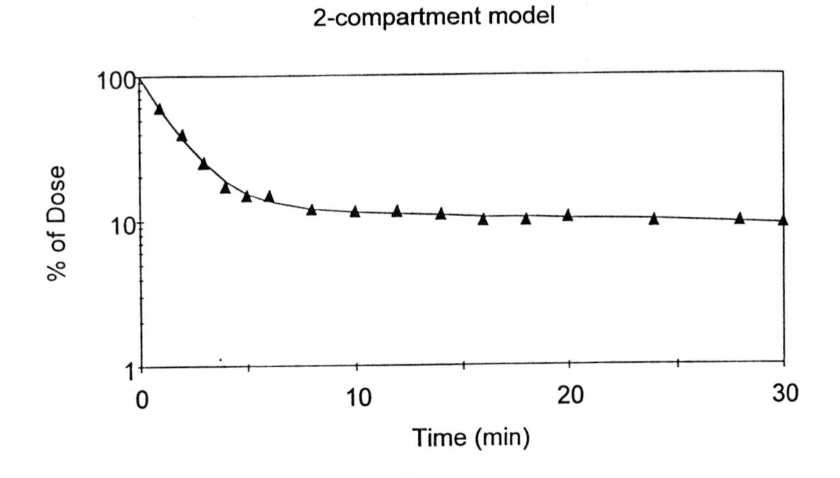

Fig 8.2b Plot of the data and model solution for a 2-compartment model. The plot was generated by the command >PLOT Q(1).

Example 3. Steady-state Solution

Problem: Determine the amount of a compound of interest in a compartment, M(I), the entry rate, U(I), and the transport rate between the compartments, R(I,J).

Problem 3a: Steady state solution for a one-compartment model.

Strategy: The relationship between tracer kinetics and the steady state solution for the tracee (or compound of interest) is shown in Fig 8.3.a.1. Note that M(I) is mass of tracee in compartment I (amount), R(I,J) is transport rate of tracee into compartment I from, compartment J, (amount/unit time) and is equivalent to: R(I,J) = M(I) * L(I,J). U(I) is rate that tracee enters the system (amount/unit time). The Text input file for calculating the solution is shown in Fig 8.3a.2. To calculate steady state values, (input rates, mass, transport rates) it is necessary to add the following to the Text input file:

- H STE heading
- An input or a mass
- Request steady state parameters of interest under H DAT (e.g., component 0)

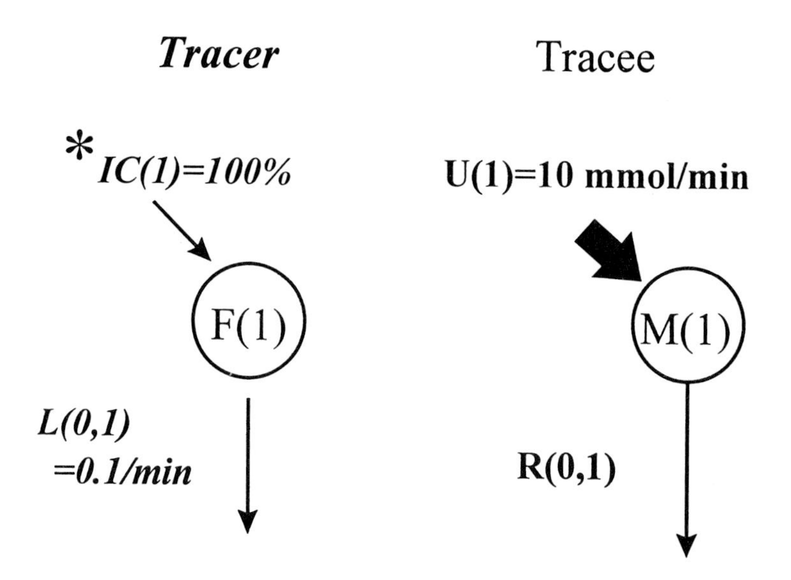

Fig 8.3a.1: Model on the left shows the tracer kinetics, where F(1) is the amount of tracer in compartment 1 at any time point. The model on the right calculates the steady state values for the tracee, or compound of interest, where U(1) is entry of tracee into compartment 1, M(1) is the mass in compartment 1 and R(0,1) is the transport rate out of compartment 1.

```
A SAAM31              CH083a.saam                    1/31/98
C EXAMPLE OF CALCULATING STEADY STATE
C ONE COMPARTMENT MODEL
H PAR
C DOSE IS %. TIME UNITS ARE MINUTES.
    IC(1)      100.
    L(0,1)     0.1
H STE
C INPUT IS 10 MMOL/MIN
    U(1)       10.
    M(1)       1.            0.             1000
H DAT
100
    U(1)
    R(0,1)
    M(1)
101F(1)
               0.
               1             90.48
               2             81.87
               3             74.08
               4             67.03
               5             60.65
               6             54.88
               7             49.65
               8             44.93
               9             40.65
               10            36.78
C ********** END OF MODEL
```

Fig 8.3a.2 Text input file for calculating the steady state parameters for a one-compartment model.

Problem 3b: Determine the input rate, compartment masses and rates of transport between compartments for the 2-compartment model shown in Example 8.2 (Fig. 8.3b)

Example 4: Equations

Problem 4a: A dose is administered and the amount of tracer in samples is determined as % dose/L. Convert the calculated units (%) to observed units (%/ml).

Strategy: Units can be converted by using an equation.

Result: The units for the 2-compartment model are converted using P(1) where P(1) is equivalent to the volume of distribution of the first pool (Fig. 8.4a).

Problem 4b: A tracer is measured as the sum of two compartments.

Strategy: Use an equation to sum the calculated values in the two compartments.

Result: The Text input file with the equation for summing the compartments, G(2) is shown in Fig 8.4b.1 The plot is shown in Fig 8.4b.2. Two examples of summing compartments are Fisher WR et al. (1997) for fitting plasma lipoproteins and Blomhoff R et al.(1989) for fitting cellular endocytosis.

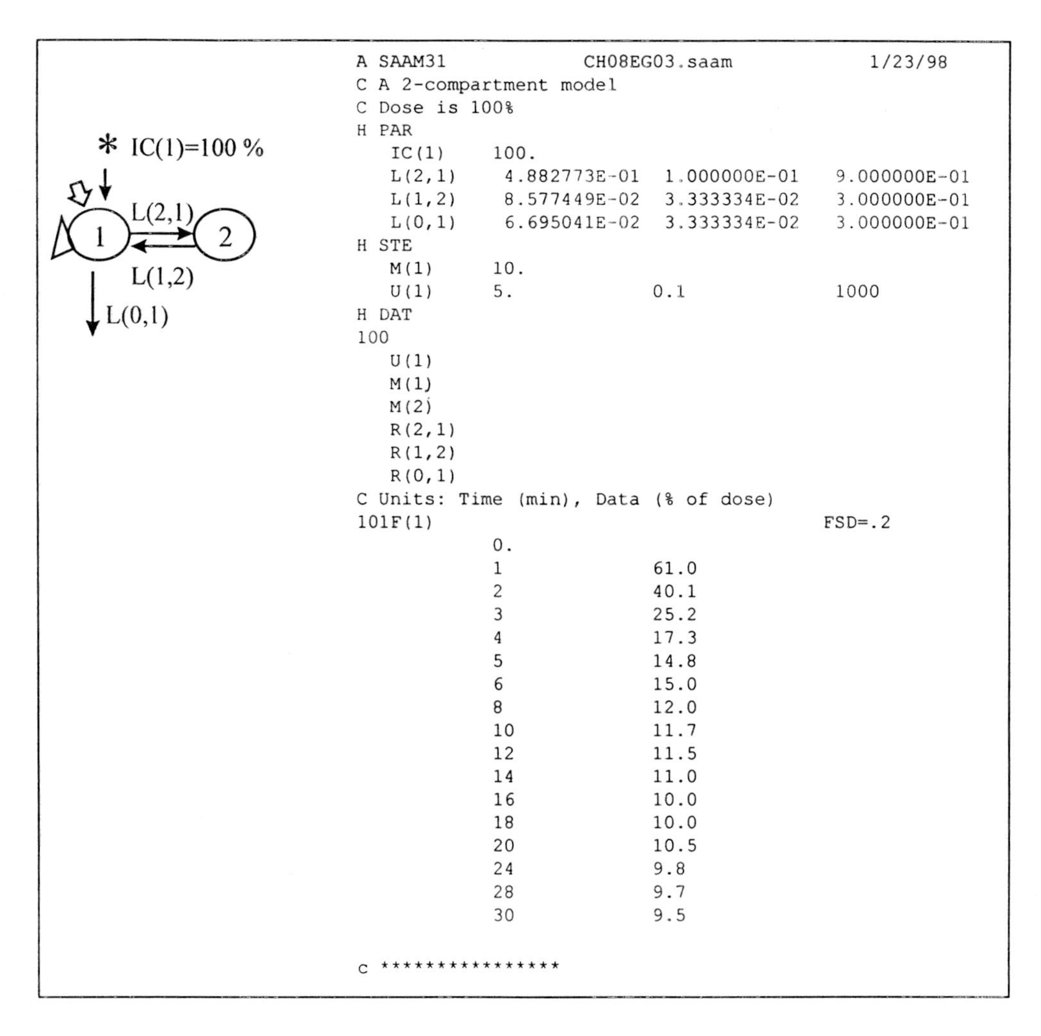


```
A SAAM31                CH08EG03.saam                1/23/98
C A 2-compartment model
C Dose is 100%
H PAR
   IC(1)     100.
   L(2,1)     4.882773E-01   1.000000E-01    9.000000E-01
   L(1,2)     8.577449E-02   3.333334E-02    3.000000E-01
   L(0,1)     6.695041E-02   3.333334E-02    3.000000E-01
H STE
   M(1)      10.
   U(1)      5.              0.1             1000
H DAT
100
   U(1)
   M(1)
   M(2)
   R(2,1)
   R(1,2)
   R(0,1)
C Units: Time (min), Data (% of dose)
101F(1)                                     FSD=.2
            0.
            1               61.0
            2               40.1
            3               25.2
            4               17.3
            5               14.8
            6               15.0
            8               12.0
            10              11.7
            12              11.5
            14              11.0
            16              10.0
            18              10.0
            20              10.5
            24              9.8
            28              9.7
            30              9.5

C ****************
```

Fig 8.3b.1 Graphic and Text input file for determining the steady state values for a compartmental model (i.e., compartment mass, transport rate, and entry rate of tracee).

CATEGORY	QC
U (1)	6.695E-01
M (1)	1.000E+01
M (2)	5.693E+01
R (2, 1)	4.883E+00
R (1, 2)	4.883E+00
R (0, 1)	6.695E-01

Fig 8.3b.2 Output listing the calculated input of tracee into compartment 1, U(1), mass in each of the compartments, M(I),and transport rates between compartments, R(i,j). The rates are mmol/min and masses are in mmol.

```
A SAAM31               CH08EG04.saam                1/23/98
C A 2-compartment model
C Dose is 100%
H PAR
   IC(1)     100.
   L(2,1)      5.156974E-01  1.000000E-01    9.000000E-01
   L(1,2)      8.496518E-02  3.333334E-02    3.000000E-01
   L(0,1)      7.333220E-02  3.333334E-02    3.000000E-01
C P(1) is the volume of distribution
   P(1)        2.332598E+03  6.666667E+02    6.000000E+03
H DAT
XG(1)=F(1)/P(1)
C Units: Time (min), Data (% of dose/ml)
101G(1)                                     FSD=.2
            0.
            1             24.0E-3
            2             16.0E-3
            3             10.0E-3
            4             6.9E-3
            5             5.9E-3
            6             6.00E-3
            8             4.80E-3
            10            4.56E-3
            12            4.6E-3
            14            4.4E-3
            16            4.0E-3
            18            4.0E-3
            20            4.2E-3
            24            3.9E-3
            28            3.88E-3
            30            3.80E-3
C ****************
```

Fig 8.4a Text input file showing how an equation is used to convert calculated units (% of dose) to observed units (% of dose/ml). Note the category field changes to G(1).

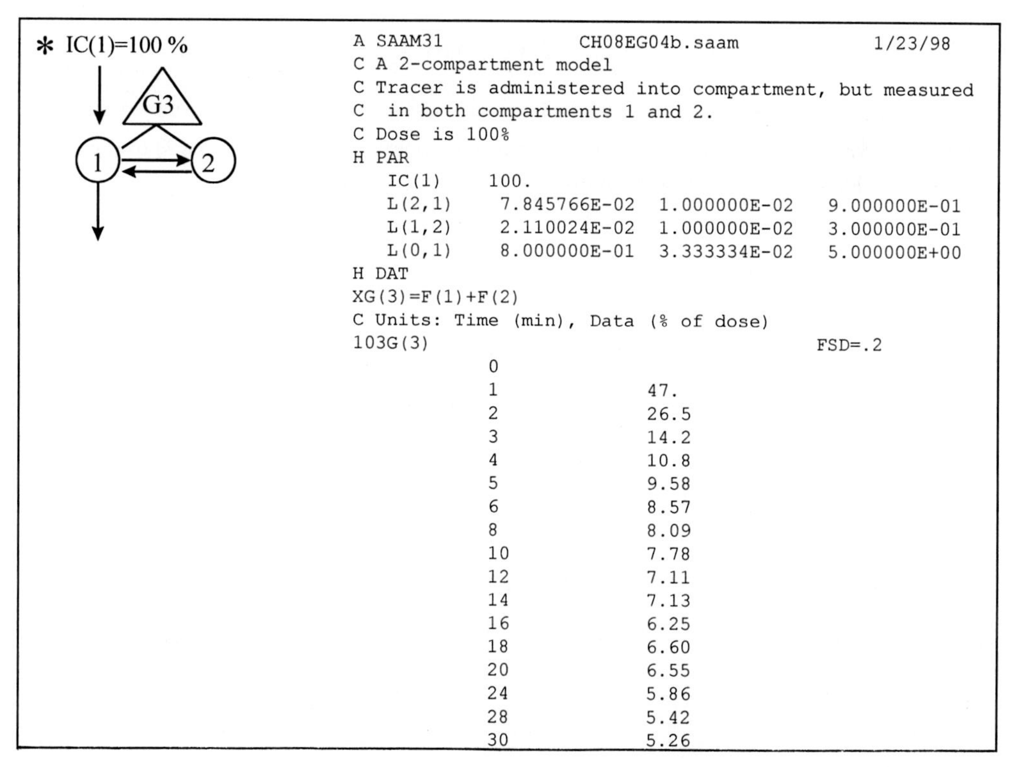


```
A SAAM31               CH08EG04b.saam                1/23/98
C A 2-compartment model
C Tracer is administered into compartment, but measured
C  in both compartments 1 and 2.
C Dose is 100%
H PAR
   IC(1)     100.
   L(2,1)      7.845766E-02  1.000000E-02    9.000000E-01
   L(1,2)      2.110024E-02  1.000000E-02    3.000000E-01
   L(0,1)      8.000000E-01  3.333334E-02    5.000000E+00
H DAT
XG(3)=F(1)+F(2)
C Units: Time (min), Data (% of dose)
103G(3)                                     FSD=.2
            0
            1             47.
            2             26.5
            3             14.2
            4             10.8
            5             9.58
            6             8.57
            8             8.09
            10            7.78
            12            7.11
            14            7.13
            16            6.25
            18            6.60
            20            6.55
            24            5.86
            28            5.42
            30            5.26
```

continued

```
C Simulate values in compartment 1 and 2 at
C one minute intervals for 30 minutes
101F(1)
                    0.
2                   1.                        30.
102F(2)
                    0.
2                   1.                        30
c ***************
```

Fig 8.4b.1 Text input file for a 2-compartment model showing how data that were obtained by sampling compartments 1 and 2 are fitted by using an equation, G(3). The file also shows how values can be simulated for compartments 1 and 2.

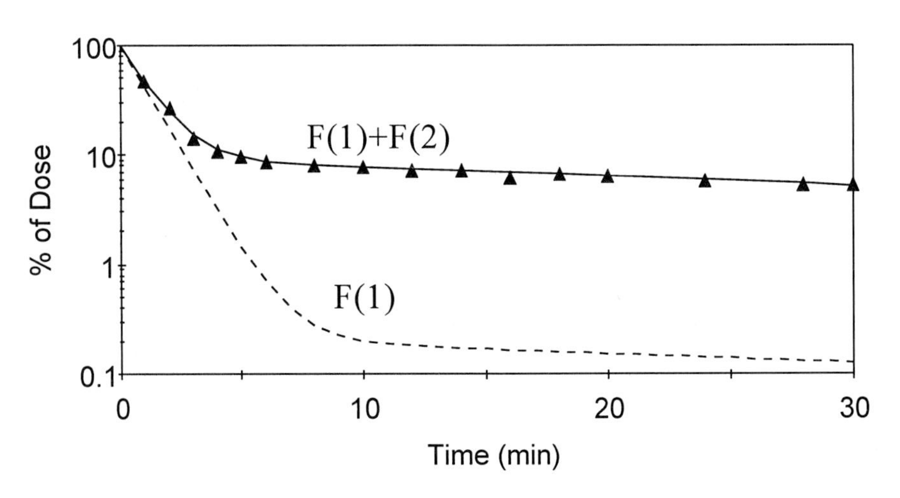

Fig 8.4b.2 Plot of model in Fig 4b.1 showing the fit generated by equation G(3) to the observed data, F(1)+F(2) and the values generated for compartment 1.

Example 5: Continuous Infusion

Problem: Tracer (or tracee) is administered by a continuous infusion.

Strategy: This is modeled by using a UF function. The function can be equated to a constant or to an equation. The model (Fig 8.5a) demonstrates the infusion of one mg/min.

Result: The plot (Fig 8.5b) shows the increase in compartment 1 resulting from a continuous infusion of 1 mg/min. See a later example for how the length of the infusion time, and also the amount infused can be set using by using the QO tool.

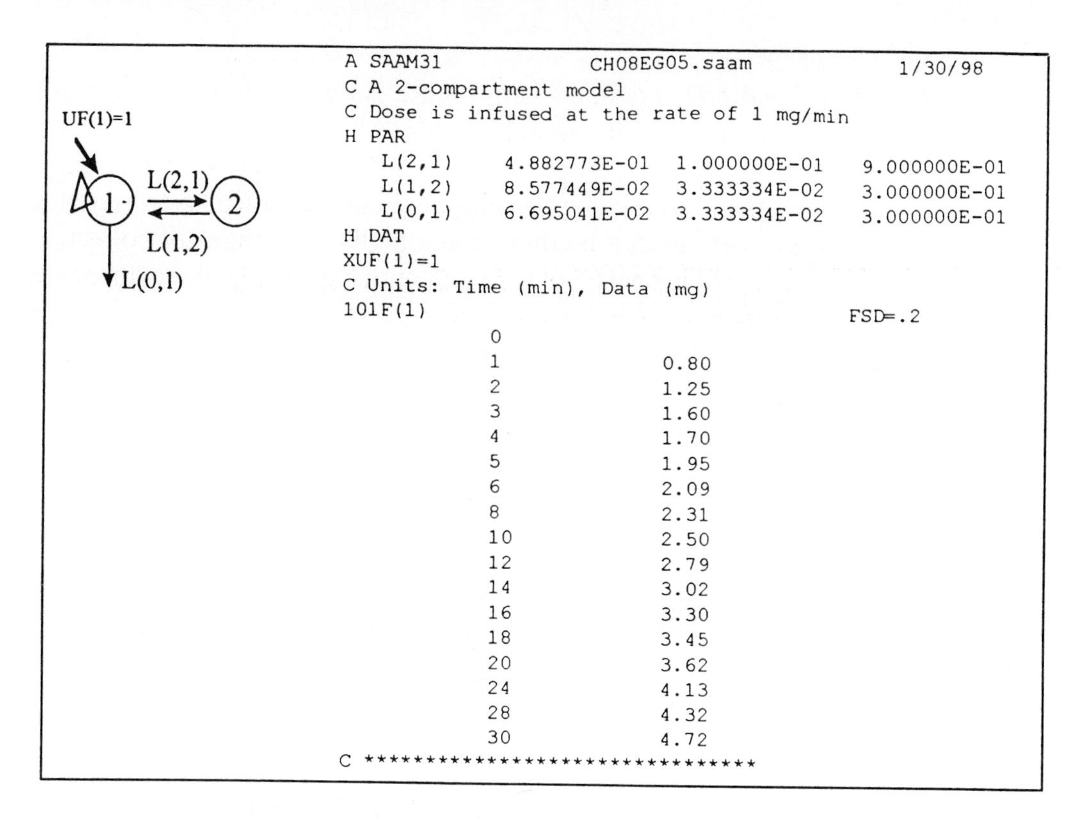


```
A SAAM31                    CH08EG05.saam                    1/30/98
C A 2-compartment model
C Dose is infused at the rate of 1 mg/min
H PAR
   L(2,1)     4.882773E-01  1.000000E-01   9.000000E-01
   L(1,2)     8.577449E-02  3.333334E-02   3.000000E-01
   L(0,1)     6.695041E-02  3.333334E-02   3.000000E-01
H DAT
XUF(1)=1
C Units: Time (min), Data (mg)
101F(1)                                        FSD=.2
           0
           1           0.80
           2           1.25
           3           1.60
           4           1.70
           5           1.95
           6           2.09
           8           2.31
          10           2.50
          12           2.79
          14           3.02
          16           3.30
          18           3.45
          20           3.62
          24           4.13
          28           4.32
          30           4.72
C ********************************
```

Fig 8.5a Model and Text input file for a continuous infusion, represented by UF(1).

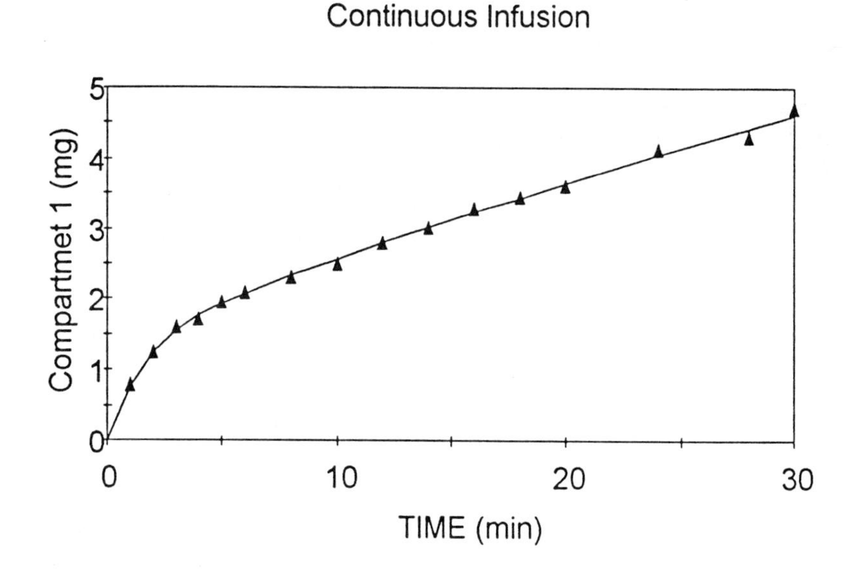

Fig 8.5b Simulation of a continuous infusion into compartment 1 (Fig 8.5a).

Example 6a: Use of QO

Problem: A tracer is administered and samples are collected from the excretion compartment (e.g., urine) at intervals.

Strategy: Collection data can be fitted by using QO, to reset the value for a compartment. The example (Fig 8.6a) shows how data collected at 30 min intervals can be fitted using a QO. The times of collection are specified under 103QO. At each time specified, the contents of compartment 3 will be reset to zero. The data are listed under component 13. Results of plotting Q(13) are shown in Fig 8.6a.2. Note that plotting Q(14) will show how material accumulates in F(3) between collections.

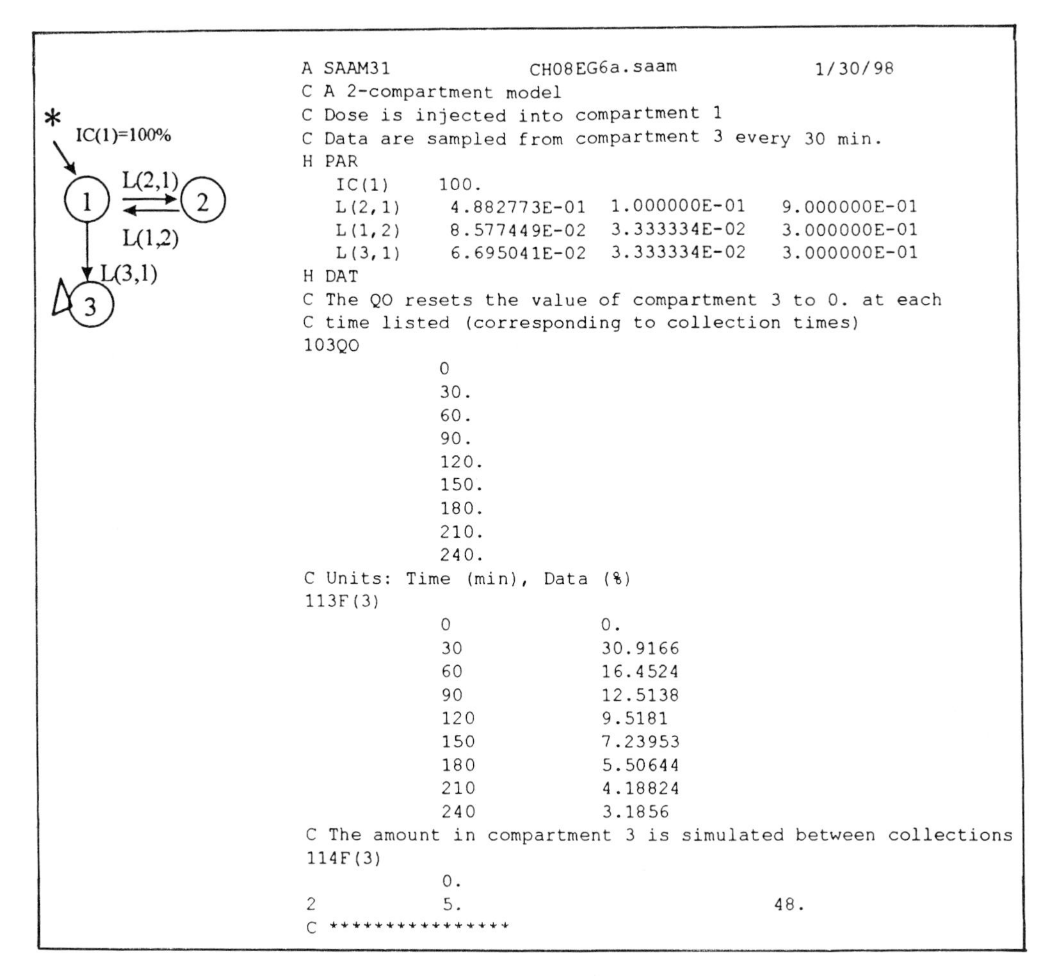


```
A SAAM31              CH08EG6a.saam              1/30/98
C A 2-compartment model
C Dose is injected into compartment 1
C Data are sampled from compartment 3 every 30 min.
H PAR
   IC(1)      100.
   L(2,1)      4.882773E-01   1.000000E-01    9.000000E-01
   L(1,2)      8.577449E-02   3.333334E-02    3.000000E-01
   L(3,1)      6.695041E-02   3.333334E-02    3.000000E-01
H DAT
C The QO resets the value of compartment 3 to 0. at each
C time listed (corresponding to collection times)
103QO
           0
           30.
           60.
           90.
           120.
           150.
           180.
           210.
           240.
C Units: Time (min), Data (%)
113F(3)
           0              0.
           30             30.9166
           60             16.4524
           90             12.5138
           120            9.5181
           150            7.23953
           180            5.50644
           210            4.18824
           240            3.1856
C The amount in compartment 3 is simulated between collections
114F(3)
           0.
2          5.                           48.
C ****************
```

Fig 8.6a.1 Model to simulate timed collections from compartment 3. The contents of compartment 3 are reset to zero after each collection by using QO. Data are entered under a different component number (here Q(13)), and equated to the calculated values for compartment 3.

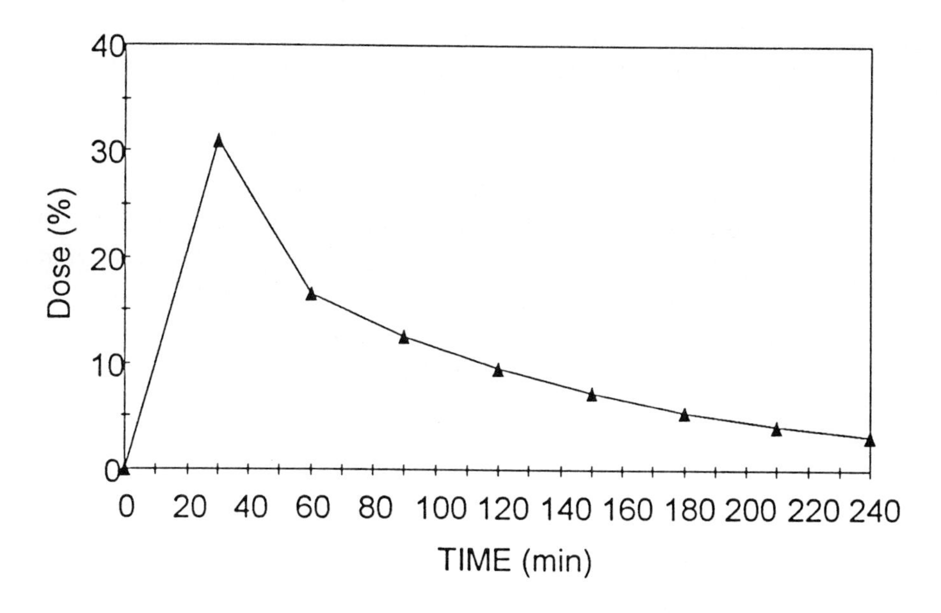

Fig 8.6a.2 Plot of Q(13), showing data collected from compartment 3 at 30 min intervals (symbols), and the calculated values (line), using the model shown in Fig 8.6a.1

Example 6b: Simulate timed and cumulative collections.

Strategy: As shown in example 6a, timed collections can be simulated using the QO tool to reset the value of a compartment. Cumulative values can be fit simultaneously by adding loss into an additional compartment, and a negative loss pathway, to balance this loss. This is demonstrated in Fig 8.6b.1. Compartment 4 is added to simulate the cumulative loss, and a loss pathway equal to -L(4,1) is is added to offset the duplicate loss pathway.

Results: The results of fitting both the timed and cumulative data are shown in Fig 8.6b.2.

Example 7: Delays

Problem: There is a delay before material appears in another compartment.

Strategy: This is modeled using the delay tools, DT (delay time) and DN (delay number). DT refers to the length of the delay and DN refers to the speed at which material is released from the delay, (a larger delay number results in a sharper release). In the demonstration (Fig 8.7a) the delay element 4 is shown as a box. The delay time refers to the time before material appears in the next compartment and, as shown by the data (Fig 8.7b), this is 60 minutes. The delay number

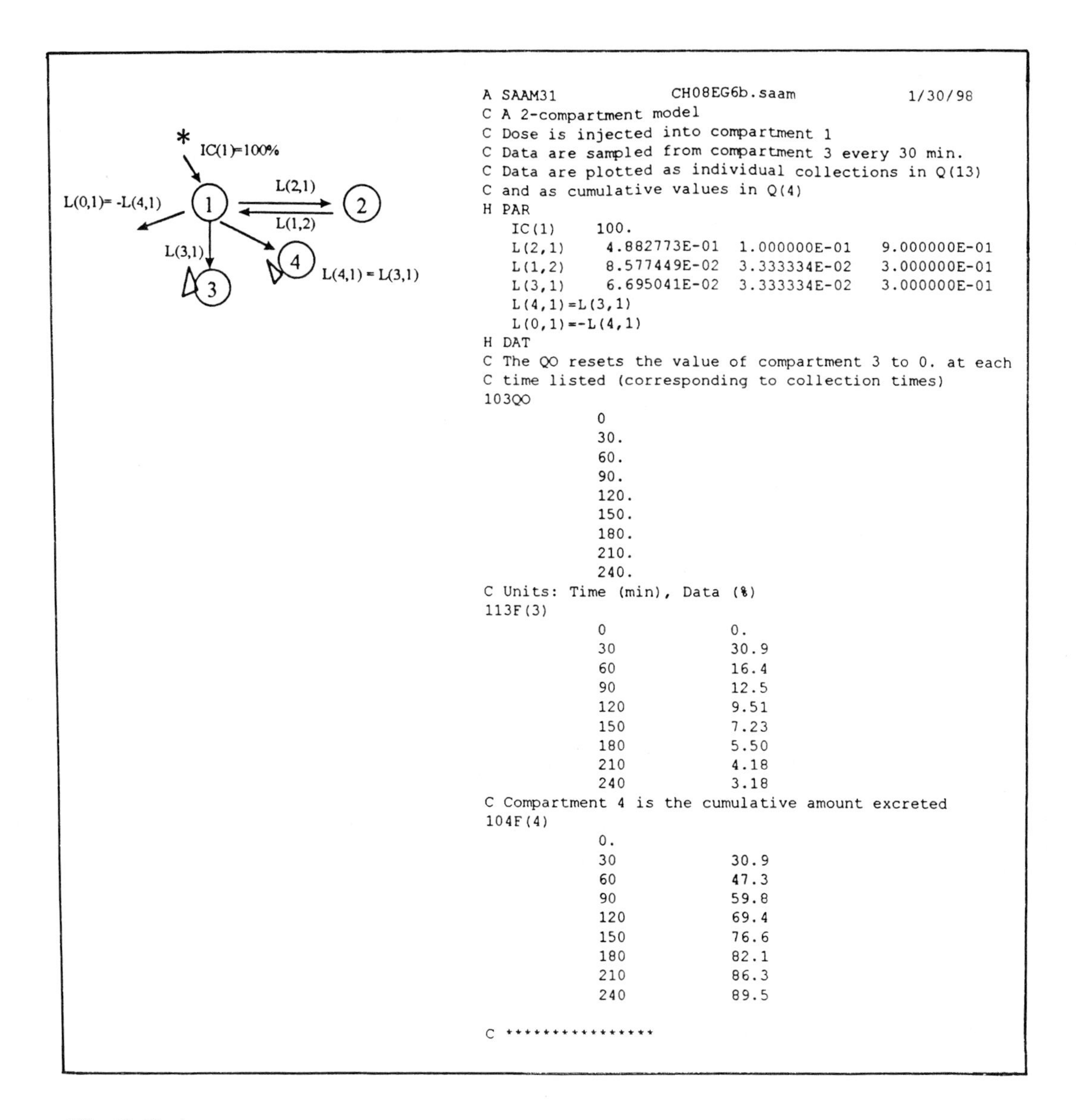


```
A SAAM31                CH08EG6b.saam                    1/30/98
C A 2-compartment model
C Dose is injected into compartment 1
C Data are sampled from compartment 3 every 30 min.
C Data are plotted as individual collections in Q(13)
C and as cumulative values in Q(4)
H PAR
  IC(1)     100.
  L(2,1)     4.882773E-01  1.000000E-01    9.000000E-01
  L(1,2)     8.577449E-02  3.333334E-02    3.000000E-01
  L(3,1)     6.695041E-02  3.333334E-02    3.000000E-01
  L(4,1)=L(3,1)
  L(0,1)=-L(4,1)
H DAT
C The QO resets the value of compartment 3 to 0. at each
C time listed (corresponding to collection times)
103QO
            0
            30.
            60.
            90.
            120.
            150.
            180.
            210.
            240.
C Units: Time (min), Data (%)
113F(3)
            0             0.
            30            30.9
            60            16.4
            90            12.5
            120           9.51
            150           7.23
            180           5.50
            210           4.18
            240           3.18
C Compartment 4 is the cumulative amount excreted
104F(4)
            0.
            30            30.9
            60            47.3
            90            59.8
            120           69.4
            150           76.6
            180           82.1
            210           86.3
            240           89.5

C ****************
```

Fig 8.6b.1 Model and Text input file demonstrating how to plot timed-collection data, Q(13), and cumulative data, Q(4), simultaneously. Two loss pathways are included: into compartment 3, which is QO'd at the collection times and into compartment 4, for the cumulative data. To balance the second loss into compartment 4, an additional loss, L(0,1) is included and set to a negative value.

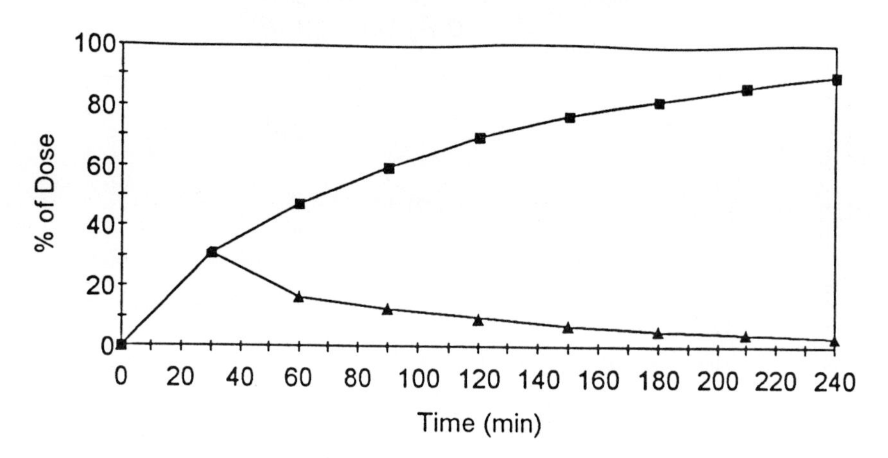

Fig 8.6b.2 Plot of timed, interval collections using the model in Fig 8.6b.1 (lower curve) and cumulative data (upper curve).

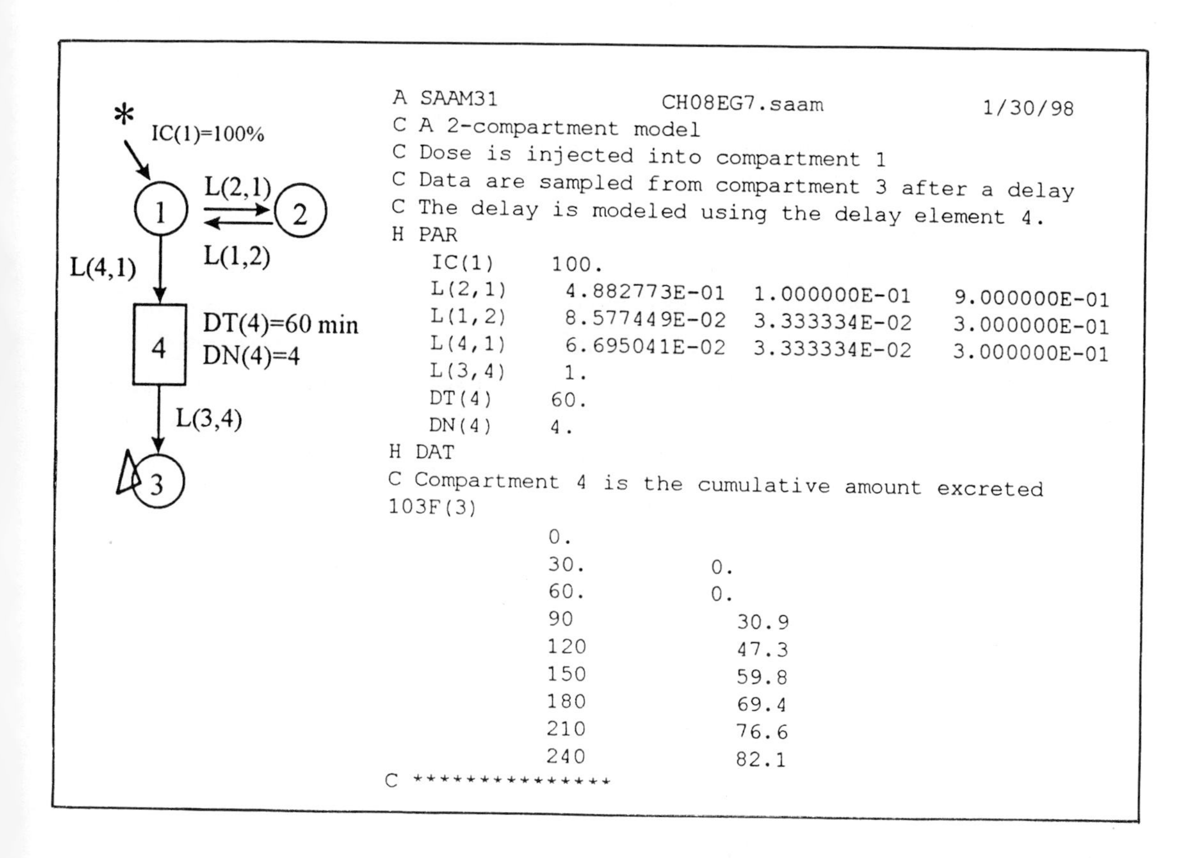


```
A SAAM31                    CH08EG7.saam                 1/30/98
C A 2-compartment model
C Dose is injected into compartment 1
C Data are sampled from compartment 3 after a delay
C The delay is modeled using the delay element 4.
H PAR
   IC(1)    100.
   L(2,1)    4.882773E-01  1.000000E-01   9.000000E-01
   L(1,2)    8.577449E-02  3.333334E-02   3.000000E-01
   L(4,1)    6.695041E-02  3.333334E-02   3.000000E-01
   L(3,4)    1.
   DT(4)    60.
   DN(4)    4.
H DAT
C Compartment 4 is the cumulative amount excreted
103F(3)
             0.
             30.          0.
             60.          0.
             90             30.9
             120            47.3
             150            59.8
             180            69.4
             210            76.6
             240            82.1
C ***************
```

Fig 8.7a Graphic and test input file for fitting material that appears in a compartment after a delay.

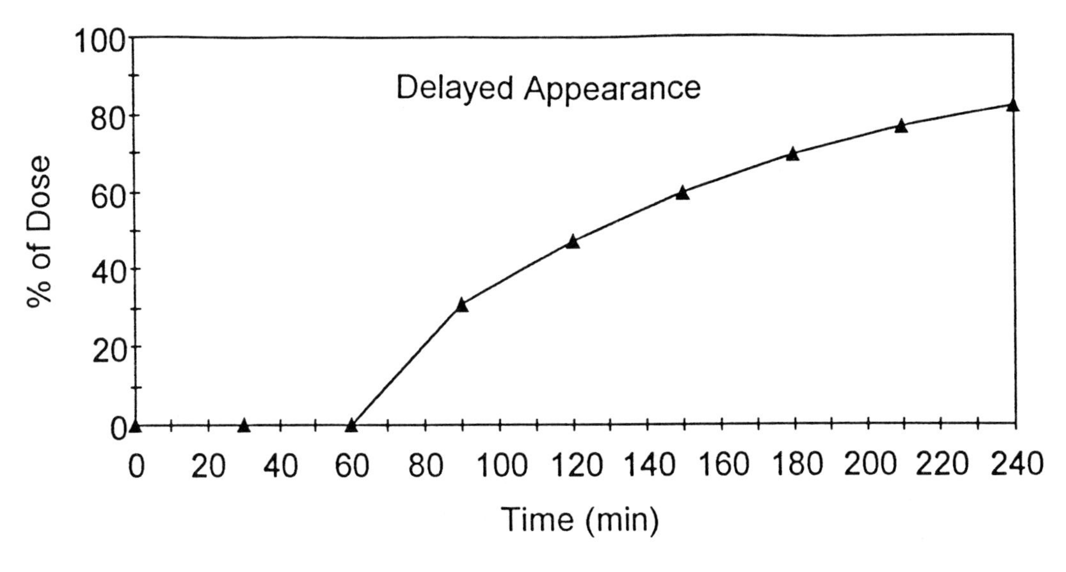

Fig 8.7b Appearance of material in a compartment after a 60 min delay.

is set at the value 4. A larger number would result in a sharper peak in the appearance of material in compartment 3. The value of L(3,4) is set at 1 and indicates that all the material in the delay goes into compartment 3. Output from a delay can go to more than one compartment. The loss pathways from the delay, indicate the proportions that go to each.

Example 8: Nonlinear System

Problem: A parameter changes in value over time (i.e., it is not constant) and therefore the system is nonlinear.

Strategy: Nonlinear parameters can be represented by equating the parameter to a function. In the example in Fig 8.8a, the rate of loss decreases inversely to the size of F(1). Parameters can be dependent upon a compartment, F(I), or function, G(I). The use of nonlinear parameters can be seen in Everts et al., (1996)

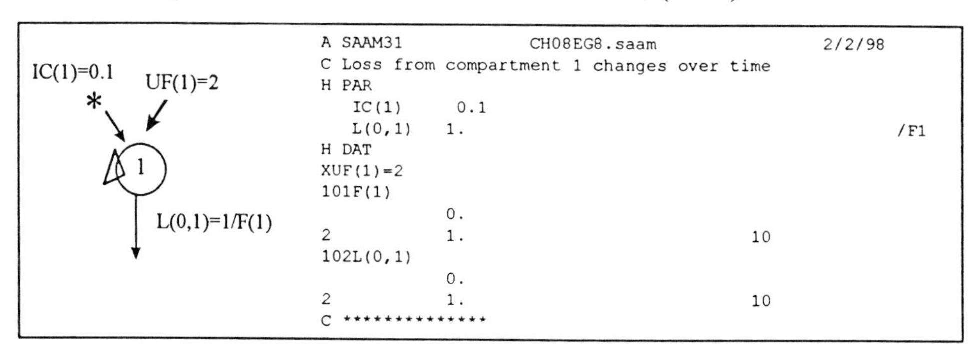

Fig 8.8a Graphic and Text input file for a nonlinear model. The value of L(0,1) is inversely related to the value of F(1).

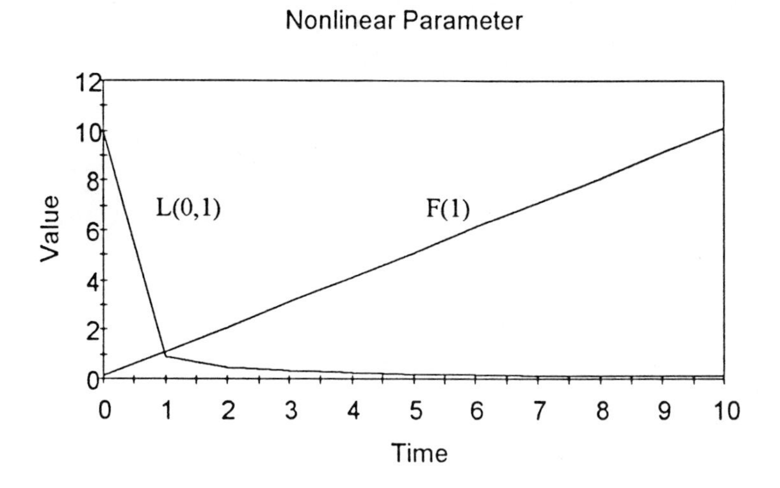

Fig 8.8b Plot of the decrease in L(0,1) with increase in F(1), generated using the text input file in Fig 8.8a.

Example 9: Modeling Tracer and Tracee

Problem: When a system is not in steady state (i.e., the pool sizes change over time), it is necessary to model both the tracer and tracee.

Strategy: Use parallel models, one for tracer and one for tracee. Initial conditions in the tracer model will be zero, except for entry of the tracer, while the tracee model will have initial conditions equal to the steady state masses of the tracer model (Fig 8.9a).

To simulate a change in tracee input, the model is modified, as shown in Fig 8.9b. Specifically, compartment 10 is added with a fast pathway to compartment 11. Compartment 10 is QO'd (i.e., the contents in compartment 10 are set to specified values which were 0 at time zero, and 100 mg at 4 hour and 6 hours). This causes an increase in tracee in compartments 11 and 12, as shown in Fig 8.9c.1. If the value of a parameter in the model changes with an increase in tracee, this can be modeled as shown in Fig 8.9c.b.2.

Note that L(2,1) is changed arbitrarily (from 2/hr to 4/hr). The system shown may represent, for example increased uptake of a compound by liver, after a meal. The change in L(2,1) could be equated to a function, such as a Michaelis-Menten function, if the nature of the process was known.

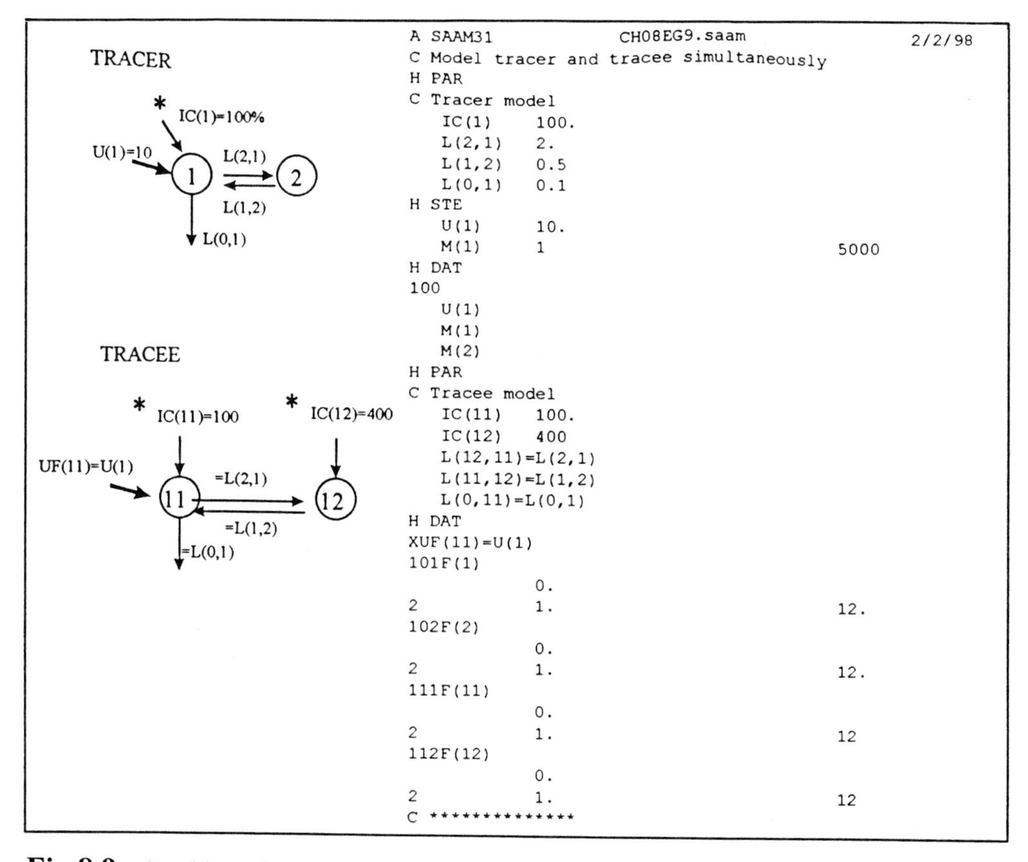


```
A SAAM31                    CH08EG9.saam                    2/2/98
C Model tracer and tracee simultaneously
H PAR
C Tracer model
    IC(1)     100.
    L(2,1)    2.
    L(1,2)    0.5
    L(0,1)    0.1
H STE
    U(1)      10.
    M(1)      1                          5000
H DAT
100
    U(1)
    M(1)
    M(2)
H PAR
C Tracee model
    IC(11)    100.
    IC(12)    400
    L(12,11)=L(2,1)
    L(11,12)=L(1,2)
    L(0,11)=L(0,1)
H DAT
XUF(11)=U(1)
101F(1)
              0.
2             1.                         12.
102F(2)
              0.
2             1.                         12.
111F(11)
              0.
2             1.                         12
112F(12)
              0.
2             1.                         12
C **************
```

Fig 8.9a Graphic and text input file for solving tracer and tracee data simultaneously. The initial conditions of the tracee model are equated to the steady state values of the tracer model.

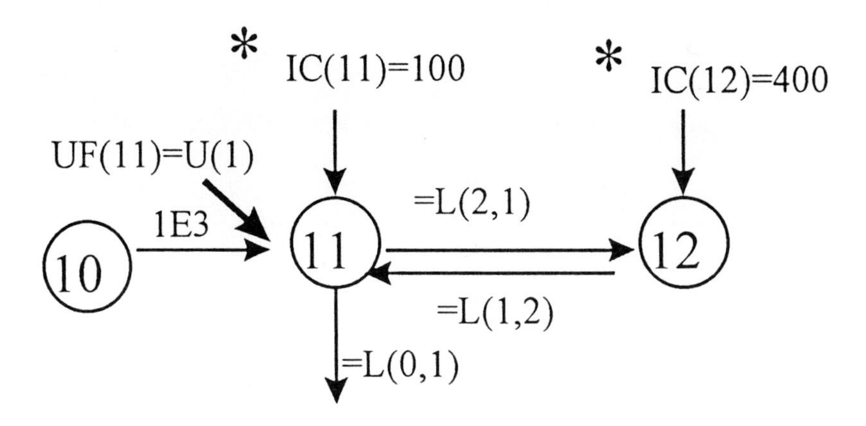

Fig 8.9b.1 Modification of tracee model in Fig 8.9a, for allowing delivery of tracee at specified times. This is performed by making Q(10) a QO, and specifying the times and values tracee is to be added.

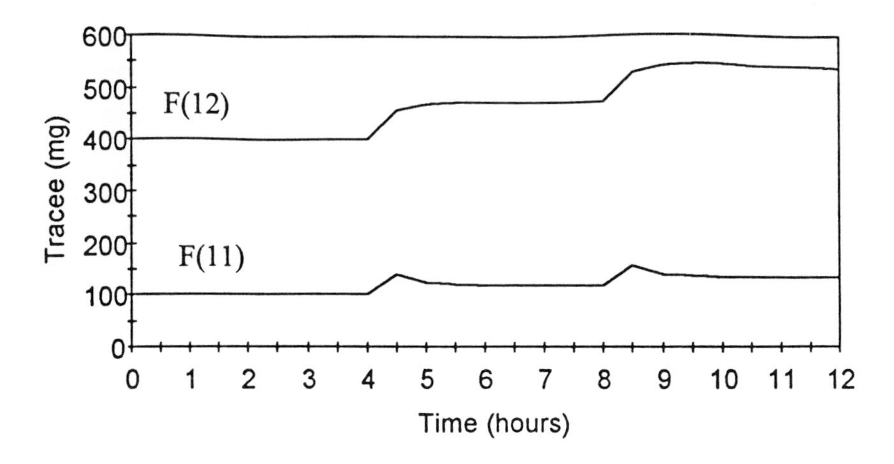

Fig 8.9b.2 Change in contents of compartments 11 and 12, when 100 mg are introduced into the tracee model at 4 and 8 hours.

```
A SAAM31                    CH08EG9c.saam                    2/2/98
C Model tracer and tracee simultaneously
C Effect of increased tracee input at 6 hours
H PAR
C Tracer model
   IC(1)      100.
   L(2,1)     1.                                                  *F21
   L(1,2)     0.5
   L(0,1)     0.1
H STE
   U(1)       10.
   M(1)       1                             5000
H DAT
100
   U(1)
   M(1)
   M(2)
H PAR
C Tracee model
   IC(11)     100.
   IC(12)     400
   L(12,11)=L(2,1)                                                *F21
   L(11,12)=L(1,2)
   L(0,11)=L(0,1)
   L(11,10) 1E3
H DAT
XUF(11)=U(1)
101F(1)
           0.
2          1.                               12.
102F(2)
           0.
2          1.                               12.
111F(11)
           0.
2          0.5                              24
112F(12)
           0.
2          0.5                              24
C COMPARTMENT 10 IS USED TO SIMULATE TRACEE ENTRY
C AT 4 AND 8 HOURS
110QO
           0.
           4.            100.
           8             100.
121QO
           0.            2.
           4.            4.
           6.            2
           8.            4
           10.           2.
C ************
```

Fig 8.9c.1 Model for simulating changes in tracee, when parameters of system change over time. The difference from the model in Fig 8.9.a is the use of compartment 10 for simulating entry of tracee at 4 and 8 hours, and compartment 21 for increasing L(2,1) for 2 hours after tracee entry increases.

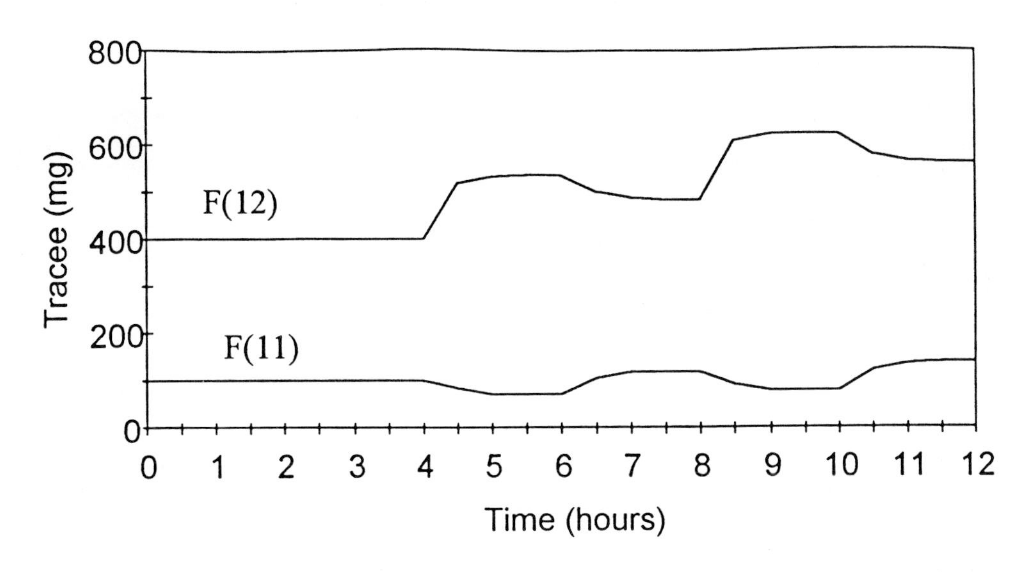

Fig 8.9c.2 Simulation of tracee in compartments 11 and 12 for the model in Fig 8.9c.1. When tracee into the system is increased at 4 and 8 hours, L(2,1) increases, resulting in a decrease in compartment 11 but higher increase in compartment 12, compared to no change in L(2,1) (i.e., the linear system, Fig 8.9b.2).

Example 10: Equivalence of Tracer and Tracee Supply

Problem: Determine the mass of compartments when the tracer is infused at a different site to entry of tracee. Determining the site of tracee entry is important for whole body metabolism of compounds such as vitamins, as described for retinol by Lewis KC et al. (1990).

Strategy: This example is included to demonstrate that the values determined for mass in a system depends on the entry site of tracee input. Duplicate models are set up (i.e., with the same structure and parameter values). In one, the tracee input is into the first compartment while in the other, tracee input is into the second compartment (Fig 8.10a).

Results: As shown in Fig 8.10b, the mass calculated for compartment 2 differs depending on site of tracee entry. The mass is larger, in this case, when input is into compartment 2. The reason for this is that when input is in to compartment 1, some is lost by L(0,1) before it reaches compartment 2. The relative masses generated by various inputs for a model can be determined by using the L-inverse (LI) tool.

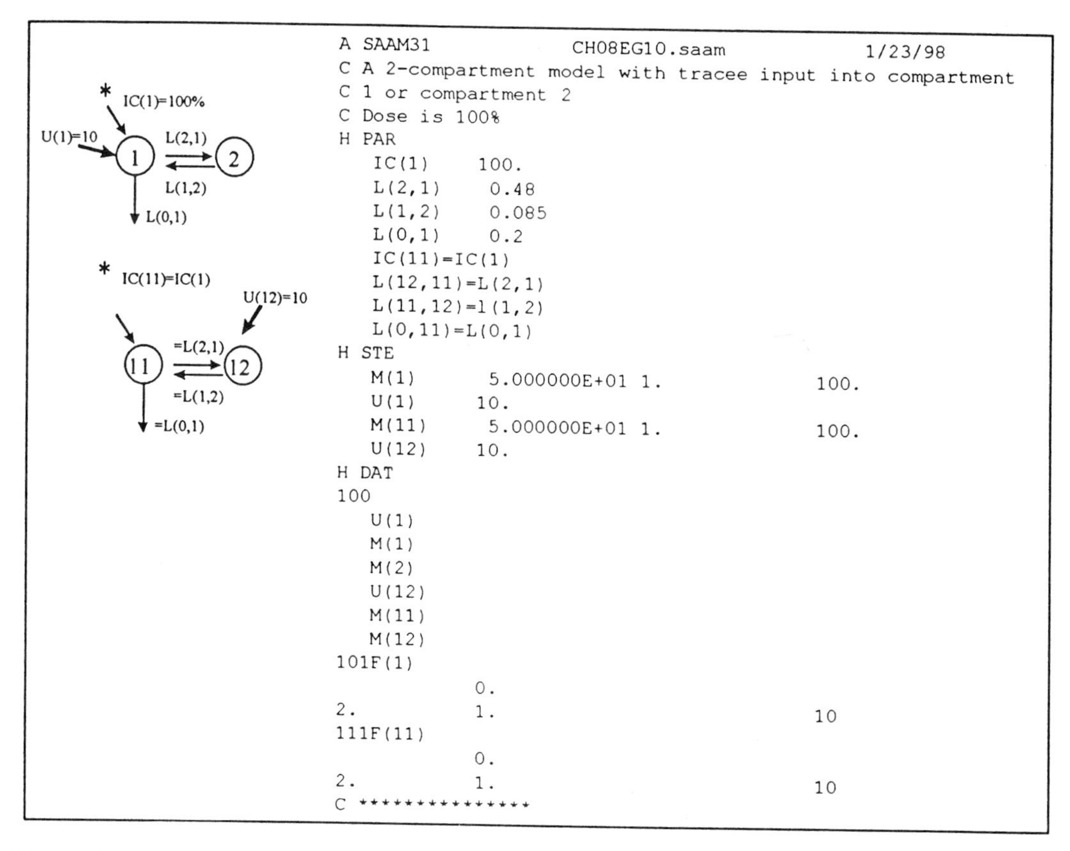


```
A SAAM31                CH08EG10.saam              1/23/98
C A 2-compartment model with tracee input into compartment
C 1 or compartment 2
C Dose is 100%
H PAR
   IC(1)      100.
   L(2,1)      0.48
   L(1,2)      0.085
   L(0,1)      0.2
   IC(11)=IC(1)
   L(12,11)=L(2,1)
   L(11,12)=l(1,2)
   L(0,11)=L(0,1)
H STE
   M(1)        5.000000E+01 1.                 100.
   U(1)       10.
   M(11)       5.000000E+01 1.                 100.
   U(12)      10.
H DAT
100
   U(1)
   M(1)
   M(2)
   U(12)
   M(11)
   M(12)
101F(1)
            0.
2.          1.                                 10
111F(11)
            0.
2.          1.                                 10
C ***************
```

Fig 8.10a Graphic showing duplicate models for calculating compartment mass with tracee entry into either compartment 1 or compartment 2 (simulated using a parallel model and using compartment 12 as compartment 2).

```
U ( 1) 1.000E+01
M ( 1) 5.000E+01
M ( 2) 2.824E+02
U (12) 1.000E+01
M (11) 5.000E+01
M (12) 4.000E+02
```

Fig 8.10b Output from model in Fig 8.10a showing that, using the same model, if tracee input is into compartment 1, mass of compartment 2 is 282, but if input is into compartment 2 (represented by a duplicate model as compartment 12), the mass is 400.

Example 11: Simulating Multiple Doses

Problem: Doses are administered at different levels and at different times.

Solution: This problem can be modeled in WinSAAM using the QO construct in conjunction with the fractional transfer function, L, and a rapidly clearing injection pool.

Demonstration: The following injection protocol is modeled using the graphic in Fig 8.11.1 and Text input file in Fig 8.11.2.

Result: The model solution is shown in Fig 8.11.3

Time	Injection Amount
Start (primer)	10
6	5
12	5
18	5

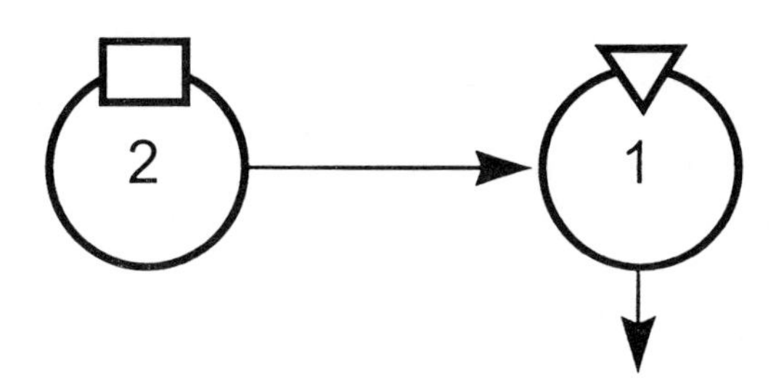

Fig 8.11.1 Model graphic for simulating multiple doses. Compartment 2 is the dosing compartment while compartment 1 is the sampled compartment.

```
 1:  A SAAM31              CH08EG11.saam
 2: H PAR
 3: C COMPT.1 IS THE SAMPLED COMPARTMENT TO WHICH
 4: C THE DRUG IS DELIVERED
 5: C COMPT. 2 IS THE DOSING COMPARTMENT
 6: C
 7:     L(1,2)    50
 8:     L(0,1)    .1
 9: C
10: H DAT
11: 102 QO
12:               0               10
13:               6               5
14:               12              5
15:               18              5
16: 101
17:               0
18: 2             .1                               240
```

Fig 8.11.2 Text input file for simulating multiple doses.

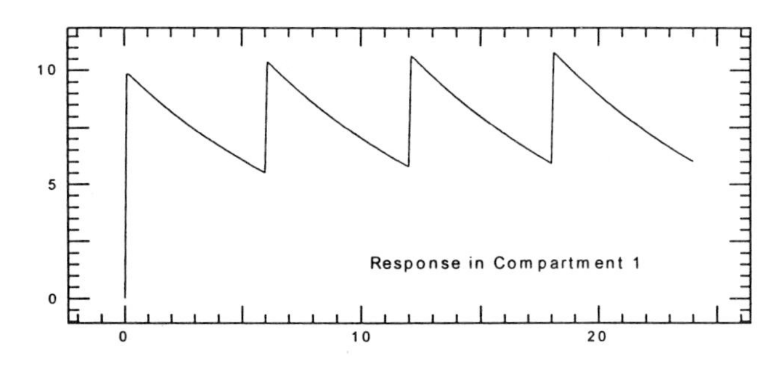

Fig 8.11.3 Plot showing response in compartment 1 over time when doses were administered at 0, 6, 12, and 18 time units.

Discussion: The QO construct sets the calculated value of compartment '2', F(2), to the 'observed' value at the indicated time point. That is F(2) is set to 10 at t=0, 5 at t=6, 5 at t=12, and 5 at t=18. Referred to as a resetting function in this context, QO emulates re-initializing F(2), as though a sequence of initial conditions were invoked. The value to which F(2) is set would remain if it weren't for the fact that compartment 2 is rapidly transferred into compartment 1. Indeed, the speed of the transfer provides the pulsatile nature of the drug application to compartment 1. If L(1,2) was set smaller, there would have been an interaction between the 'dumping' process and the elimination process from compartment 1, as it is, the elimination rate is just 0.2% of the transfer rate.

The question arises as to why the dosing compartment is needed. Why couldn't the QO have been applied to compartment 1? The reasons for this are 1) from the principle of linear superposition drug accumulates in the sampled compartment and the application of a new bolus adds to the amount of drug residing there and 2) the QO function totally resets the value of the compartment to which it is applied.

The figure (Fig 8.11.4) contrasts responses where a) dose dumping is applied using the dosing compartment (multiplied by 1.1 for clarity) versus b) simply resetting the sampled compartment. Following the priming dose in case b, the sampled compartment response is actually lowered after the administration of the second dose and therefore is not correct. An example can be seen in Jackson and Zech (1991).

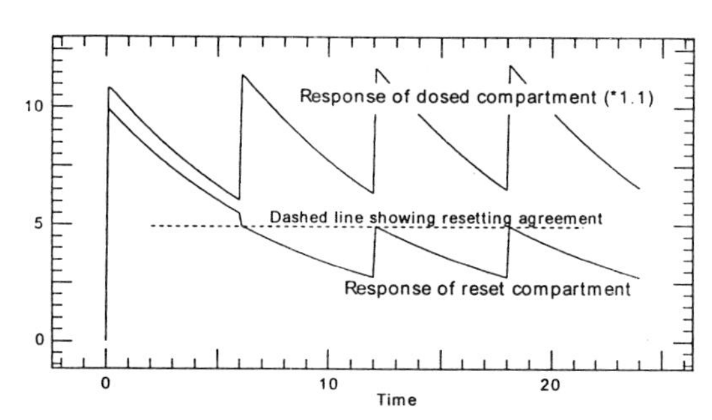

Fig 8.11.4 Response of resetting compartment 1 using QO (reset compartment) vs. dosing compartment (Compartment 2).

Example 12: Simulating Different Experimental Conditions by Using T-Interrupts (e.g., switching continuous infusions on and off)

Problem: Wagner (1993) presents a large number of models to portray system responses to various drugs. One is the response of 'an open compartment model to a zero-order input'. The goal is to represent the continuous infusion of an antibiotic, lincomycin, into a subject for a predefined time and then follow the elimination of the drug from the subject. The turnover, or action space, of the subject is represented as a single compartment.

Solution: Multiple experiments can be simulated in WinSAAM using the 'time block' (TC) tool to reflect the system changes in terms of model changes. For example, to simulate a zero-order drug input, the input into the model would be simulated for a period of time and then turned off.

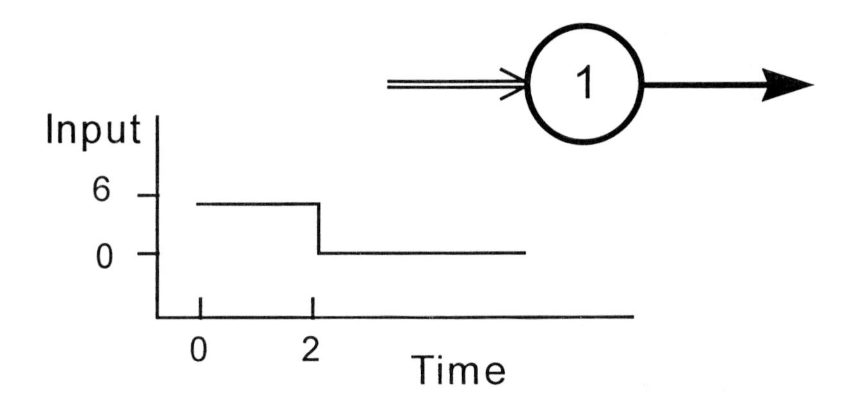

Fig 8.12.1 Graphic used to simulate an infusion into compartment 1 for 2 time units.

Demonstration: An infusion for a period of time can be simulated using the model in Fig 8.12.1

Model solutions: The solution is plotted in Fig 8.12.3

```
1: A SAAM31             CH08EG12.saam
2: C
3: C DEMONSTRATION FROM P. 6 WAGNER
4: C
5: C
6: C  UF(1)=KO/V INFUSION  [UG/ML/H]
7: C  P(2)=K
8: C
9: H PAR
```

```
10:      P(1)       5.77
11:      P(2)       0.221
12:   L(0,1)=P(2)
13:      UF(1)      5.77
14:  H DAT
15:  101
16:                 0
17:  2              .1                                20
18:  H DAT                          TC(1)
19:  101
20:                 2.1
21:  2              .1                                100
22:  H PCC                          TC(1)
23:   UF(1)=0
```

Fig 8.12.2 Text input file to simulate an infusion for 2 hours (Time block 0). The response in compartment 1 is simulated for an additional 10 hours (Time block 1).

Discussion: There are two time blocks in the WinSAAM lexical model, time block 0 and time block 1. Time block '0' is implied in models if no other time block is explicitly specified or if other domains are specified but some of the data is not included in the specified domains. Note that time block '0' runs from t=0, to t=2.0, and that time block '1' runs from t=2.1, to t=12.1, (Fig 8.12.3). Note that UF(1) is 5.77 during time block 0. Then, at time block '1', UF(1) is assigned the value zero. This is the input scheme described by Wagner and the plots of the results are identical to those of Wagner (3, 4).

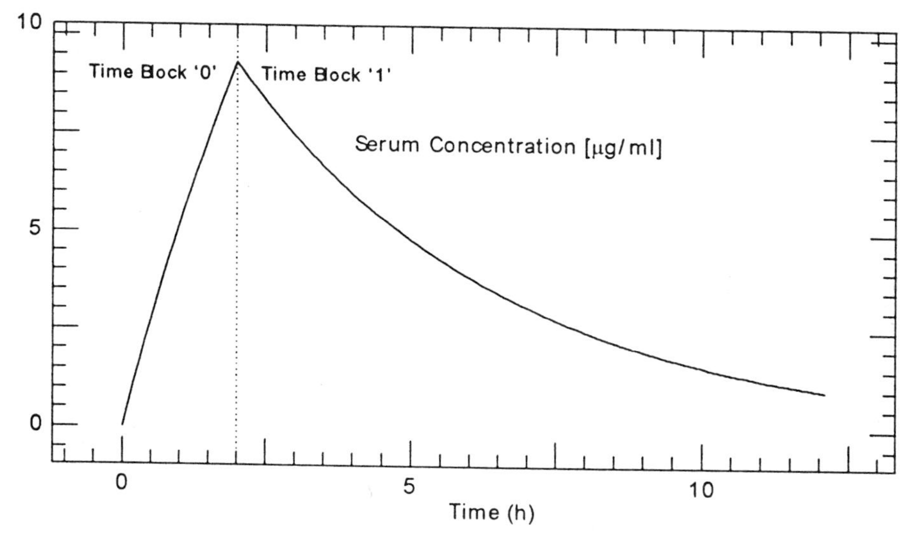

Fig 8.12.3 Response of compartment 1 during an infusion (0-2 hours) and following the infusion (2-12 hours).

Example 13a: Simulating Different Experimental Conditions Using QO

Problem: An alternate technique to model the problem in Example 12.

Solution: The switching in problem 11 simulating a sequence of varying drug injections, was accomplished using the QO function. For problem 12 where the switching was needed to achieve a cessation of an infusion the time block (TC) tool was used. To demonstrate the similarity of these two tools, problem 12 will be resolved using the QO function.

Demonstration: The model graphic (Fig 8.13a.1) is represented in the Text input file (Fig 8.13a.2).

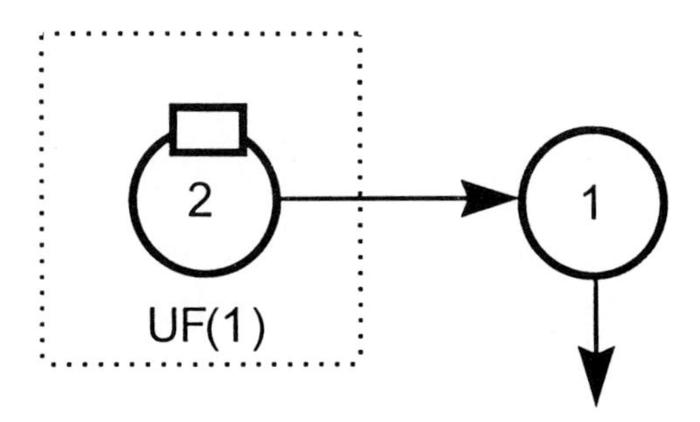

Fig 8.13a.1 Graphic to simulate a timed infusion using a QO. Compartment 2 acts as a switch to turn the infusion on and off by setting the value as either 1 or 0 using a QO.

```
 1: A SAAM31          CH08EG13.saam
 2: C
 3: C DEMONSTRATION FROM P. 6 WAGNER
 4: C
 5: C
 6: C  P(1)=K0
 7: C  P(2)=K
 8: C  P(3)=V
 9: C
10: H PAR
11:    P(1)       150000
12:    P(2)       0.221
13:    P(3)       26000
14:  L(0,1)=P(2)
15: H DAT
16: X UF(1)=F(2)*P(1)/P(3)
17: 102 QO
18:               0              1
19:               2              0
20: 101
21:               0
22: 2             .1                              120
```

Fig 8.13a.2 Text input file to switch an input on and off by using a QO on compartment 2.

Model solutions: The result of a timed infusion into compartment 1 is shown in Fig 8.13a.3.

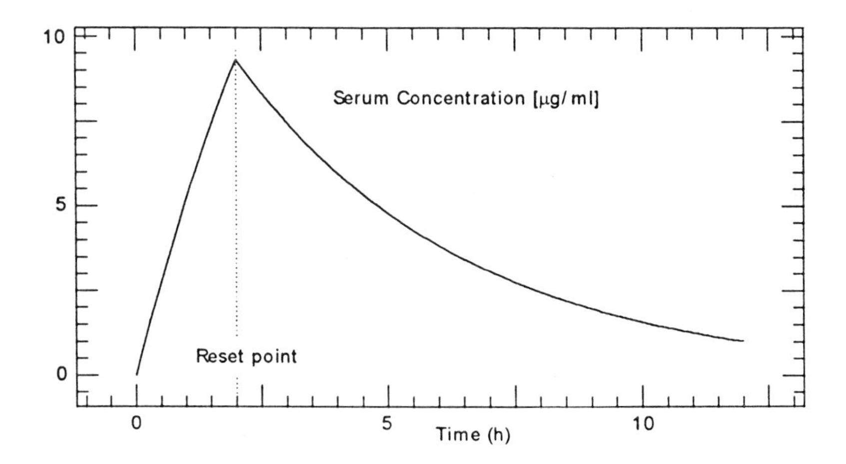

Fig 8.13a.3 Plot of response in compartment 1 to a timed infusion.

Discussion: QO was used in example 11 to achieve a rapid injection while in this example it was used for an infusion. For Example 11 the QO was used to reset F(2) and this was rapidly transferred into compartment 1, simulating a bolus application. Here the QO was used to switch the level of an infusion (UF(1)) causing a change to the delivery of a drug.
The switching process (F(2) unity and zero) is decoupled from the object needing to be switched (UF(1)), which is switched, between 5.77 and 0 [mg/ml.hr]. The parameters (P(1) and P(3)) are used to correct for the units. The advantage of decoupling is that the switch range can be changed by changing the parameters. F(2) has the value 1 from t=0 to t=2, at which time it is set to zero. F(2) is never changed beyond t=2. The data generation sequence calls for values for F(1) to be calculated for t=0 to t=12 at intervals of 0.1 units of time.

Example 13b: Simulating Different Experimental Conditions Using Function Dependence

Problem: A third alternative for Example 12. It will be demonstrated using a function-dependent form of UF although it could also be solved using a function-dependent L.

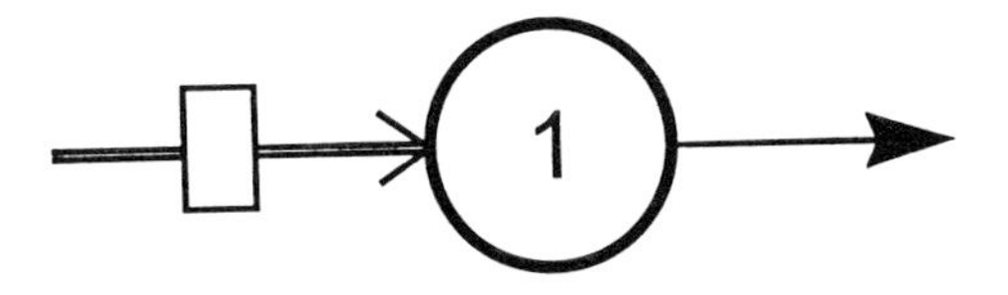

Fig 8.13b.1 Graphic used to demonstrate input switching using a function-dependent UF.

Solution: Inputs can be switched using function dependent parameters. Only two of the WinSAAM parameters (L and UF) can be dependent (on either G or F functions) and either parameter can be switched to provide the input.

Demonstration: The graphic is shown in Fig 8.13b.1 and the Text input file is shown in Fig 8.13b.2.

```
 1: A SAAM31              CH08EG13b.saam
 2: C
 3: C DEMONSTRATION FROM P. 6 WAGNER
 4: C
 5: C
 6: C  P(1)=K0
 7: C  P(2)=K
 8: C  P(3)=V
 9: C
10: H PAR
11:    P(1)        150000
12:    P(2)        0.221
13:    P(3)        26000
14:  L(0,1)=P(2)
15:  UF(1)=P(1)                                   8G 1
16: H DAT
17: X G(1)=(2-T)/P(3)/ABS(2-T)
18: 101
19:                0
20: 2              .1                            100
```

Fig 8.13b.2 Text input file to simulate a timed infusion using a function dependent UF.

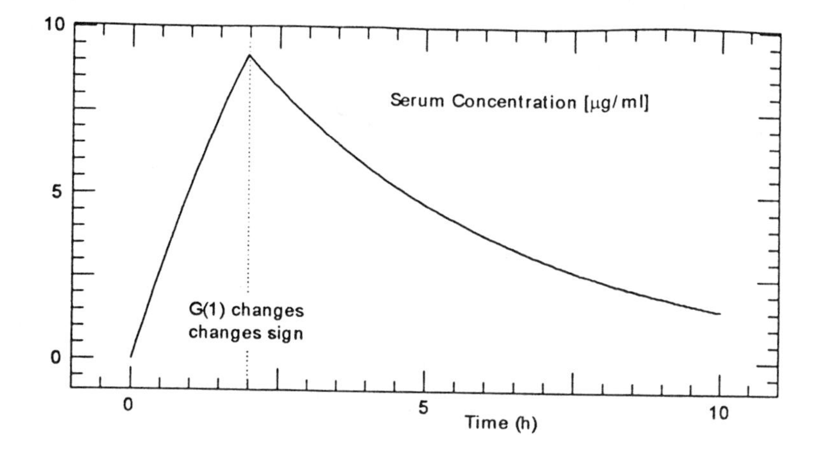

Fig 8.13b.3 Plot of response of compartment 1 to a timed infusion using a function dependent UF.

Model Solutions: The solution is shown in Fig 8.13b.3

Discussion: The input, UF(1) is function dependent on a G-function, G(1), which switches between 1 and -1 as t passes through the value 2. The '8' modification on the UF function dependent line ensures that if G(1) is greater than zero UF(1) equals the product of P(1) and G(1), i.e. P(1), however, the instant G(1) becomes less than or equal to zero UF(1) becomes zero. Mathematically,

$$UF(1) = P(1) \textit{ if } G(1) > 0$$
$$\textit{else } 0$$

Example 14: Solving Equations and Parameter Sensitivity

Problem: Gabrielsson and Wiener (1994, hereafter referred to as GW) present a large number of pharmacokinetic and pharmacodynamic problems and discuss their solutions. One problem uses a sigmoidal model to describe the changing 'effect' of a drug with concentration. The sigmoidal model has the form (symbols of GW)

$$\text{Effect} = E_0 - E_{max}.C_p^{\,n}/(EC^n_{50} + C^n_p) \qquad (8.1)$$

where,

Effect is the drug effect measure
C_p is the drug concentration
E_{max} is the 'maximum' drug affect
$EC^n_{50}$is a drug concentration for a 50% drug effect
E_0 is a baseline.
Data are presented relating the drug effect measurements and the goal of this demonstration is to use this data to estimate EC^n_{50}.

Solution: This problem is modeled in WinSAAM using a G-function because Eq 8.1 is both explicit and algebraic.

Demonstration: The Text input file is shown in Fig 8.14.1.

Model solutions: The plot is shown in Fig 8.14.2 and the calculated values for G(1) available in the output spreadsheet are given in Fig 8.14.3.

```
1: A SAAM31
2: 2        10
3: C
4: C DEMNSTRATION OF SIMPLE NONLINEAR FITTING
5: C FIT TO SIGMOIDAL PROCESS ... P. 426 A & W
6: C
7: H PAR
8: C
9: C EMAX = P(1) = 35
```

```
10: C EC50 = P(2)  = 143
11: C N    = P(3)  = 2
12: C E0   = P(4)  = 175
13: C
14:     P(1)        3.551020E+01  0.000000E+00   1.000000E+02
15:     P(2)        1.426883E+02  0.000000E+00   1.000000E+03
16:     P(3)        1.955074E+00  0.000000E+00   1.000000E+01
17:     P(4)        1.714526E+02  0.000000E+00   1.000000E+03
18: C
19:  P(5)=P(4)-P(1)
20: H DAT
21: X G(1)  = P(4)-P(1)/(1+(P(2)/T)**P(3))
22: 102
23:     P(5)
24: 101 G(1)                                     SD=10
25:               10            173
26:               25            168
27:               50            169
28:               75            162
29:               90            160
30:               120           160
31:               150           153
32:               200           146
33:               300           143
34:               500           139
```

Fig 8.14.1 Text input file to model a sigmoidal response.

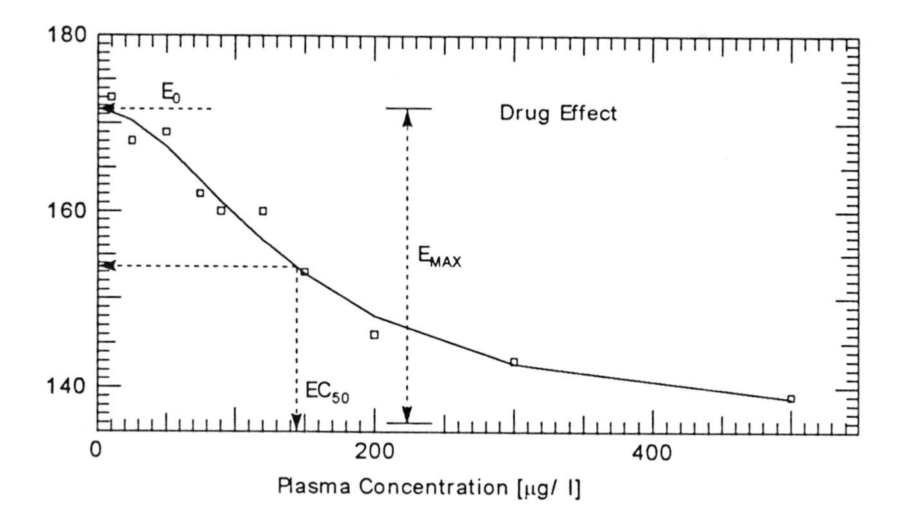

Fig 8.14.2 Plot of the response G(1), the solution to Equation 8.1.

The calculated values (QC) are within 1.5% of the observed values (Q0). Furthermore, the computed value of the difference between the baseline and the maximum effect is approximately 136. The fitted parameter values from the WinSAAM output spreadsheet are shown in Fig 8.14.4 P(1) is E_{MAX}, P(2) is EC_{50}, P(3) is n, and P(4) is E_0. All parameters are resolved to within 30% and P(4) appears somewhat better resolved than the others.

Component	Category	Time	QC	QO
1	G(1)	10	171.263	173
1	G(1)	25	170.315	168
1	G(1)	50	167.402	169
1	G(1)	75	163.588	162
1	G(1)	90	161.193	160
1	G(1)	120	156.674	160
1	G(1)	150	152.831	153
1	G(1)	200	148.043	146
1	G(1)	300	142.675	143
1	G(1)	500	138.758	139
2	P(5)	0	135.934	0

Fig 8.14.3 Output spreadsheet listing the value of G(1), the solution to Equation 8.1.

Category	Form	CurrentValue	Minimum	Maximum	FSD
P(1)	A	35.5252	0	100	0.150666
P(2)	A	142.698	0	1000	0.159538
P(3)	A	1.95347	0	10	0.275781
P(4)	A	171.459	0	1000	0.011349

Fig 8.14.4 Parameter values for the model Equation 8.1.

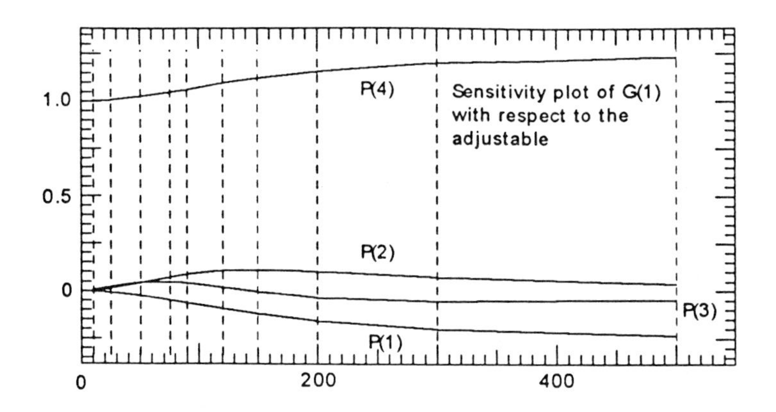

Fig 8.14.5 Sensitivity of the parameters. The vertical dashed lines correspond to abscissa observation points. G(1) is universally most sensitive to P(4), then to P(1), then P(2) followed by P(3).

Discussion: WinSAAM is able to fit linear and nonlinear explicit functions to data. Furthermore, dependent nonlinear functions can also be fit directly provided the system is reducible to explicit forms.

P(4) appears to be better resolved than the other parameters based on the FSD. Also, by inspecting the sensitivity profiles of the fitted response to the adjustable parameters, well-resolved parameters can be characterized by 1) a substantial temporal domain in which they contribute in a 'unique' way to the response, 2) large sensitivity values in this particular domain, and 3) substantial data for model fitting associated with this domain. The sensitivity of G(1) with respect to parameters P(1), P(2), P(3), and P(4) are shown in Fig 8.14.5.

The order of precision of the parameters estimates is P(4), P(1), P(2), and then P(3), i.e., the precision of the parameters is in direct relation to the sensitivity of G(1), their fitted response. If data had been clustered around abscissa at 175 or 320 there may have been disproportionately better resolution of P(2) and P(3) respectively, but the flatness of the sensitivity profiles probably limits the likelihood of this.

There are two model features which support identifying P(4); the value of G(1) (and presumably data values) for small abscissa, and values of G(1) for extremely large abscissa value. For small abscissa the expected value of G(1) is indeed P(4), and for large abscissa the expected value of G(1) is P(4)-P(1). Indeed, P(2) and P(3) only contribute to G(1) by resolving from the data the rapidity with which G(1) changes from P(4) to P(4)-P(1) and the abscissa point at which this takes place.

Notice that in the Text input file SD type data weighting are assigned uniformly to the observations. This causes the larger data values to influence the fit more significantly than the smaller ones. In fact, with WinSAAM's SD data weighting, the weight to each individual observation is in inverse proportion to the SD value (in this case uniformly set to 10);

$$W_i \approx 1/SD_i^2 \quad \text{and} \quad \Sigma W_i = N$$

where N is the total number of data points.

If the observation range spans 2 orders of magnitude (and presumably the model of the data also spans 2 orders of magnitude) then a residual associated with the lower value end of the response with a value of 10% of the observation (a realistic fitting condition) will in turn be just 10% of its equi-accurate counterpart at the upper end of the response value range. This means that SD type weighting for data fitting, with no change to data SD value assignment across the data, will lead to these two equal proportioned differences effecting the fit by a ratio of 10,000, with the larger value having the greater weight. The situation for the data in this example covers a range of 1.24 and so that higher points would have about a 50% greater influence on fitting for residuals of the same magnitude.

Returning to the discussion on the precision of the parameter estimates, it can be concluded that in addition to sensitivity support for estimation of P(4), the observation value range in conjunction with the selected data weighting scheme further supported estimation of P(4). With FSD type data weighting the following results would have been obtained (Fig 8.14.6)

PARAMETER	VALUE	ERROR	FSD
P (4, 0)	1.715E+02	2.040E+00	1.190E-02
P (1, 0)	3.551E+01	5.049E+00	1.422E-01
P (2, 0)	1.427E+02	2.090E+01	1.465E-01
P (3, 0)	1.965E+00	5.210E-01	2.651E-01

Fig 8.14.6 Parameter values for Eq. 8.1 obtain using FSD weighting.

Slightly more precise estimates of all parameters are obtained because of the movement of our fitting emphasis away from large to small numbers. The reason that a greater shift is not obtained on the precision estimation is because P(4) benefits from both high and low observation weights. The simplicity of this model permits the exploration of some interesting model features without getting into complex accounts relating to the model *per se*. Other features of a system that would need to be included are: the mean response level, the area under the fitted curve, the mean value of the independent variable, the profile of the cumulative form of the response, and the profile of the derivative of the response
The WinSAAM model to calculate each of those items is shown in Fig 8.14.7. In general, the line, UF(I) = G(K), sets the derivative of F(I) to G(K) and thus F(I) is just the integral of G(K). This property is used to determine the area under G(1) and the area under T*G(1). Plotting F(5) presents the cumulative profile of G(1) and dividing the value of F(5) at the appropriate end point by the integration gives the average value of G(1) over the indicated range (Role's Theorem). Dividing the integral of T*G(1) by the integral of G(1), where integrals are taken over the entire T range gives the average value of T. Note that the integrals are taken from t=1 to t=500, as opposed to starting at t=0. The reason for this is that there is a singularity in G(1) at t=0. A more complete examination of integration in association with the pole would be warranted if this approximation were found to be inadequate. The cumulative plot, F(5), is shown in Fig 8.14.8. The derivative of G(1) is shown in Fig 8.14.9.

```
 1: A SAAM31              CH08EG14c.saam
 2: 2       10
 3: C
 4: C DEMONSTRATION OF SIMPLE NONLINEAR FITTING
 5: C FIT TO SIGMOIDAL PROCESS ... P. 426 A & W
 6: C
 7: H PAR
 8: C
 9: C EMAX = P(1) = 35
10: C EC50 = P(2) = 143
11: C N    = P(3) = 2
12: C E0   = P(4) = 175
13: C
14:    P(1)      3.551020E+01  0.000000E+00   1.000000E+02
15:    P(2)      1.426883E+02  0.000000E+00   1.000000E+03
16:    P(3)      1.955074E+00  0.000000E+00   1.000000E+01
17:    P(4)      1.714526E+02  0.000000E+00   1.000000E+03
18:  P(5)=P(4)-P(1)/2
19: H DAT
20: C
21: C G(3) IS THE NUMERICAL DERIVATIVE OF G(1)
22: C F(5) IS THE UN-NORMALIZED CUMULATIVE FORM OF G(1)
23: C F(6) IS THE INTEGRAL OF T*G(1)
24: C G(6) IS THE EXPECTED VALUE OF T
25: C G(7) IS THE AVERAGE VALUE OF G(1) FROM ROLE'S THEOREM
26: C
27: X G(1) = P(4)-P(1)/(1+(P(2)/T)**P(3))
28: X G(2) = P(4)-P(1)/(1+(P(2)/(T+.01))**P(3))
29: X G(3)=(G(2)-G(1))/.01
30: X UF(5)=G(2)
31: X UF(6)=T*G(2)
32: X G(6)=F(6)/F(5)
33: X G(7)=F(5)/500
34: 101 G(1)
35:            1
36: 2          5                        100
37: 103 G(3)
38:            1
39: 2          5                        100
40: 105 F(5)
41:            1
42: 2          5                        100
43: 106 F(6)
44:            500
45: 106 G(6)
46:            500
47: 106 G(7)
48:            500
```

Fig 8.14.7 Model to calculate properties of a system.

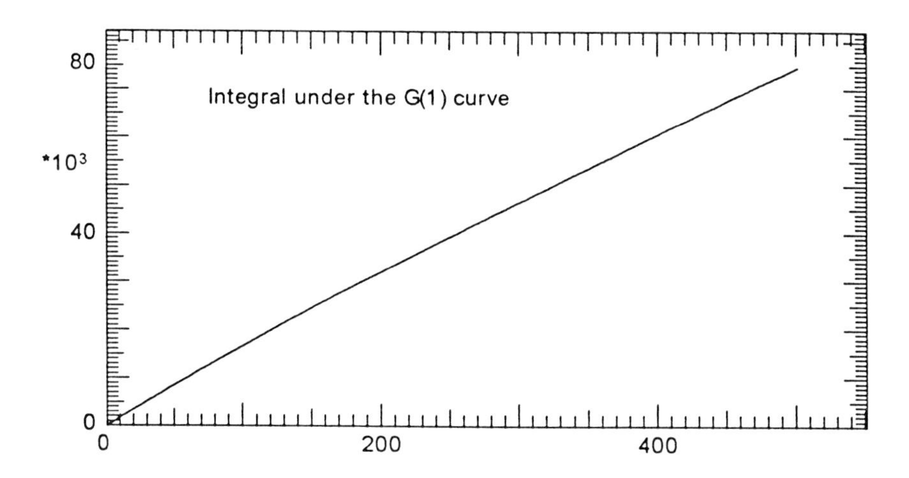

Fig 8.14.8 Plot of F(5) for model in Fig 8.14.7.

The remaining results from WinSAAM are as follows:
The mean value of T=G (6) 2.407E+02.
Note that if G(1) were a linear function the mean value of T would have been 250. The mean value of G(1) from Role's theorem is:
G (7) 1.491E+02
and this compares very closely with an EC50 based value for G(1), P(5); P (5) 1.537E+02
The reason for the difference between G(7) and P(5) can again be attributed to the degree of curvilinearity of G(1).

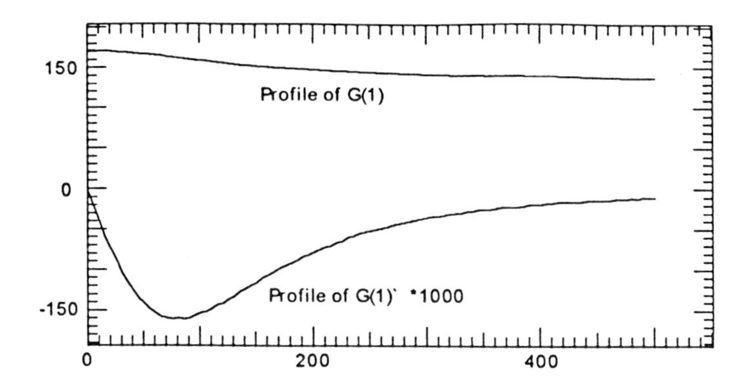

Fig 8.14.9 Derivative of G(1) for model in Fig 8.14.7.

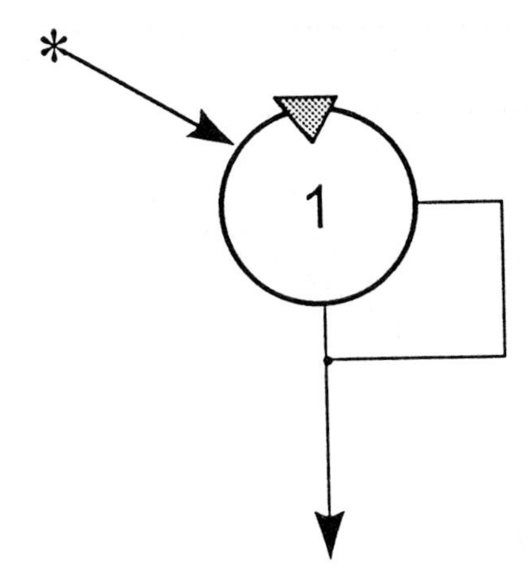

Fig 8.15.1 Model graphic for fitting Michaelis-Menten kinetics. The line joining compartment 1 and the elimination process from compartment 1 implies that elimination from compartment 1 is facilitated by the level of metabolite in compartment 1. The asterisked arrow to compartment 1 implies an initial condition to compartment 1, and the triangle on compartment 1 implies that compartment 1 is sampled.

Example 15: Michaelis-Menten Kinetics

Problem: Gabrielsson and Weiner (4) demonstrate the effect of a new drug by modeling its elimination for two subjects dosed with different amounts of the drug. The elimination profiles were characterized using Michaelis- Menten kinetics.

Solution: Michaelis-Menten kinetics can be modeled in WinSAAM using function dependent fractional turnover rates. G-functions will be used to define the function dependency of the L's and unassigned nonlinear adjustable parameters (P(J)'s) to set the Km and Vmax values.

Demonstration: The model is shown in Fig 8.15.1 and Fig 8.15.2.

Model Solutions: Doses given to the two subjects were 25 and 100 mg respectively. The Vmax in each subject was 42,804 ng/l/h but Km's were about 330 ng/l and 1000 ng/l respectively (Fig. 8.15.3).

```
 1: A SAAM31                 CH08EG15.saam
 2: 2         10
 3: C
 4: C DEMONSTRATION OF NONLINEAR KINETICS FROM G & W P. 380
 5: C
 6: H PAR
 7: C
 8: C P(1)  = VD = 45
 9: C P(2)  = VMAX = 42.804
10: C P(3)  = KM1 = 33
```

```
11: C P(4)  = KM2 = 100
12: C P(5)  = DOSE UNIT CONVERSION,1000
13: C P(6)  = KM UNIT CONVERSION, 10
14: C
15:   IC(1)=25000/P(1)
16:   IC(2)=100000/P(1)
17:   P(25)=P(2)*P(5)
18:     L(0,1)    1                                                  *G 1
19:     L(0,2)    1                                                  *G 2
20:     P(1)        4.775117E+01   0.000000E+00    1.000000E+02
21:     P(2)        4.619469E+01   0.000000E+00    1.000000E+03
22:     P(3)        3.332329E+01   0.000000E+00    1.000000E+03
23:     P(4)        1.170177E+02   0.000000E+00    1.000000E+04
24:   P(5)=1000
25:   P(6)=10
26: H DAT
27: X G(1)=P(25)/(P(3)*P(6)+F(1))/P(1)
28: X G(2)=P(25)/(P(4)*P(6)+F(2))/P(1)
29: 101                                                RQO=.4
30:                 .08             554.4
31:                 .25             358.1
32:                 .5              272.9
33:                 .75             155
34:                 1               102.4
35:                 1.5             28.8
36:                 2               13.1
37: 102                                                RQO=.4
38:                 .08             2290
39:                 .25             1876
40:                 .5              1776
41:                 .75             1531
42:                 1               1395
43:                 1.5             1279
44:                 2               1038
45:                 4               391
46:                 6               79.5
47:                 8               11.4
```

Fig 8.15.2 Text input file to simulate Michaelis-Menten elimination of a metabolite.

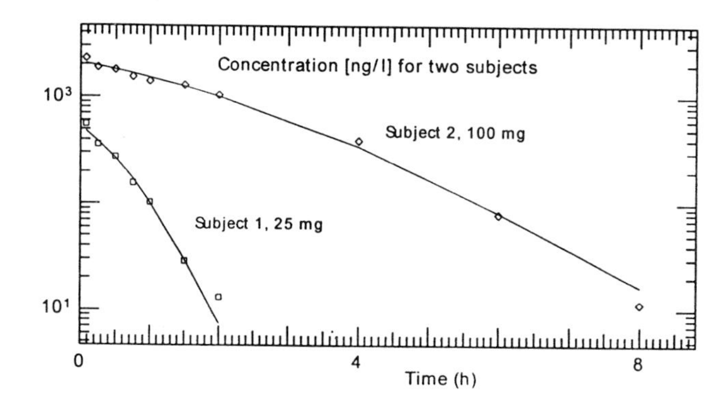

Fig 8.15.3 Plot of the elimination of a metabolite from two subjects given different doses, and with different Km's.

Discussion: Two points concerning modeling Michaelis-Menten type processes. Firstly, either the function dependent L or the UF approach can be used (see below). Function dependency is the preferred option. The following lines in the input file

```
L(0,1)  1                    *G 1
X G(1)=P(25)/(P(3)*P(6)+F(1))/P(1)
```

establish the differential equation;

$$F(1)'=-P(25)*F(1)/((P(3)*P(6)+F(1))/P(1))$$

which is the format required. The nonlinear variation of the Michaelis- Menten process is reflected in terms of a minor modification to the ordinary linear L(0,1). Specifically, by referencing L(0,1), the F(1) factor is applied automatically to the 'modified' transfer process.

When the transfer cannot simply be written in the form

$$L(0,1) = F(1)*h(P(K),F(J))$$

where F(J) does not involve F(1) it may be more appropriate to use the UF form to establish the differential system. If h(P(K), F(J)) cannot be specified in the form:

$$h(P(K), F(J)) = I(P(L).F(P))/(P(M)+F(Q))$$

then there may be initialization problems. In particular, poles due to the zero initial values of some of the F(J) may cause h to be singular. To circumvent this problem a small positive constant can be added to the denominator of h. It is important to ask the question 'what are the consequences of the technique in regard to the accuracy of the final solutions?' The effect of different magnitudes of small constants added to the denominator of h should be investigated, and also different approaches should be tried to see if they lead to similar problem solutions.

This is not so abstract as it may appear. A variation of this problem was mentioned with Example 14. There non-invertability of G(1) at t=0 was circumvented by starting solutions at t=1. The justification for this was the following; the earliest data point coincided with an independent variable value of '10', therefore it was not an initialization problem, and the range of the independent variable was from 10 to 500, i.e. dropping the domain 0 to 1 was probably of little consequence.

To achieve the same results for this problem using UF, as opposed to function dependent L strategy the following variations would be made to the input file (Fig 8.15.4)

```
15:  IC(1)=25000/P(1)
16:  IC(2)=100000/P(1)
17:  P(25)=P(2)*P(5)
18:    P(1)        4.775117E+01  0.000000E+00   1.000000E+02
19:    P(2)        4.619469E+01  0.000000E+00   1.000000E+03
20:    P(3)        3.332329E+01  0.000000E+00   1.000000E+03
21:    P(4)        1.170177E+02  0.000000E+00   1.000000E+04
22:  P(5)=1000
23:  P(6)=10
```

```
24: H DAT
25: X UF(1)=-P(25)*F(1)/(P(3)*P(6)+F(1))/P(1)
26: X UF(2)=-P(25)*F(2)/(P(4)*P(6)+F(2))/P(1)
```

Fig 8.15.4 Use of UF's for modeling Michaelis Menten kinetics. The initial conditions have been retained, the L(0,1) and L(0,2) lines have been dropped, G(1) and G(2) have been replaced with UF(1) and UF(2) respectively, and the former G lines have been modified by a) negating them, and b) adding the F(J) factor, F(1) to UF(1) and F(2) to UF(2).

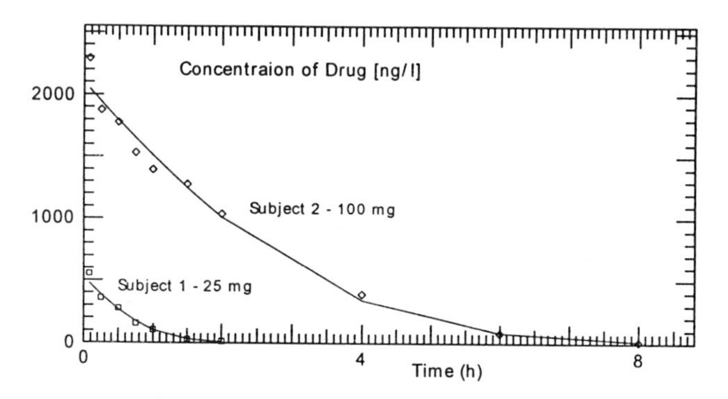

Fig 8.15.5 Plot of Michaelis Menten elimination using UF's.

The line

25: X UF(1)=-P(25)*F(1)/(P(3)*P(6)+F(1))/P(1)

leads to the establishment of the differential equation

F(1)′=-P(25)*F(1)/(P(3)*P(6)+F(1))/P(1)

which is identical to the one derived above for the case of the function dependent L. To see that the new approach indeed worked, the results are plotted in Fig 8.15.5.

Note the difference in plotting in semi-logarithmic versus linear fashion. The first solution plot of this model and observations presented in semi-logarithmic style showed better agreement between the calculated and observed values than the linear plot. This is because in viewing plots the importance of small values diminish compared to large values. A logarithmic perspective suppresses large fluctuations (statisticians tend to use logarithmic transformation when data ranges span several cycles) and hence users are often more satisfied with logarithmic perspectives of the same results than linear perspectives. The point is that this may be purely and aberration. It always pays to examine model fits from both semilogarithmic and linear perspectives because the former highlights the concordance of observed and calculated values in the region of low values and the latter highlights the same concordance on the high value region. Fits showing good linear agreement in the high value range and good semilogarithmic agreement in the low value range indicate good uniform response coverage by the model.

Reference: Gabrielsson and Weiner (1994).

Example 16: Multiple Dosing Regimens

Problem: Drugs are often administered by multiple routes such as intravenous infusion (also referred to as rapid IV) and extravascular injection. The IV application primes the system, and the extravascular application provides a steady delivery of drug.

Solution: Drug administration regimens will be modeled using the following model (Fig 8.16.1) and some data published in Gabrielsson and Weiner, (1994), (Fig 8.16.2).

Note the following regarding this model:

1. the model essentially duplicates the actual system, compartments 1 and 2 represent the plasma compartment respectively pertaining to the extravascular injection and the intravenous drug infusions, and compartments 3 and 4 are the peripheral compartment for the two drug applications.
2. the initial conditions in all compartments set 'baseline' values (in the discussion presented by Gabrielsson and Weiner, 1994)
3. the entire system is modeled using UF's so that readers can compare the system and results directly with AW. The system could have been represented using L's and the reader is encouraged to do this.
4. FSD data weighting scheme is used. The data straddles two cycles in value range and small values will only have importance if this data weighting scheme is used.

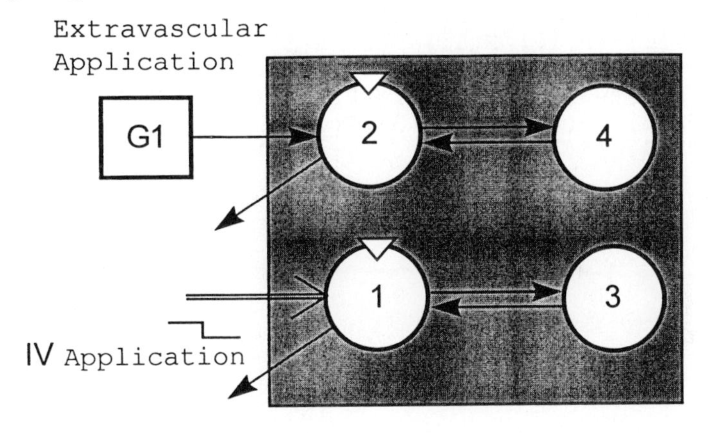

Fig 8.16.1 Model for simulating different dosing regimens.

Model solutions: The solution is shown in Fig 8.16.3 and the fitted parameter values are shown in Fig 8.16.4.

```
1: A SAAM31                    CH08EG16.saam
  2: 2        10
  3: C
  4: C DEMONSTRATION FROM P. 360 FROM GABRIELSSON AND WEINER
  5: C
  6: C VC = P(1) = 82
```

```
 7: C CL = P(2) = 0.7
 8: C CLD = P(3) = 1.5
 9: C VT = P(4) = 43
10: C KA = P(5) = 0.5
11: C BIO = P(6) = 0.6
12: C DOSE1 = P(7) = 1000
13: C CHANGING PART OF DOSE2 = P(10)
14: C DOSE2 = P(10) = 700,T <= TINF ELSE 0
15: C TINF  = P(11) = 2
16: C
17: H PAR
18:     IC(1)    2.45
19:     IC(2)    2.8
20:     IC(3)    2.45
21:     IC(4)    2.8
22: C
23:     P(1)       3.089583E+01  0.000000E+00   1.000000E+02
24:     P(2)       1.068236E+00  0.000000E+00   1.000000E+01
25:     P(3)       6.308818E+00  0.000000E+00   1.000000E+01
26:     P(4)       1.784599E+02  0.000000E+00   1.000000E+03
27:     P(5)       8.259688E-02  0.000000E+00   1.000000E+01
28:     P(6)       6.346210E-01  0.000000E+00   1.000000E+01
29:     P(7)     1000
30:     P(10)    700
31:     P(11)    2
32: H DAT
33: X G(1)=P(6)*P(7)*P(5)*EXP(-P(5)*T)
34: X UF(1)=(P(10)/P(11)-P(2)*F(1)-P(3)*F(1)+
35:    P(3)*F(3))/P(1)
36: X UF(2)=(G(1)-P(2)*F(2)-P(3)*F(2)+P(3)*F(4))/P(1)
37: X UF(3)=(P(3)*F(1)-P(3)*F(3))/P(4)
38: X UF(4)=(P(3)*F(2)-P(3)*F(4))/P(4)
39: C
40: 101                                             FSD=.4
41:              0            2.45
42: 102                                             FSD=.4
43:              0            2.8
44:              2            4.95
45: H DAT                     TC(2)
46: 101                                             FSD=.4
47:              4            13.95
48:              6            10.15
49:              8            7.8
50:              8            7.9
51:              10           6.2
52:              12           5.55
53:              17           4.5
54:              22           4.45
55:              27           4.050
56:              32           4.25
57:              42           4.2
58:              52           4.05
59:              62           3.9
60:              92           3.15
61:              122          2.7
62:              182          2
```

```
63:                 242         1.5
64:                 302         1.35
65:                 362         1
66: 102                                     FSD=.4
67:                 4           5.55
68:                 6           7.2
69:                 8           6.85
70:                 10          7.45
71:                 15          6.5
72:                 20          5.4
73:                 25          4.8
74:                 30          5.15
75:                 40          4.1
76:                 50          4.05
77:                 60          3.85
78:                 90          3.6
79:                 120         3.25
80:                 180         2.4
81:                 240         1.4
82:                 300         1.3
83: H PCC                       TC(2)
84:   P(10)=0
```

Fig 8.16.2 Text input file for simulating multiple dosing regimes.

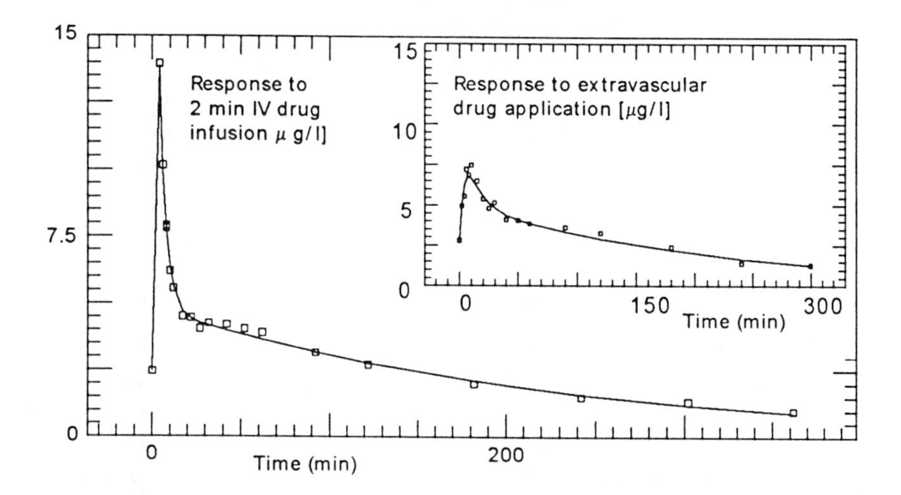

Fig 8.16.3 Plots of appearance of a drug after an IV or extravascular infusion.

```
* P ( 1)   3.078E+01(31.06)  FSD( 5)   1.415E-01
* P ( 2)   1.068E+00(1.061)  FSD( 3)   3.185E-02
* P ( 3)   6.309E+00(6,236)  FSD( 4)   6.056E-02
* P ( 4)   1.782E+02(179.7)  FSD( 6)   4.334E-02
* P ( 5)   8.260E-02(0.082)  FSD( 2)   1.220E-01
* P ( 6)   6.338E-01(0.647)  FSD( 1)   4.675E-02
```

Fig 8.16.4 Parameter values obtained by fitting the model in Fig 8.16.2 to data in Fig 18.6.3. Values in parentheses are those published in Gabrielsson and Weiner, 1994.

Discussion: This two-part study has been set up in WinSAAM using two parallel 2-compartment models with two time interrupts. The time interrupts allow the intravenous drug infusion to run for 2 minutes and then be turned off for the remainder of the study. This approach allows both sets of data to define the parameter values. Both sets of data were weighted equally for the data fitting (FSD=0.4) but it is likely that one data set would be better than the other for estimation of the parameters. In particular the IV response data may be more appropriate for estimating vascular transport parameters than the data relating to the extravascular application. The relative confidence in the data sets could be reflected by using different FSD value assignments, i.e., if the IV response were twice as accurate in quantifying the parameters as the extravascular drug application response data the FSD assignment for these data could be set to 0.2 vs. FSD=0.4 for the extravascular response data. Of course the actual standard deviation of the data is the best vehicle whereby WinSAAM generates data weights.

It is unlikely that two apparently unconnected studies would have been combined in this fashion. The studies appear to be unconnected because if they were connected temporally, i.e. concurrent drug applications to the same system, then separating the study model into two non-interacting sub-systems would not have been justified because this would be in violation of the principle of superposition. However, there is evidence of the same subject being the host as the baseline values for the drug levels (i.e. levels prior to data collection), therefore they are identical for the two subsystems. If the drug applications were concurrent to the same system the two sources of data could have served equally well using a single model for the estimation of all parameters simply by using a time interrupt operational unit.

Example 17: Pharmacokinetics

Problem: Describe drug metabolism when both drug and metabolite data are available. Data were obtained following a 100 mg IV drug bolus and by monitoring the plasma concentrations of the drug and its conjugate metabolite (Gabrielsson and Weiner, 1994).

Solution: Two WinSAAM models will be used to describe these data, one replicating the approach of Gabrielsson and Weiner (1994), (Fig 8.17.1) and one showing how a more mechanistic representation of the system can be established more easily using WinSAAM.

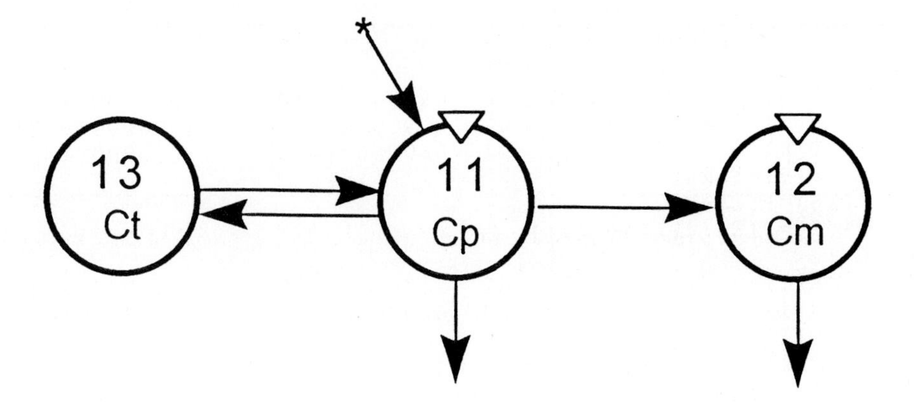

Cp = Compt. 11 = Plasma Compartment
Cm = Compt. 12 = Metabolite Compartment
Ct = Compt. 13 = Peripheral Compartment

Fig 8.17.1 Model for fitting a drug in plasma and metabolite data.

```
 1: A SAAM31              CH08EG17.saam
 2: C
 3: C THE ANALYSIS OF DRUG AND METABOLITE DATA
 4: C FOLLOWING GABRIELSSON AND WEINER. P. 326
 5: C
 6: C
 7: C THE PROBLEM IS SOLVED USING TWO METHODS ...
 8: C THE APPROACH BY G & W IS PRESENTED FIRST
 9: C THE APPROACH BY THE AUTHOR IS PRESENTED NEXT
10: C BOTH APPROACH YIELD THE SAME ANSWER
11: C
12: C
13: C NOTE THAT WE HAVE IN THIS EXAMPLE USE DATA WEIGHTS
14: C TO ENSURE THAT WE GET THE SAME ANSWER FOR EACH METHOD
15: C
16: 2      10
17: H PAR
18: C CLR = P(7) = 1.24
19: C VC = P(1) = 12
20: C VT = P(2) = 20
21: C CLD = P(3) = 1
22: C CLM = P(4) = 1.7
23: C CLME = P(5) = 3
24: C VM = P(6) = 15
25: C
26:    IC(1)=100/P(1)
27:    P(1)      1.247886E+01  0.000000E+00   1.000000E+02
28:    P(2)      9.588226E+00  0.000000E+00   1.000000E+02
29:    P(3)      1.094478E+00  0.000000E+00   1.000000E+02
```

```
30:    P(4)        1.564092E+00  0.000000E+00    1.000000E+02
31:    P(5)        3.447858E+00  0.000000E+00    1.000000E+02
32:    P(6)        1.018815E+01  0.000000E+00    1.000000E+02
33:    P(7)=1.24
34:  P(8)=P(1)+P(2)
35:  P(9)=P(8)/(P(4)+P(7))
36: C
37:    K(11)       8.013529E-02  0.000000E+00    1.000000E+03
38:    K(12)       9.815320E-02  0.000000E+00    1.000000E+03
39:    P(11)=1/K(11)
40:    P(16)=1/K(12)
41:    P(14)=L(12,11)/K(11)
42:    P(15)=L(0,12)/K(12)
43:    P(12)*L(11,13)=L(13,11)*P(11)
44:    P(17)=1.24
45:    P(13)=P(12)*L(11,13)
46:    L(0,11)=P(17)*K(11)
47:    L(13,11)  8.770733E-02  0.000000E+00    1.000000E+02
48:    L(12,11)  1.253410E-01  0.000000E+00    1.000000E+02
49:    L(0,12)   3.384279E-01  0.000000E+00    1.000000E+02
50:    P(12)     9.588136E+00  0.000000E+00    1.000000E+02
51:    IC(11)    100
52:  G(1)=P(1)-P(11)
53:  G(2)=P(2)-P(12)
54:  G(3)=P(3)-P(13)
55:  G(4)=P(4)-P(14)
56:  G(5)=P(5)-P(15)
57:  G(6)=P(6)-P(16)
58: H DAT
59: X UF(1)=(-F(1)*(P(4)+P(7)+P(3))+F(3)*P(3))/P(1)
60: X UF(2)=(F(1)*P(4)-P(5)*F(2))/P(6)
61: X UF(3)=P(3)*(F(1)-F(3))/P(2)
62: 120
63:    G(1)                      0              .50
64:    G(2)                      0              .50
65:    G(3)                      0              .50
66:    G(4)                      0              .50
67:    G(5)                      0              .50
68:    G(6)                      0              .50
69: 110
70:    P(8)
71:    P(9)
72: 101                                         SD=1
73:               0              8
74:               .1             7.76
75:               .25            7.41
76:               .5             6.87
77:               1              5.91
78:               2              4.41
79:               3              3.33
80:               5              1.98
```

```
 81:          8              1.04
 82:          12             .558
 83:          16             .363
 84:          24             .19
 85:          36             .079
 86:          48             .033
 87: 102                                     SD=1
 88:          0              0
 89:          .1             .173
 90:          .25            .395
 91:          .5             .682
 92:          1              1.04
 93:          2              1.28
 94:          3              1.27
 95:          5              1.09
 96:          8              1.04
 97:          12             .524
 98:          16             .363
 99:          24             .19
100:          36             .079
101:          48             .033
102: 111                                     SD=1
103:          0              8
104:          .1             7.76
105:          .25            7.41
106:          .5             6.87
107:          1              5.91
108:          2              4.41
109:          3              3.33
110:          5              1.98
111:          8              1.04
112:          12             .558
113:          16             .363
114:          24             .19
115:          36             .079
116:          48             .033
117: 112                                     SD=1
118:          0              0
119:          .1             .173
120:          .25            .395
121:          .5             .682
122:          1              1.04
123:          2              1.28
124:          3              1.27
125:          5              1.09
126:          8              1.04
127:          12             .524
128:          16             .363
129:          24             .19
130:          36             .079
131:          48             .033
```

Fig 8.17.2 Text input file for analyzing drug and metabolite data.

The Text input file (Fig 8.17.2) is discussed below with reference to the numbered lines.

Lines 26-35 and 59-61 define the model in the terms used by GW. There are 6 adjustable parameters, P(1) to P(6) and 3 dependent parameters, P(7) to P(9). The three differential equations are defined by UF(1) - UF(3). The system is linear.

Lines 37-38 and 46-51 define the model in terms of fractional transfers (L(I,J)'s) as well as linear parameters (K(J)'s). Again there are in all 6 adjustable parameters. However, in this version of the model, two parameters are linear adjustable (the K(J)) and four are nonlinear adjustable (the L(I,J)). This means that fitting of the compartmental form of the model will be easier computationally than fitting using the differential definitions (UF(J)).

Another important distinction between the models is that whereas the distribution spaces are estimated directly in the differential model, they are estimated indirectly in the 'compartmental' model via the K(J)'s.

This is often an issue examined at great length by the proponents of identifiability. The concern is whether the data at hand provides a basis for dilution space size estimation given the model to be fitted. Clearly, where dilution space factoring permits the scaling of the differential solution, the issue of identifiability is trivial. On the other hand, for nonlinear systems where the measurements are not scalably related to the differential system solutions, identifiability is an important concern (discussed later).

Lines 39-45 reflect a set of parameter dependencies which transform the results of the fitted compartmental model to their counterparts for the differential model such that P(J) from the differential model can be compared to P(J+10) from the compartmental model.

Lines 52-57 together with lines 63-68 weight the fitting process in favor of producing very nearly the same estimates of P(J) and P(J+10) for the two models. This technique is useful when fitting two data sets to a particular model when one set of data has more noise than the other. Although, such a step can only be used as a corroborative mechanism because if there were genuinely different processes present in one data set, coupling their fitting in this fashion would literally throwaway the extra information.

Lines 72-101 reflect our organization of the data for fitting the differential model while lines 102-131 reflect our organization of the data for fitting the compartment model.

PARAMETER	VALUE	ERROR	FSD
P (4, 0)	1.564E+00	1.042E-01	6.662E-02
P (3, 0)	1.094E+00	1.063E-01	9.709E-02
P (1, 0)	1.248E+01	4.630E-02	3.711E-03
P (5, 0)	3.448E+00	2.377E-01	6.893E-02
P (6, 0)	1.019E+01	7.767E-01	7.623E-02
P (2, 0)	9.588E+00	2.595E+00	2.707E-01

```
L (12,11)   1.253E-01   8.457E-03   6.748E-02
L ( 0,12)   3.384E-01   2.058E-02   6.082E-02
P (12, 0)   9.588E+00   2.595E+00   2.707E-01
L (13,11)   8.771E-02   8.605E-03   9.812E-02
K (11, 0)   8.014E-02   4.370E-04   5.453E-03
K (12, 0)   9.815E-02   8.001E-03   8.152E-02
P(  1, 0)   1.2479E+01 A       P( 11, 0)   1.2479E+01  D
P(  2, 0)   9.5882E+00 A       P( 12, 0)   9.5879E+00  A
P(  3, 0)   1.0945E+00 A       P( 13, 0)   1.0945E+00  D
P(  4, 0)   1.5641E+00 A       P( 14, 0)   1.5641E+00  D
P(  5, 0)   3.4479E+00 A       P( 15, 0)   3.4480E+00  D
P(  6, 0)   1.0188E+01 A       P( 16, 0)   1.0188E+01  D
P(  7, 0)   1.2400E+00  D P( 17, 0)  1.2400E+00  D
P(  8, 0)   2.2067E+01  D
P(  9, 0)   7.8696E+00  D
```

Fig 8.17.3 Parameter values showing the agreement between the two fits. Agreement between the two model forms was not entirely left to chance or optimal selection of data weights.

Results: The final parameter estimates for the two models are shown in Fig 8.17.3 and the plots are shown in Fig 8.17.4.

Discussion: The linear plot shows that the high points are fitting quite well but the semi-logarithmic plot shows only fair agreement between the model and the observations for the low values. It is not surprising that the low values fit poorly because with SD type data weighting the emphasis is on the high values.

In drug turnover and metabolism studies it is usually more important to focus on the high value range than the low value range of the data. The reason for this is that if data are modeled with say, 10% error uniformly distributed across the entire data range, then the high values could be poorly estimated in absolute

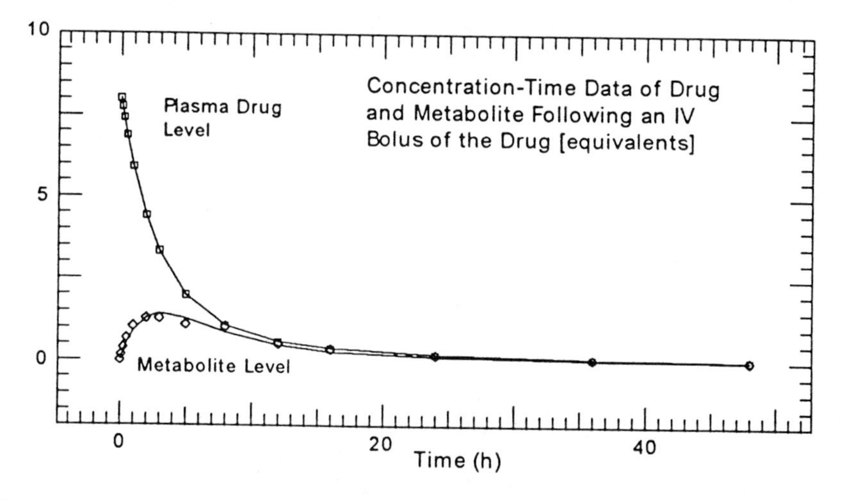

Fig 8.17.4a Linear plot of the drug and metabolite data.

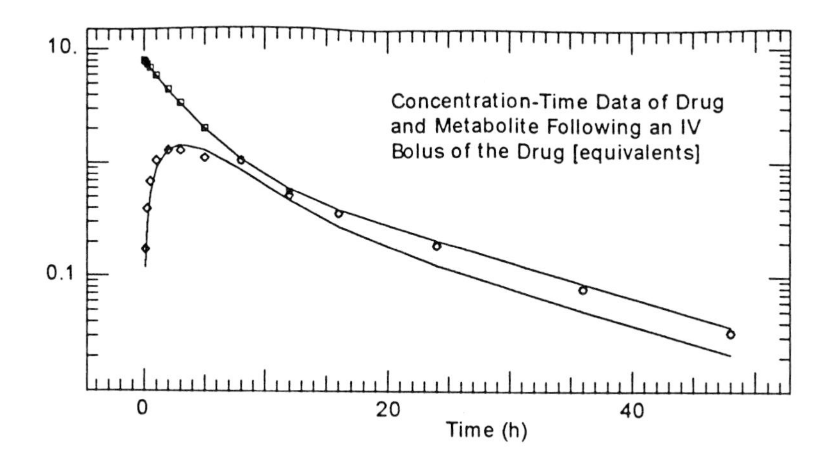

Fig 8.17.4b Semi-logarithmic plot of the model fit to the drug and metabolite data.

terms in contrast to the small values. If the model is used as a basis for predicting drug delivery then such errors in the high value range could produce significant mal-predictions leading potentially to toxic dosing of a subject.
The solutions produced for compartment 20, which monitors the differences between the P(J) values for the two models is shown in Fig 8.17.5.

#	COMP	TC	CATEGORY	QC
1	20	0	G (1)	-4.005E-05
2	20	0	G (2)	2.956E-04
3	20	0	G (3)	-1.156E-05
4	20	0	G (4)	-2.551E-05
5	20	0	G (5)	-1.054E-04
6	20	0	G (6)	9.537E-07

Fig 8.17.5 Differences between parameters for fitting two models to the same set of data. The technique used to force the two model fits to agree has indeed worked because all differences between the parameters estimates are less than 0.0003 which is smaller than the magnitude of any parameter.

Example 18: Determining Areas Under Curves (AUC) and Forcing Functions

Problem: Metabolic studies often call for areas under curves (AUC) and whereas these can be inferred from fits, it is consoling to see the concordance between model based AUC's and data based AUC's. In this study WinSAAM is used to derive model based and data based AUC's.

Solution: The AUC for first order elimination of a drug will be estimated from the data (data from GW) and from the model. Plasma AUC will also be estimated for the drug using the QL tool with the model in Fig 8.18.1

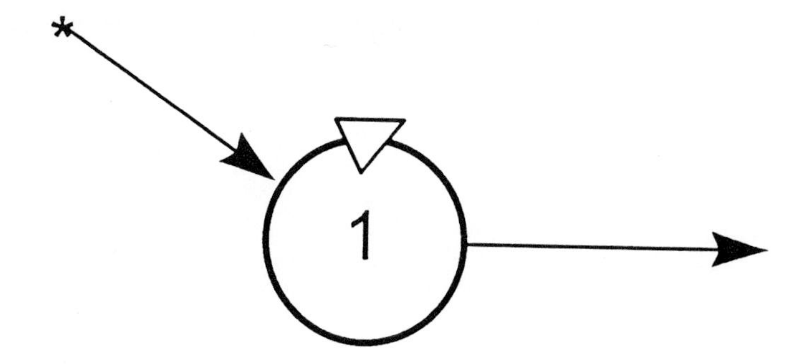

Fig 8.18.1 Graphic of model for determining AUC.

```
 1: A SAAM31              CH08EG18.saam
 2: C
 3: C ANALYSIS OF ONE COMPARTMENT RESPONSE FROM GW P. 120
 4: C
 5: H PAR
 6:     IC(1)      10000
 7:     L(0,1)      1.985634E-02  0.000000E+00   1.000000E+00
 8:     K(1)        1.007050E-01  0.000000E+00   1.000000E+03
 9:  P(1)=1/K(1)
10:  P(2)=1/L(0,1)
11:  IC(4)=IC(1)*K(1)
12: C
13: C L(0,1) IS THE ELIMINATION RATE CONSTANT
14: C K(1) IS A DOSE DILUTION FACTOR
15: C P(1) IS THE DISTRIBUTION SPACE
16: C P(2) IS THE RESIDENCE TIME
17: C IC(4) INITIALISES THE FORCING FUNCTION COMPARTMENT
18: C
19: H DAT
20: C
21: C F(2) IS THE AREA UNDER THE FITTED CURVE
22: C F(3) IS THE AREA UNDER THE LINE SEGMENTED DATA
23: C FF(4) IS A FORCING FUNCTION BUILT TO REPRESENT THE DATA
24: C F(5) IS THE AREA UNDER THE TIME WEIGHTED CALCULATIONS
25: C G(1) IS THE MEAN TRANSIT TIME
26: C
27: X UF(2)=F(1)*K(1)
28: X UF(3)=FF(4)
29: X UF(5)=T*F(1)
30: X G(1)=F(5)*K(1)/F(2)
31: X FF(4)=F(4)
32: 101                                            FSD=.4
33:              10            769
34:              20            710
35:              30            585
36:              40            472
37:              50            363
38:              60            300
```

```
39:              70          256
40:              90          170
41:              90          170
42:              110         109
43:              150         52
44: 102 F(2)
45:              150
46: 102 F(3)
47:              150
48: 102 G(1)
49:              1500
50: 104 QL
51:              0           1000
52:              10          769
53:              20          710
54:              30          585
55:              40          472
56:              50          363
57:              60          300
58:              70          256
59:              90          170
60:              110         109
61:              150         52
62: 104 FF(4)
63:              10          769
64:              20          710
65:              30          585
66:              40          472
67:              50          363
68:              60          300
69:              70          256
70:              90          170
71:              110         109
72:              150         52
```

Fig 8.18.2 Text input file for calculating AUC from data and from the model.

Lines 6-8 represent the model, the initial condition (IC), a fractional turnover (L), and a scale factor (K). The scale factor is not part of the model but rather represents an amount by which the calculations will be scaled to bring the differential solutions into alignment with the data. This scaling is to attach, or change, units to those of the solution.

Lines 9-10 transform model features to system attributes. P(1) transforms K(1) into the distribution space, and P(2) transforms L(0,1) into the system residence time. Note that each of these transforms simply inverts the estimated parameter. This may indicate that modeling is performed in the wrong dimensions! Line 11 initializes compartment 4 in preparation for data integration.

Of the functions generated in lines 27-31;

UF(2) integrates QC(2) providing one estimate of the AUC.

UF(3) integrates under FF(4), which is a line segmented

representation of the data generated in compartment 4 as a result of the QL operational unit applied to it. Thus F(3) is a direct AUC derived from the data.

UF(5) integrates t*F(1), allowing G(1)=F(5)*/F(2) to provide an estimate of the mean transit time.

The assignment of FF(4) to F(4) simply exteriorizes FF(4) allowing it to be subsequently referenced (see lines 28 and 60).

Solutions: The solutions are plotted in Fig 8.18.3.

The computed values 'stored' in component 2 are as follows:

#	COMP	CATEGORY	T	QC
31	2	F (2)	1.500E+02	4.813E+04
32	2	F (3)	1.500E+02	4.854E+04
35	2	G (1)	1.500E+03	5.030E+01

F(2) and F(3) are within 1% of one another suggesting that integrating the fitted model and integrating under the formed line segmented data using QL provides similar estimates of the AUC. Displaying the calculated parameters shows that there is agreement between the area under the normalized, t-weighted response, G(1), and the reciprocal of the fractional turnover rate, P(2), suggesting that either may be used for residence time (also in this model mean transit time) estimation for this model.

P(1, 0) 9.9300E+00 D
P(2, 0) 5.0362E+01 D

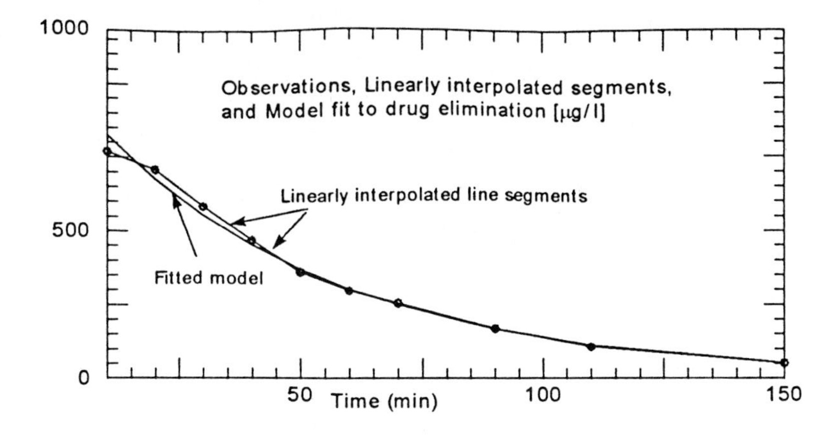

Fig 8.18.3 Solution to model shown in Fig 8.18.2.

Example 19: Forcing Functions

Problem: Data, as opposed to calculated responses, are required to drive a sub-system.

Solution: Consider the following synthetic data:

T	Observation
0	1
1	2
2	3
3	4
4	5

The goal is to construct a WinSAAM function so that the data are treated as though they were computed values. The model (Fig 8.18.3) creates a forcing function, FF(1) which drives the adjoining compartment and a G-function, G(2), is used to contain the predicted solutions. If the QL causes an FF(1) which joins the data in component '1' to be generated then the response of compartment 2 should be predicted on the basis of a smooth function representing the data. This is done recognizing that

$$F(2)' = L(2,1).FF(1)$$
$$= L(2,1).T$$

and replacing F(2) by G(2) gives

$$G(2) = L(2,1)*(T^2/2 + T)$$

```
 1: A SAAM31                    CH08EG19.saam
 2: C
 3: C QL DEMONSTRATION
 4: C RCB
 5: C
 6: H PAR
 7:      IC(1)       1
 8:      L(2,1)      .1
 9: H DAT
10: X G(2)=L(2,1)*(T+2)*T/2
11: 101 QL
12:                  0             1
13:                  1             2
14:                  2             3
15:                  3             4
16:                  4             5
17: 102
18:                  0
19:                  1
20:                  2
21:                  3
```

```
22:                4
23: 102 G(2)
24:                0
25:                1
26:                2
27:                3
28:                4
29: C
30: C THIS IS A DEMONSTRATION OF THE QL RESETTING FUNCTION
31: C THE PURPOSE OF QL IS TO SET FF(I) TO QO(I), OR ITS
32: C LINEARLY INTERPOLATED VALUE AT THE NOMINATED TIME POINT.
33: C
34: C IN THIS EXAMPLE FF(1) = T + 1 INFERRED FROM THE QO VALUES
35: C AND THUS F2` = L(2,1) * (T + 1) OR INTEGRATING
36: C              F2 = .1 * (T+2) * T/2
37: 103 FF(1)
38:                0
39:                1
40:                2
41:                3
42:                4
```

Fig 8.19.1 Text input file to demonstrate QL function in WinSAAM.

Component	Category	Time	QC	Weight
2	F(2)	0	0	1
2	F(2)	1	0.15	1
2	F(2)	2	0.4	1
2	F(2)	3	0.75	1
2	F(2)	4	1.2	1
3	FF(1)	0	1	1
3	FF(1)	1	2	1
3	FF(1)	2	3	1
3	FF(1)	3	4	1
3	FF(1)	4	5	1
2	G(2)	0	0	1
2	G(2)	1	0.15	1
2	G(2)	2	0.4	1
2	G(2)	3	0.75	1
2	G(2)	4	1.2	1
1	QL(1)	0	1	1
1	QL(1)	1	0.904837	1
1	QL(1)	2	0.818731	1
1	QL(1)	3	0.740818	1
1	QL(1)	4	0.67032 1	

Fig 8.19.2 Output confirming that F(2) and G(2) are equivalent

To interpret Fig 8.19.2 remember;

1. Component denotes the structural unit including an associated set of categories and observations. For example, in component 1 the QL resetting function is associated with the observations at time points 1, 2, 3, and 4 leading to the generation of FF(1). By contrast, in component 3 the FF(1) calculations are not associated with actual data. Thus we see that the data area of a component can be blank. The calculated values area of a component can also be blank, however if both the data segment and calculated values area are blank then of course the component need not exist.
2. Category, usually denotes the name of the calculated values in the component. Thus in component 2 there are two sets of calculated values F(2) (the solution to the equation F(2)`=L(2,1)*FF(1)), and G(2) (explicitly given by G(2)=L(2,1)*$(T^2/2+T)$. In component 3, is FF(2). The assignments associated with component 1 are less obvious because of their implicit as opposed to explicit nature. The only category associated with component '1' is the QL function. However, whereas this achieves the generation of FF(1) the determinations (or calculations) associated with component '1' are those for F(1). Here of course F(1) represents the solution to F(1)`=-L(2,1)*F(1). The values deriving from the QL operation must be explicitly assembled in a component, such as '3', before they can be accessed.
3. Accessing calculated values deriving from resetting operations such as QL, QO, and QF is most easily achieved using WinSAAM's data segment name utility. Here a name is set up for the data segment which embraces calculated values as well and then the named data segment is plotted or printed with the appropriate plotting or printing path. The syntax of the name utility is:

   name name = c:#1<category,tc(#2)>, ...

   where, #1 denotes the component encapsulating the calculated values, and #2 denotes the time block values to be accessed. For example to retrieve FF(1) from component '3' in the above demonstration use the name following name command:

   name ff30 = c:3<FF(1),TC(0)>

   or

   name ff30 = c:3<FF(1)>

   The specification of the TC block is optional if none exists, or TC(0) (time block zero) is implied. A name for a data segment may be up to 4 characters long. The name command extends to enable the name association of up to five data segments.

   To associate all the calculations in the above WinSAAM Text input file into a single data segment use a name command such as

   name all = c:2<f(2)>,c:2<g(2)>,c:3<ff(1)>

   and then to plot or print these values

   *plot qc(*all*)*

   *prin qc(*all*)*

Example 20: Function Dependencies

Problem: Clarify the purpose of the '+' function dependency of a fractional transfer. Recall that a fractional transfer, L(I,J), is by default linear and represents the rate of transfer of material in compartment J to compartment I. If the specification of the L(I,J) is modified to reflect an additive dependency on some function G(K), viz:

L(I,J),TAB>val(L(I,J))<3 TABS>+G K

where <TAB> represents tabulation to facilitate indicated insertion. Then the differential form generated by this line may be one of the following:

F(J)'=-val(L(I,J))*F(J) + G(K)

F(J)'=-(val(L(I,J)) + G(K))*F(J)

F(J)'=-val(L(I,J) + G(K)*F(J)

Solution: To determine which of the above applies, though clearly the first two seem more likely than the third, construct a simple model that will unequivocally provide the answer. The test problem needs to be assembled appropriately and also the solution to confirm the correct understanding of the '+' function dependency.

```
 1: A SAAM31                 CH08EG20.saam
 2: H PAR
    C 345678 1 2345678 2 2345678 3 2345678 4 2345678 5 23456789
 3:     L(0,1)    .1                                        +G 1
 4:     IC(1)     100
 5:     P(1)      .1
 6: H DAT
 7: X G(1)=P(1)
 8: X G(2)=IC(1)*EXP(-(L(0,1)+P(1))*T)
 9: 101
10:               1
11:               2
12:               4
13:               8
14: 101 G(2)
15:               1
16:               2
17:               4
18:               8
19: 101 L(0,1)
20:               1
21:               2
22:               4
23:               8
```

Fig 8.20.1a Text input file to show function dependency.

The respective solutions to the output categories if our understanding of the '+' transfer modifier is correct are as follows:

F(1)=IC(1)*EXP(-(L(0,1)+P(1))*T)
G(2)=IC(1)*EXP(-(L(0,1)+P(1))*T)
L(0,1)=val(L(0,1))

The model solutions obtained for two P(1) values are as follows:

```
> deck
* DECK BEING PROCESSED
***ALL WEIGHTS=1.*2*
PRE-PROCESSING TIME :        0.160 SECS
> new
* NEW DEVICE  30 OPENED
> switch storage
* AUTOMATIC SOLUTION STORAGE (STOR ON)
> p(i)
P(  1, 0)  1.0000E-01   F
> solve
*** MODEL CODE  8 SOLUTION
***NO ADJUSTABLE PARAMETERS**67*
SOLUTION TIME :        0.051 SECS
* SOLUTION SAVED ON DEVICE 30 AS RECORD  1
> p(1)=-.1
> solve
*** MODEL CODE  8 SOLUTION
***NO ADJUSTABLE PARAMETERS**67*
SOLUTION TIME :        0.000 SECS
* SOLUTION SAVED ON DEVICE 30 AS RECORD  2
> restore 1
* RECORD   1  RESTORED
* STORED AT 12:46:08     ON 07/20/1997   USING SOLV COMMAND
> print qc(1)
-----------------------------------------------------------------
*** NAME :    1 CURRENT KOMN
 #  COMP TC  CATEGORY          T            QC
  1    1   0  F ( 1)      1.000E+00    8.187E+01
  4    1   0  F ( 1)      2.000E+00    6.703E+01
  7    1   0  F ( 1)      4.000E+00    4.493E+01
 10    1   0  F ( 1)      8.000E+00    2.019E+01
  2    1   0  G ( 2)      1.000E+00    8.187E+01
  5    1   0  G ( 2)      2.000E+00    6.703E+01
  8    1   0  G ( 2)      4.000E+00    4.493E+01
 11    1   0  G ( 2)      8.000E+00    2.019E+01
  3    1   0  L ( 0, 1)   1.000E+00    2.000E-01
  6    1   0  L ( 0, 1)   2.000E+00    2.000E-01
  9    1   0  L ( 0, 1)   4.000E+00    2.000E-01
 12    1   0  L ( 0, 1)   8.000E+00    2.000E-01
> restore 2
* RECORD   2  RESTORED
* STORED AT 12:46:21     ON 07/20/1997   USING SOLV COMMAND
> print
------------------------------------------------------------------
*** NAME :    1 CURRENT KOMN
 #  COMP TC  CATEGORY          T            QC
  1    1   0  F ( 1)      1.000E+00    1.000E+02
  4    1   0  F ( 1)      2.000E+00    1.000E+02
  7    1   0  F ( 1)      4.000E+00    1.000E+02
 10    1   0  F ( 1)      8.000E+00    1.000E+02
  2    1   0  G ( 2)      1.000E+00    1.000E+02
```

```
  5    1    0  G ( 2)       2.000E+00     1.000E+02
  8    1    0  G ( 2)       4.000E+00     1.000E+02
 11    1    0  G ( 2)       8.000E+00     1.000E+02
  3    1    0  L ( 0,  1)   1.000E+00     0.000E+00
  6    1    0  L ( 0,  1)   2.000E+00     0.000E+00
  9    1    0  L ( 0,  1)   4.000E+00     0.000E+00
 12    1    0  L ( 0,  1)   8.000E+00     0.000E+00
```

For the two P(1)'s used, the values of G(2) and F(1) agree indicating that our understanding of the operational significance of the '+' transfer modifier is correct, i.e.

F(J)′=-(val(L(I,J)) + G(K))*F(J)

Discussion: The commands used (in **bold** text) and aspects of their purpose are explained;

> **deck**

Compile the WinSAAM input checking for errors and building data structures

> **new**

Starts up a new solution storage facility

> **switch storage**

Ensures that subsequent solutions are automatically stored

> **p(i)**

Displays the values of all P(J) parameters

> **solve**

Solves the model for the current (initial) parameter values

SOLUTION TIME : 0.051 SECS
* SOLUTION SAVED ON DEVICE 30 AS RECORD 1

Shows that the first solution was automatically stored

> **p(1)=-.1**

Sets the value of P(1) to -.1. Note that this means the L(0,1) is now zero

> **solve**

Solves the model for the new parameter values

SOLUTION TIME : 0.000 SECS
* SOLUTION SAVED ON DEVICE 30 AS RECORD 2

Shows that the solution was automatically stored

> **restore 1**

Requests restoration of solution 1

* RECORD 1 RESTORED
* STORED AT 12:46:08 ON 07/20/1997 USING SOLV COMMAND

Shows solution 1 was automatically stored

> **print qc(1)**

Prints solutions for component '1'

> **restore 2**

Requests restoration of solution 2

* RECORD 2 RESTORED
* STORED AT 12:46:21 ON 07/20/1997 USING SOLV COMMAND

Shows solution 2 was automatically stored

> **print**

Prints solutions for component '1'

L(0,1) can be inserted in component's category field, meaning that its value for the indicated time points become part of component's solution area. Most WinSAAM modeling constructs can be inserted into the category field of a component which means that it is possible to fit almost any construct. The strengths of this demonstration are that it is both simple and precise.

A simple model has been used to explore a complex aspect of WinSAAM. The simplicity meant that a mathematical solution could be obtained for the system to test an assumption and clarify understanding of a particular construct. WinSAAM supports the following function dependency operators '+, -, *, /, 5, 6, 7, 8, 9, A, B, C, D, and E' and only the first '+' has been demonstrated. The role of others can be investigated using the table in Chapter 6, Section 5.3 and the demonstration below (Fig 8.20.1b).

In Fig 8.20.1a, x = val(L(I,J)) and z = indexed function. For the specification line

3: L(0,1) 2.3 +G 1

x = 2.3, and z = G(1) leading to $\underline{L(0,1)}$ = 2.3 + G(1), $\underline{L(0,1)}$ is underlined to show that it is a function dependent transfer. From the modeling perspective remember that elimination from compartment '1' follows the differential form F(1)′=-2.3*G(1)*F(1). The F(1) is implied by the '1' in L(0,1). Whereas $\underline{L(0,1)}$ has a function dependent form, if L(0,1) were used in conjunction with any other system functions in the model it would have val(L(0,1)) substituted for its place in the system function. That is the function dependent form $\underline{L(0,1)}$ is not directly available for use in the formation of other functions. It is available indirectly via the parts being accessible for creating other functions. A demonstration of the A-E function dependency set and the model is given in Fig 8.20.1b.

```
A SAAM31                 CH08EG20.saam
H PAR
   L(0,1)    2                                              AG 1
   L(2,1)    1                                              BG 1
   L(3,1)    3                                              CG 1
   L(4,1)    4                                              DG 1
   L(5,1)    5                                              EG 1
```

```
H DAT
X G(1)=-SIN(2*T+.1)
101 L(0,1)
                0
2               .1                              100
101 L(2,1)
                0
2               .1                              100
101 L(3,1)
                0
2               .1                              100
101 L(4,1)
                0
2               .1                              100
101 L(5,1)
                0
2               .1                              100
```

Fig 8.20.1b Example of function dependency.

The plots of the D and E operations are shown in Figs 8.20.2. In each case the driving function, G(1), is represented as the oscillating line and the function dependent L by the discontinuous line.

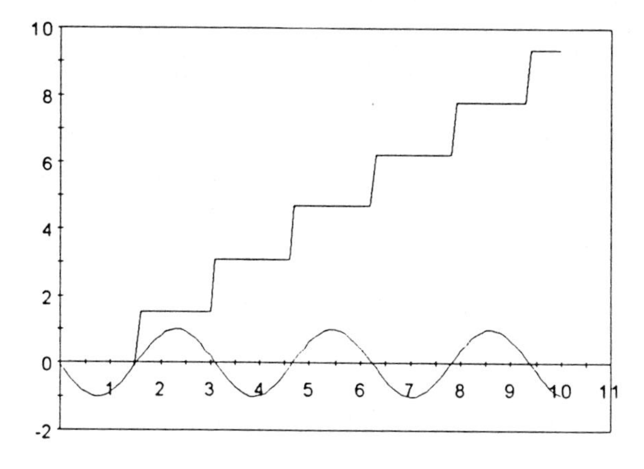

Fig 8.20.2a The D operator

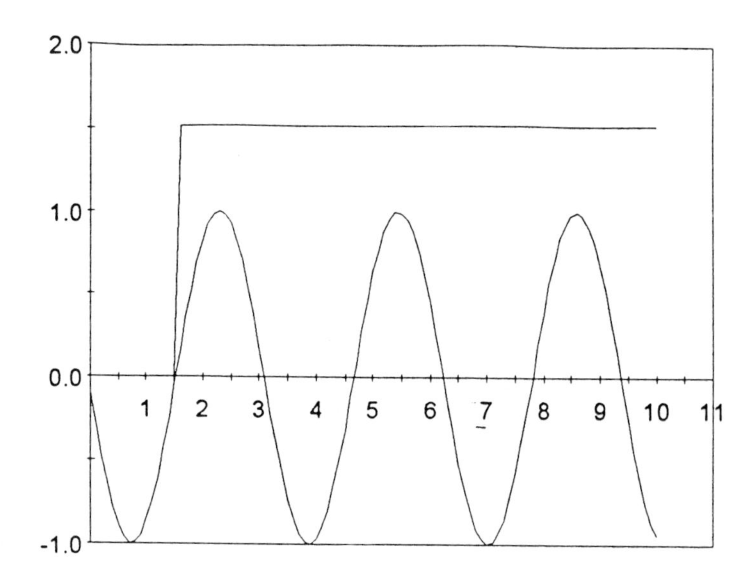

Fig 8.20.2b The E operator

REFERENCES

1. Blomhoff R et al. 1989. A multicompartmental model of fluid-phase endocytosis in rabbit liver parenchymal cells. Biochem J 262: 605-610.
2. Everts, H.B., Jang, H-K, Boston, R.C. and Canolty, N.L. 1996. A compartmental model predicts that dietary potassium affects lithium dynamics in rats. J. Nutr. 126: 1445-1454.
3. Fisher WR et al. 1997. The ^{3}H-leucine tracer: Its use in kinetic studies of plasma lipoproteins. Metab. 46:333-342.
4. Gabrielsson, J., and D. Weiner. 1994. Pharmacokinetic and Pharmacodynamic Data Analysis: concepts and applications. Swedish Pharmaceutical Press, Stockholm.
5. Jackson, A. J., and L. A. Zech. 1991. Easy and practical utilization of CONSAM for simulation, analysis, and optimization of complex dosing regimens. *J Pharm Sci.* 80:317-320.
6. Lewis KC et al. Retinol metabolism in rats with low vitamin A status: a compartmental model. J. Lipid Res. 31:1535-1548, 1990
7. Wagner, J. 1993. Pharmaceutics for the Pharmaceutical Scientist. Technomic Publication., Lancaster, PA.
8. Wagner, J., and C. D. Alway. 1964. Serum levels of 'Lincomycin' from single dose Serum levels when Lincomycin (as the hydorchloride) was administered by constant rate infusion. *Nature,*. 201:1101-1103.

SECTION IV

Strategies for Modeling Biological Systems

9

EXPERIMENTAL DESIGN AND DATA COLLECTION

To determine properties of biological systems, such as rates of uptake, movement or loss of material, pool turnover, and pool size, it is necessary to perform kinetic studies where changes in the system are measured over time. Data obtained from these studies can be used for example, to determine differences between healthy vs. sick populations for therapeutic strategies or between normal and null mutations to determine the function of a gene product. The first step in designing a kinetic study is to establish the purpose for the study. The next step is to state the corresponding hypothesis. Lastly, design the experiments necessary to test the hypothesis. The design of the experiments will often determine the approach to be used for data analysis and the amount of information that will be obtained from the study. This chapter explores some theoretical issues to consider when designing experiments such as kinetics vs. dynamics, tracer vs. tracee and steady state vs. non-steady state. Then it considers practical issues relating to tracer administration, data collection, the importance of pilot studies, and the number of subjects to be studied.

I. Theoretical Considerations

1.1. Kinetics vs Dynamics

Most biological systems are dynamic, or nonlinear, because parameter values of the system change over time. These changes occur because most systems contain zeroeth-, second- or higher-order processes (such as enzyme reactions and receptor-mediated binding processes). Dynamic systems, however, can be treated as linear (or kinetic), with constant parameter values if they are studied in steady state (described below) using tracers (1). The dynamics of the system can be defined by multiple kinetic studies, i.e., by studying a system in one steady state and then performing another study after it has been perturbed to a new steady state. Non-linear processes can be identified by parameters that change between the two states. (If parameters do not change between the two states, the system is linear). This approach was used to study zinc kinetics in humans during normal zinc intake and when intake was increased by 10-fold (13). By modeling both the tracer and tracee (material being traced) data simultaneously, the transition period between two steady states could be used to provide information on the nature of the nonlinear processes. Likewise, copper metabolism was studied during pregnancy and lactation in dairy cows and by modeling tracer and tracee simultaneously, changes due to growth of the fetus were described in the pool sizes of copper (4, 5). The published papers provide detailed descriptions of the methods used. Dynamic analysis is harder than modeling linear systems, as it is more difficult to predict the outcome of non-linear interactions.

> Tip: Tracer experiments permit the observation of complex non-linear systems as linear systems (1).

1.2. Tracer vs. Tracee

It is necessary to tag systems that change over time in order to measure them. For example, to determine the flow of a river, one approach would be to throw a stick into the river and to measure how quickly it passes between two points. The stick in this case would represent the tag (or tracer) while the water would represent the tracee (or unlabelled compound of interest). Note that the stick should be small enough that it does not impede the flow of the river, but heavy enough that it flows at the same rate as the water and is not blown back upstream. Tracers therefore must act like the substance being traced and must not perturb the system. Examples of tracers are dyes, fluorescent markers, and isotopes such as radioactive or stable isotopes.

Some systems cannot be studied using tracers because there are no suitable tracers available (see Chapter 3 for discussion of tracers). For example, lithium does not have a suitable tracer. Studies of lithium metabolism have been performed dynamically (or in the non-steady state) by administering different doses of lithium and a model was developed based on changes of tracee levels over time (6).

> Tip: A tracer is a tag used to measure the substance of interest (or tracee) in a system

1.3. Steady state vs Non-steady State

When interpreting kinetic data it is important to know whether the system is in steady state. Steady state means that the amount of tracee in the system is constant during the study. If a system is in steady state, information on fractional transfer rates determined from analyzing the tracer data can be used to determine synthesis rates, pool sizes, and transport rates of the tracee. Initial studies of a system should be performed, if possible, while the system is in steady state. If the system is not in steady state (e.g., the system is perturbed by administration of a drug, or the level of intake of some nutrient is increased) rate constants may be changing during the course of the study. In the example discussed above on determining the flow of a river, it would be important to know that the river was at a constant level during the measurement i.e., it was not increasing due to a dam being released above the site of measurement, or getting lower because water was being taken off for irrigation. Either change would affect the flow. Likewise, if a system is not in steady state during a kinetic study, it is necessary to model both the tracer data and tracee data simultaneously (see later section).

It is important to consider the time scale of any perturbation relative to the length of a study. For instance, while meals perturb the metabolism of nutrients such as glucose and amino acids for several hours after a meal, these small perturbations would not affect the analysis of the data for a study conducted over weeks to months. For a study of several hours, however, meals could have a large impact on the analysis and interpretation of tracer data.

> Tip: If system is not in steady state during a kinetic study it is necessary to model both the tracer data and tracee data simultaneously

II. Practical Considerations

2.1. Tracer administration

The type of tracer to use is an important consideration in designing a tracer-based study. Radioactive isotopes are measured by counting particles they emit. Advantages of radioisotopes are that they are generally inexpensive, samples generally require minimal preparation before counting, and they have insignificant mass and therefore do not perturb systems. Some disadvantages of radioisotopes are safety concerns due to exposure to radiation, that their use in humans is limited, they may decay rapidly, and their disposal is highly regulated. There is renewed interest in some radioisotopes, however, because they can be detected at very low levels by using Accelerator Mass Spectrometry.

Stable isotopes occur naturally and are measured by mass difference. Advantages of using stable isotopes are that they do not decay and so samples can be stored for long periods before analysis and there are no safety concerns (e.g., they are used in preterm infants). Disadvantages are that they are expensive, the samples require chemical purification before analysis, and they contribute mass. Therefore, application of a stable-isotope may perturb the system. Moreover, because more than one isotope is necessary for mass determination (as these are based on ratios between two isotopes), only compounds with more than one isotope can be studied. Generally, the form of isotope that is most highly enriched (and therefore will contribute the lowest mass) is chosen. For example, selenium has several stable isotopes: ^{74}Se is the least abundant, consisting of 0.087% of natural selenium but ^{76}Se, the next least abundant stable isotope consists 9.02%. If ^{76}Se were used as a tracer it would be less detectable than ^{74}Se because of the difference in natural abundance. However in some situations, such as oral feeding of the isotope, the tracer can replace part of the normal daily intake and therefore the mass of the tracer will not perturb the system. When calculating the amount of stable isotope in samples, a correction may be needed because administered stable isotopes of high natural abundance can add significantly to the total mass of each sample (10, 11).

The form of labeling is also important. For some compounds, e.g. minerals and trace elements, the tracer will behave in exactly the same way as the compound under study. Other factors may be important however. Will the tracer be administered in the fasting state, or with food? Will it be presented as an inorganic chemical, or incorporated organically into food? For some nutrients, such as zinc, the form does not seem to affect absorption and metabolism. However, for others such as calcium and selenium there are large differences in solubility and absorption between the different forms (8, 12). As a general rule, the tracer should be introduced in as natural conditions as possible. For example, when tracer is to be administered intravenously (or orally in infant formula), the tracer should be equilibrated with plasma (or the formula), by mixing it several hours before administration to ensure that the tracer is bound to the species in the media in the same ratio as the tracee, so that it will be metabolized in the same way as the natural compounds.

For other compounds, like glucose, the site of the labeling on the molecule affects the kinetics. Glucose can be labeled at the carbon or hydrogen sites.

Because the carbon is recycled through other compounds while the hydrogen is lost as water, the kinetics will appear to differ with the hydrogen label disappearing faster than the carbon label (15). Furthermore, individual carbon atoms will be metabolized differently and a model has been developed to show how the various carbon atoms are metabolized (7). An even more complicated system to label is lipoprotein metabolism. There, it is necessary to label each of the separate moieties (cholesterol, free fatty acids, proteins) in order to define fully the kinetics of the lipoproteins (3).

The site of tracer administration can provide information on different aspects of the system. For example, administering tracer orally to subjects can provide information on absorption, while administering the tracer intravenously provides information on how rapidly tissues take up the compound from blood. Administration of two tracers simultaneously, but with the same sampling schedule, can often provide information on absorption and distribution with minimal increase in cost for sample collection or analysis.

Tracer can be administered as a bolus, an infusion, by a combination of these approaches, or by multiple dosing. The type of administration can be readily simulated using modeling software, so the choice of protocol is decided based on achieving adequate levels of tracer for measuring the parts of the system that are to be sampled. It is critically important that the amount of tracer administered be known precisely. A sample of the dose should always be analyzed to test for chemical purity and activity (for radioisotopes) and enrichment (for stable isotopes).

2.2. When and where to collect data

The best rule for kinetic studies is that there can never be too much data. The number of samples obtained is normally limited by criteria such as clinical or analytical considerations, the total amount of blood which can be drawn per subject, or the cost of sample analysis. If the kinetics of the system are known, models can be used to determine the times at which data will provide the most information on the system, i.e., how useful each datum will be in defining a parameter. In WinSAAM this is done using the INFO command (2).

The length of data collection depends on the purpose of the study. For example, if the purpose is to study absorption of calcium in humans, then a study of 8-12 hr will be sufficient. However, if the purpose is to study calcium deposition in bone, then the study needs to be at least 3 wk (9). A shorter study will overestimate bone deposition.

The site of sample collection is important with respect to tracer administration when interpreting the tracer data. One example is the interpretation of tracer levels in feces after tracer is administered intravenously. In this case, appearance of tracer in feces is usually equated to endogenous excretion (i.e., material secreted from plasma into the intestine and then excreted). The amount of tracer appearing in feces, however, is related to the amount in blood. When calcium kinetics were compared in adolescent girls and young women, more of the intravenous tracer was excreted in feces by the young women, suggesting that endogenous excretion was higher in this group (14). However, when the fecal data were analyzed with the serum data, it could be seen that the reason that more i.v tracer was excreted in feces was because tracer levels were higher in serum in the young women, while the rates of endogenous excretion were the same in the two groups. If only fecal data had been sampled, it would have been concluded that young women excreted more calcium than adolescent girls.

2.3. Pilot studies

Pilot studies are essential before proceeding to a main study. Due to the expense of most kinetic studies and because of the number of samples to be collected and analyzed are large, pilot studies can ensure that in the main study, sufficient tracer will be administered. This will ensure that that the tracer can be reliably measured for the length of the study at all sites of interest, that the data obtained are sufficient to define parameters of interest, and that critical data (sites or time points) are not being missed. Analysis of the pilot study will provide information on the variability of the data; this along with the error estimates on the parameters can be used as a basis for determining the number of subjects to be studied.

2.4. Number of Subjects or Experiments

If previous studies have been performed, power analysis can provide some estimate of the number of studies required to determine a parameter of interest. However, in the absence of data, there is no method for determining the number of studies to perform. If the aim of the study is to determine the pathophysiology of a system, it is generally better to study fewer subjects in more detail than a large number of subjects with fewer data per subject.

If the purpose of the study is to compare kinetics under two conditions, the power of the study will be increased if each subject acts as their own control. Carryover of tracer from one study to the next can be accounted for in the modeling (see Time-Block tool in Chapter 8). Animal studies usually can provide more data than studies on humans due to the greater opportunity for sampling tissues. However, data at each time point may include different animals, and this adds error to the data. To account for this, generally several animals are sacrificed at each time point.

Historically, kinetic studies are performed on fewer subjects than using other methods. This is due to the large amount of data obtained per subject. For example, balance studies that determine absorption as the difference between dietary intake and fecal excretion require the collection and analysis of two samples per subject (diet and feces). Kinetic studies to determine absorption require samples from several sites (diet, urine, blood and feces) over a period of several days. Because absorption is determined more accurately using kinetics, fewer subjects would be required for a kinetic study. Once a model has been determined, power analysis can be performed to determine the number of subjects required to detect a difference in one or more parameters for a system. During model development, however, information on model variance may not be available and so the number of subjects to be studied must be decided on an *ad hoc* basis depending on the resources available. The information obtained from a few well-designed kinetic studies is often greater than a large number of studies where less data are obtained per subject. Due to the amount of data obtained per subject, published kinetic studies are often based on a small number ($n<10$) subjects.

REFERENCES

1. Berman, M. 1969. Kinetic modeling in physiology. *FEBS Letters*. 2:S56-S57.
2. Berman, M., W. F. Beltz, P. C. Greif, R. Chabay, and R. C. Boston. 1983. CONSAM User's Guide. *In* DHEW Publication No 1983-421-123:3279. US Govt Printing Office, Washington, DC.
3. Berman, M., S. M. Grundy, and B. V. Howard. 1982. Lipoprotein kinetics and modeling. Academic Press, New York.
4. Buckley, W. T. 1991. A kinetic model of copper metabolism in lactating dairy cows. *Can. J. Anim. Sci.* 71:155-166.
5. Buckley, W. T. 1995. Copper metabolism in dairy cows: development of a model based on a stable isotope tracer. *In* Kinetic Models of Trace Element and Mineral Metabolism During Development. K. N. Siva Subramanian and M. E. Wastney, editors. CRC Press, Boca Raton, FL. 37-52.
6. Everts, H. B., H.-Y. Jang, R. C. Boston, and N. L. Canolty. 1996. A compartmental model predicts that dietary potassium affects lithium dynamics in rats. *J. Nutr.* 126:1445-1454.
7. Goebel, R., M. Berman, and D. Foster. 1982. Mathematical model for the distribution of isotopic carbon atoms through the tricarboxylic acid cycle. *Fed Proc*. 41:96-103.
8. Heaney, R. P., C. M. Weaver, and M. J. Barger-Lux. 1995. Food factors affecting calcium availability. *In* Nutritional Aspects of Osteoporosis '94. Vol. 7. P. Burckhardt and R. Heaney, editors. Ares-Serono Symposia Publications.
9. Jung, A., P. Bartholdi, B. Mermillod, J. Reeve, and R. Neer. 1978. Critical analysis of methods for analysing human calcium kinetics. *J. Theor. Biol.* 73:131-157.
10. Liu, Y.-M., P. Neal, J. Ernst, C. Weaver, K. Rickard, D. L. Smith, and J. Lemons. 1989. Absorption of calcium and magnesium from fortified human milk by very low birth weight infants. *Pediatr Res*. 25:496-502.
11. Lowe, N. M., D. M. Shames, L. R. Woodhouse, J. S. Matel, R. Roehl, M. P. Saccomani, G. Toffolo, C. Cobelli, and J. C. King. 1997. A compartmental model of zinc metabolism in healthy women using oral and intravenous stable isotope tracers. *Am J Clin Nutr*. 65:1810-1819.
12. Patterson, B. H., L. A. Zech, C. A. Swanson, and O. A. Levander. 1995. An overview of selenium kinetics in humans. *In* Kinetic Models of Trace Element and Mineral Metabolism During Development. K. N. Siva Subramanian and M. E. Wastney, editors. CRC Press, Boca Raton.
13. Wastney, M. E., R. L. Aamodt , W. F. Rumble, and R. I. Henkin. 1986. Kinetic analysis of zinc metabolism and its regulation in normal humans. *Am J Physiol*. 251:R398-R408.
14. Wastney, M. E., J. Ng, D. Smith, B. R. Martin, M. Peacock, and C. M. Weaver. 1996. Differences in calcium kinetics between adolescent girls and young women. *Am. J. Physiol.* 271:R208-R216.
15. Wastney, M. E., J. E. Wolff, R. Bickerstaffe, C. F. Ramberg, and M. Berman. 1983. Kinetics of glucose metabolism in sheep. *Aust. J. Biol. Sci.*:463-474.

10

STARTING MODELING AND DEVELOPING A MODEL

This chapter presents a strategy for developing mechanistic compartmental models. It covers the approach for choosing a starting model, identifying the initial conditions, identifying the inputs into the model, issues relating to data units, how to fit a linear model, how to calculate steady state data, fitting data obtained under two conditions, calculating functions such as absorption, solving inconsistencies, comparing data from two steady states, fitting nonlinear models, and fitting tracer and tracee data simultaneously. As stated by Finklestein and Carlson (5), modeling cannot be reduced to a set of recipes, but requires insight and imagination. This chapter demonstrates how models have been developed using several systems as examples.

I. Choosing a starting model

The starting model should be the simplest model or hypothesis that is consistent with known features of the system, the nature of the experiment, and the qualitative features in the data (1). If the proposed model is too simple, it will not fit all the data. If the model is too complex, there will be a lack of confidence in the estimates of its parameter values.

The first model to try is any that is in the literature for the same or a similar system. Some published models are available through a model library on the Internet (see Chapter 22). These models can be downloaded in formats compatible with several modeling packages. It is important to begin with available models because many different models can provide an acceptable fit to a particular set of data. New features of a system will only become apparent when the current models are compared against new data. If the current models do not fit the new data, the differences will need to be explained by adding new features to the model.

If models have not been published previously for the system under study, a general approach is to identify known (or expected) compartments in the system. For example, if it is an enzyme reaction, start with the reactants and products, if it is a cellular study, start with extracellular and intracellular pools, if it is a whole body study, start with the sampled compartments. The arrangement of compartments is to some degree arbitrary in that mammillary and series models will often fit the same data. The arrangement may be determined by physiological aspects of the system. Known pathways between the compartments and losses from the system can be added. A *de novo* approach to modeling selenium metabolism in humans is described in Patterson and Zech (13)

It is important to 'model the experiment', i.e., to model the study in the same order in which it was performed. For example, if the enzyme and substrate were mixed several minutes before a catalyst was added to start the reaction, it should be modeled this way, as some substrate may be bound to enzyme during the mixing phase. Similarly, if a tracer is administered orally, and a second tracer is given intravenously some time later, it is necessary to model it this way.

We will use as an example of model development a study of calcium metabolism in adolescent girls (18). The purpose of this study was to compare and contrast calcium metabolism in girls and women, while on the same calcium intake, to identify differences in metabolism that could be used for increasing bone deposition of calcium in girls. This model was adapted from the model proposed for adults by Neer et al. (11). For the purpose of this example, we will walk though the decisions made in analyzing the data following intravenous tracer administration. (Each subject was studied following administration of an intravenous stable isotope tracer and oral administration of a second stable isotope tracer. Fitting oral and iv data simultaneously will be described below, in Section VII).

When beginning a modeling project, it is important to visually inspect the data as this will often provide clues as to the type of model required. Data show that the serum curve is multiexponential because it is not a straight line (Fig 10.1), implying that more than one compartment is required, and that tracer is excreted in urine and feces. A starting model is shown in Fig 10.2

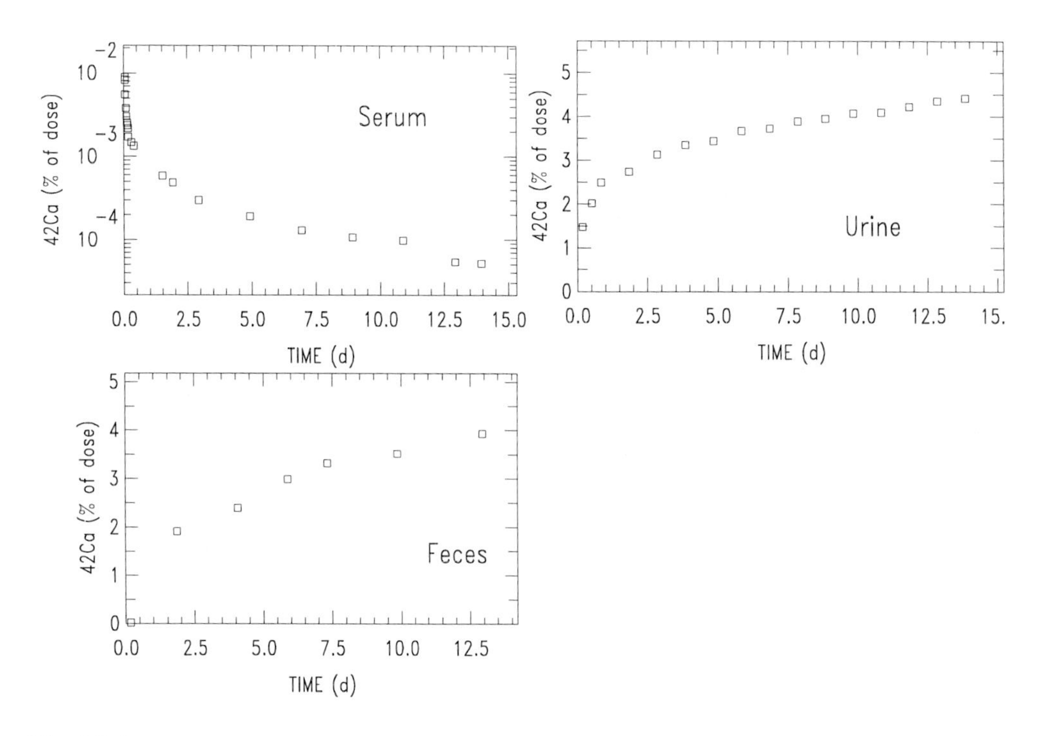

Fig 10.1 Calcium kinetic data from humans.

II. Identifying initial conditions

The initial conditions indicate the amount of material of interest in each compartment at the start of a study. For a tracer study, the initial condition will indicate the site and the rate of tracer administration. Examples of how to model an oral dose, intravenous dose, or infused dose and doses at specific times during a study are described in Chapter 8 using the tools IC, UF, and QO. In the

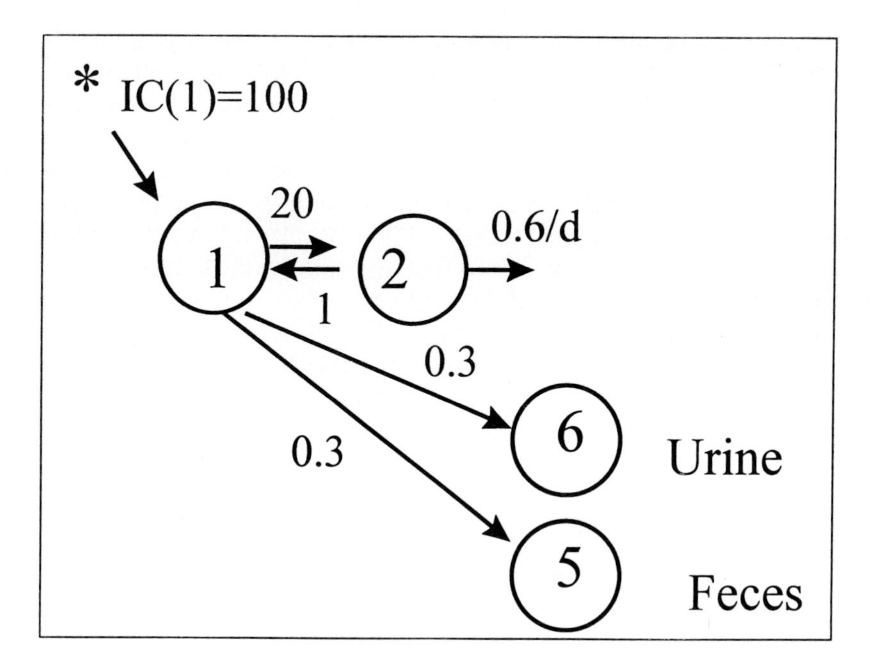

Fig 10.2 Starting model for calcium kinetics.

calcium example tracer was given by intravenous injection. As the infusion was rapid (over 30 sec) it could be considered to be instantaneous and the initial condition for the first compartment was set at 100% of the dose.

III. Identifying inputs

Inputs include entry of tracer and tracee into the system. For the calcium example, tracer was given once intravenously at the beginning of the study. In pharmacokinetics, drugs may be administered by various routes and at various times. An example showing how these various inputs can be modeled using WinSAAM has been described (9). For tracee models, inputs represent endogenous production of the compound of interest (e.g., if glucose was being studied, inputs would include release of glucose by liver and absorption of glucose from the diet).

IV. Data units

To fit the model to observed data, the units need to be consistent between the calculated values and the observed data. The calculated values will have the same units as the initial conditions. For example, if the initial condition is 100% of the dose, the values calculated for each compartment will also be in % of dose. If the data are expressed in units other than % of dose, the units of the calculated values will need to be changed to match the data by using an equation. For example, if the initial condition is in % of dose but observed serum data are expressed as % dose/ml of serum, it is necessary to divide the calculated values for serum, (% of dose), by a parameter representing the serum volume in ml (e.g., G(1)=F(1)/P(1), where F(1) is serum tracer in % of dose and P(1) is the estimate of serum volume). Serum volume can be estimated by the model if tracer is administered into blood. The intercept at time zero for the blood data will represent 100% of dose/total ml. Sometimes this value is larger than blood volume because the compound of interest exchanges rapidly with extravascular fluid, or binds to receptors or tissues. The volume is then referred to as the 'initial volume of distribution. Alternatively, the blood volume can be fixed from other studies. In humans, blood volume is about 7% of body weight and plasma is 4% of body weight.

It is also important to keep the time units consistent throughout the problem. If the data are expressed in hours, then the parameters will have the units of /hr. Radioactive data are sometimes expressed as specific activity (tracer/mass in the compartment). WinSAAM has an option of converting calculated units to specific activity (see Tools, Chapter 8). In the calcium example, units for the dose are expressed as %, the units for fecal and urine excretion are cumulative % dose, and the units for serum are % dose/L.

Tip: Calculated values will have the same units as the initial conditions. They need to be adjusted using equations to match units of the observed data.

V. Fitting a linear model

Before the model can be solved, it is necessary to enter initial values for the parameters. By inspecting the serum curve (Fig 10.1), the initial slope appears to decay by 10/0.5. Therefore the initial loss, L(2,1) appears to be 20/d. Tracer reappears in serum at 1 day, therefore the fractional rate of return to plasma, L(1,2) is set initially to 1/d. The final slope appears to be 10/15, therefore L(0,2) is set to 0.6/d. Note that L(0,2) is the loss to the outside the system. In the case of calcium, it refers to calcium that does not exchange during the time course of the study, and therefore calcium deposited into bone. The loss to urine is 4%/12 d or 0.33/d and the loss to feces was also about 4% over 12 d, and was also set to 0.33/d. These values form the initial parameter values (Fig 10.2). The fit is shown in Fig 10.3

It is apparent that the second serum slope is too rapid (i.e., material needs to be returned from the second pool). Therefore L(1,2) needs to be increased. The fit when L(1,2) is set to 4/d is shown in Fig 10.4. While this change improves the fit to the initial slope of the serum curve, it results in an overestimate in the amount of tracer excreted in urine and feces. Therefore L(5,1) and L(6,1) are both reduced to 0.1/d. This fit is shown in Fig 10.5

It is apparent that the tail of the serum curve is underestimated. This implies that there is a third compartment in the body (i.e., not all of the tracer is lost irreversibly from compartment 2, but some returns to serum). The tracer appears in serum after about 7 days, therefore the initial estimate for L(2,3) is 1/7 or 0.14/d. L(0,3) is set at approximately 0.1/d. The fit using this set of parameter values is shown in Fig 10.6.

The fit is close enough so that it is appropriate to finish the fitting using the iterative feature of WinSAAM. To do this, first parameters need to be made adjustable, i.e. upper and lower limits are assigned to the L(I,J). Second, weights need to be assigned to the data. For fitting biological data, a fractional standard deviation (FSD) of 20% is reasonable (see Chapter 15 for detailed information on assigning weights to data). Note that the first point in the fecal data is left unweighted. This datum has a low value because little tracer has been excreted at

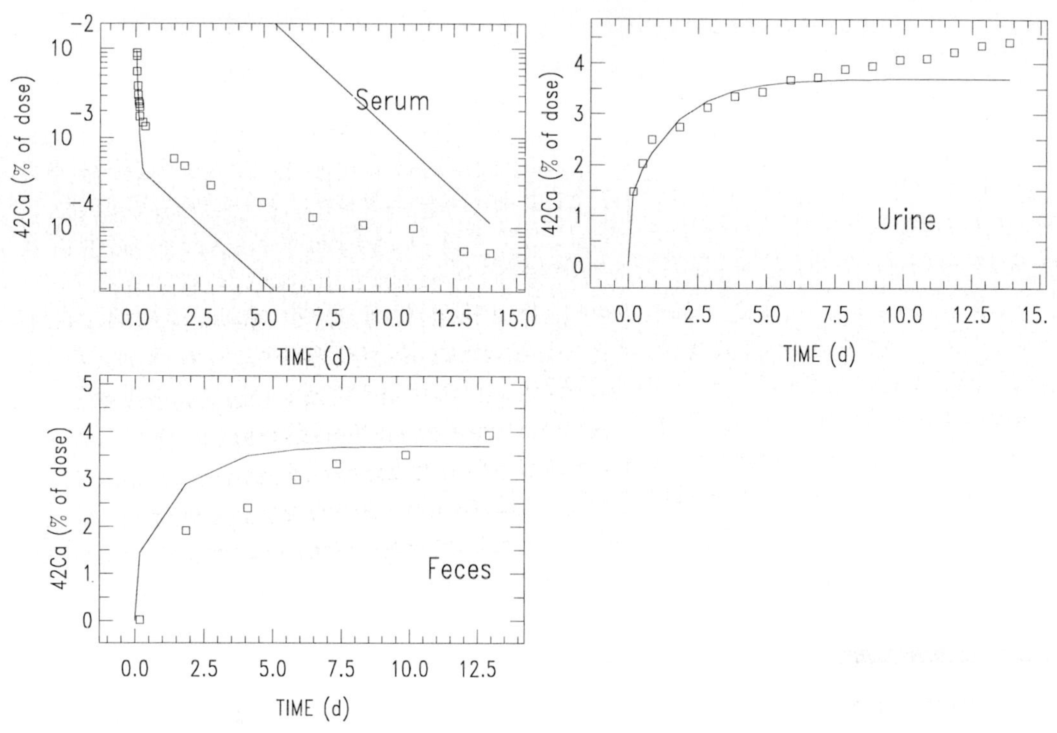

Fig 10.3 Fit of initial model for calcium kinetics. Note the calculated value for serum (line) wraps around below the observed data (symbols) as it is plotted on a log scale.

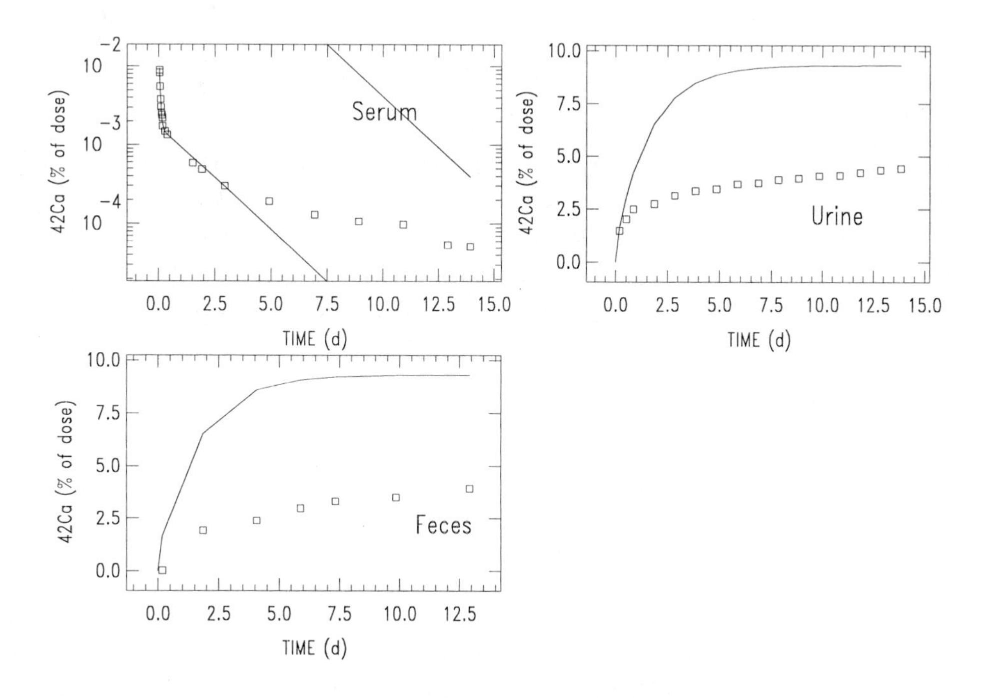

Fig 10.4 Fit of second model for calcium kinetics. Note the calculated value for serum (line) wraps around below the observed data (symbols) as it is plotted on a log scale.

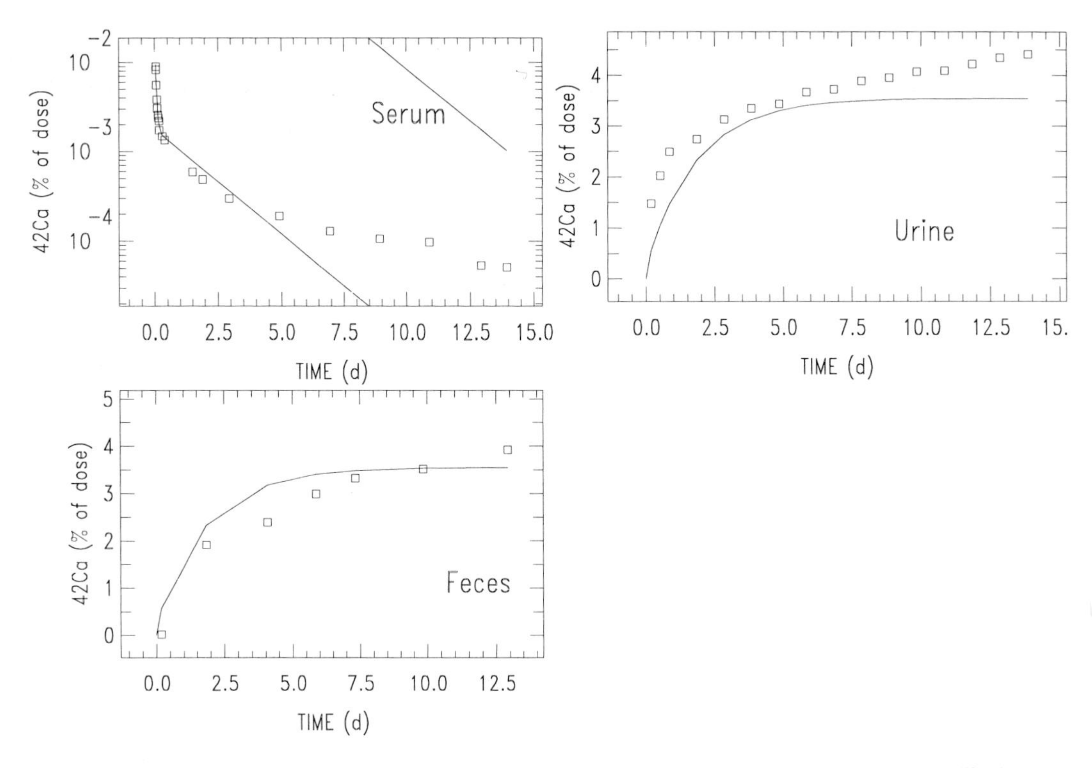

Fig 10.5 Third fit of model to calcium kinetic data. Note the calculated value for serum (line) wraps around below the observed data (symbols) as it is plotted on a log scale.

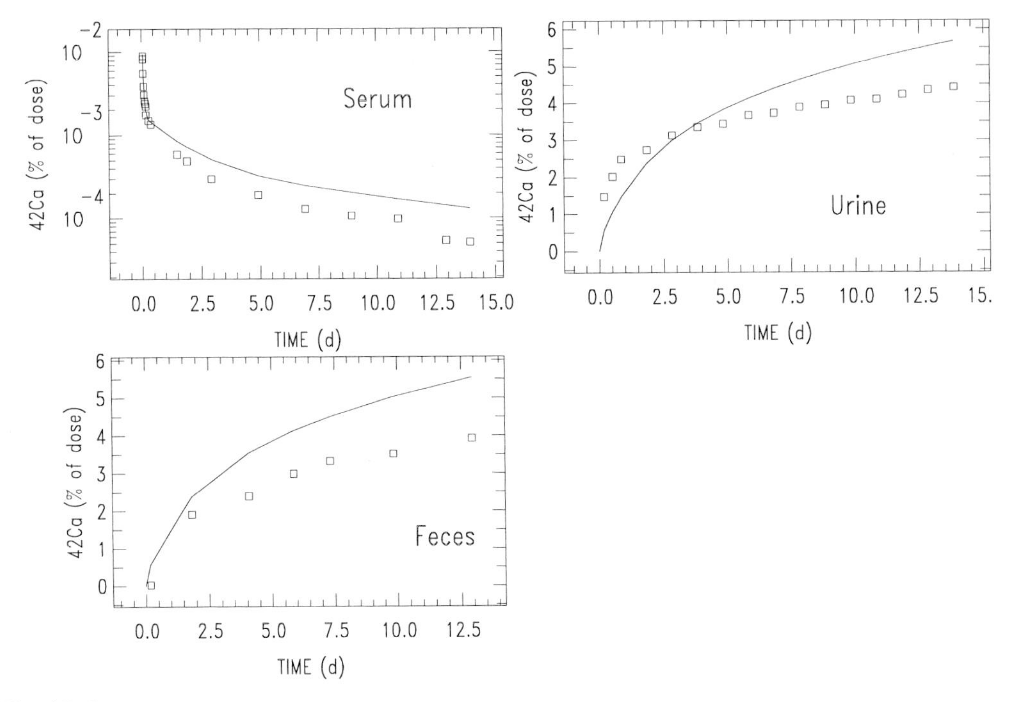

Fig 10.6 Fourth fit of model to calcium kinetic data.

this time point. The FSD weighting scheme would thus have assigned a large weight to this value, and this would have affected the fit. The fit after iterating is shown in Fig 10.7.

Although the fit is close, there is a deviation at the start of the urine data. It appears that some tracer is excreted rapidly into urine. This can be tested by assigning a small initial condition to compartment 6, by equating IC(6) to P(6), putting limits on P(6), and iterating. The final fit after iterating is shown in Fig 10.8 and the model in Fig 10.9. This example simplifies the modeling process. In reality, many approaches may be tried before finding an approach that fits the data.

VI. Fitting steady state data

Steady state data refers to the amount of tracee or mass in a compartment, the inputs of this unlabeled material into the system, and the transport rates between compartments.
The tracer data is used to determine the connectivity of the system and the fractional flows between compartments. Once these values have been determined, the steady state values for the system can be calculated by specifying one steady state value for the system, such as the mass of a compartment (e.g., mass in blood), an input (e.g., intake in the diet), or a loss (rate of urinary excretion). Note that the mass calculated for a compartment depends on where the flow of material into the system occurs (see example on equivalence of tracer-tracee supply in Chapter 8). Therefore, in addition to specifying one steady state value, it is also necessary to specify in the model where the entry of unlabelled material into the system occurs.

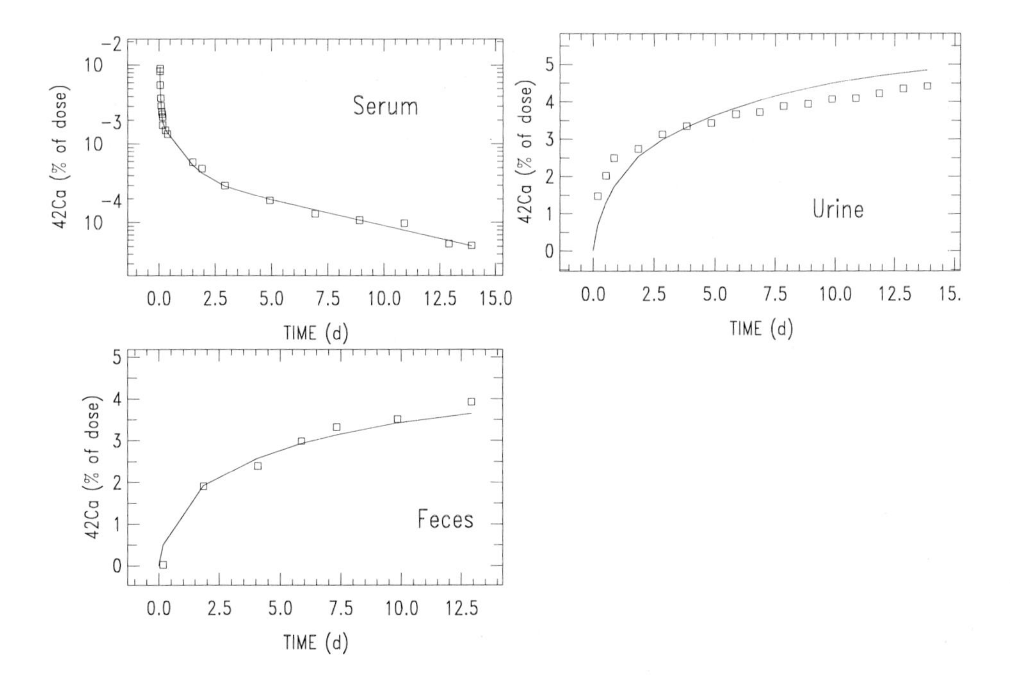

Fig 10.7 Iterated fit of model to calcium kinetic data.

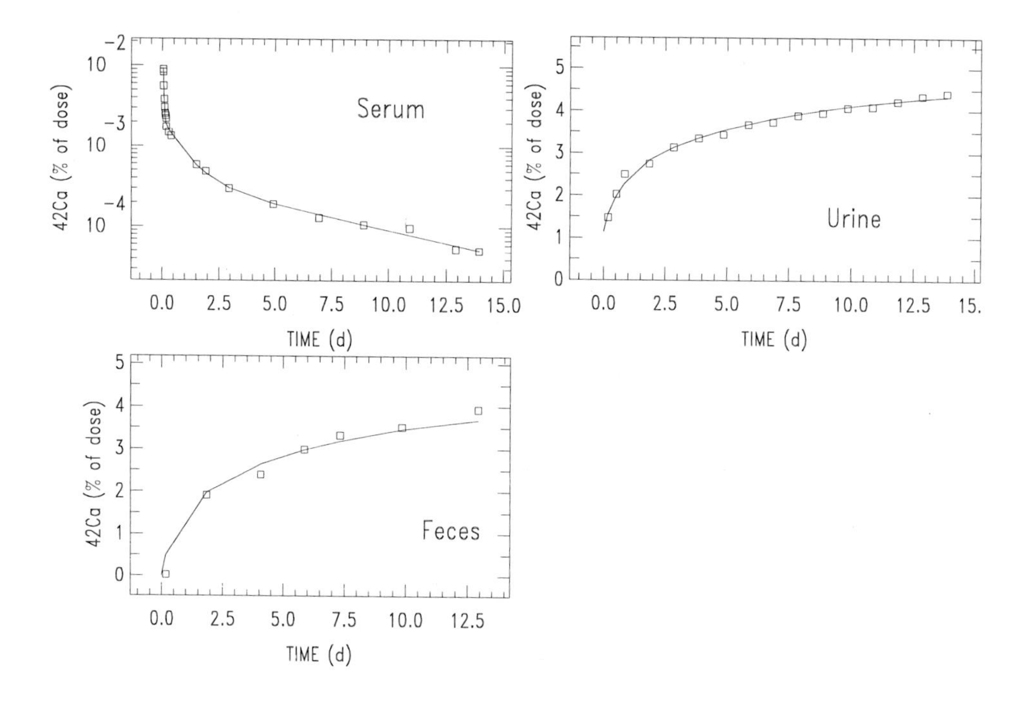

Fig 10.8 Final fit of the model to calcium kinetic data.

If there is additional steady state information on the system (e.g., other compartment sizes are known, rates of loss are known), these values can be entered as data and weighted. These data will be fitted during the iterative process. The weight assigned is important. If a fractional weight is assigned and the steady state values are large numerically compared to the tracer values, they may be assigned a large weight (see Chapter 15). It is important therefore to list the weights of each compartment to compare the relative weights before iterating, so that one compartment is not weighted significantly more than another as a result of larger data units.

In the example above, we know there is a dietary input of calcium (1300 mg/d). In addition, it is known that calcium is released from bone. Some calcium "lost" from the system is actually deposited in bone and reenters serum during bone remodeling. Entry of calcium from dietary sources as well as bone resorption is represented by an input of tracee into compartment 1, U(1), and we let this parameter adjust. We know the serum concentration of calcium and this is specified as M(1), using an equation to convert the concentration in serum to the mass in compartment 1 by multiplying the concentration in serum by serum volume. Compartment 1 represents a space 10-fold larger than serum (i.e., called the initial volume of distribution, or serum-equivalent volume). WinSAAM calculates the mass of each compartment and the transport rates, as shown in Fig 10.9.

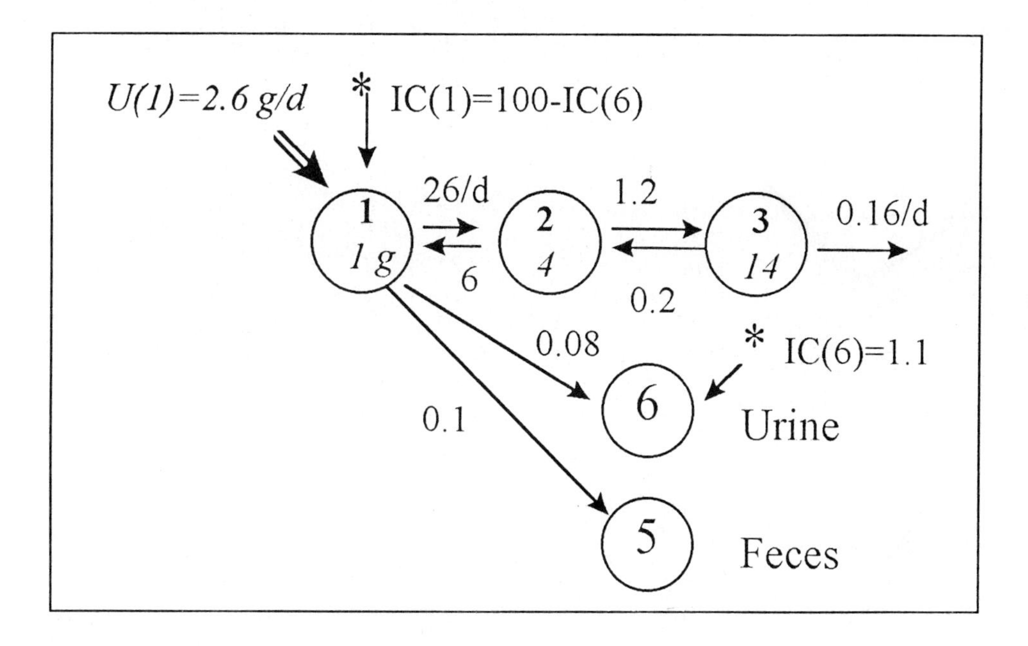

Fig 10.9 Final model for calcium kinetics.

VII. Fitting data obtained under two conditions

Often studies are performed under several conditions; for example, with two (or more) enzyme concentrations or with two (or more) tracers. All data should be modeled simultaneously because each study provides information about different aspects of the system. For instance, an intravenous tracer will provide information on the distribution of a compound from blood, while an oral tracer provides information on absorption. There are two methodological approaches to fitting data obtained under different conditions: by using duplicate models and by using time-interrupts (time-blocks, TC). With the duplicate model approach, parameter values, initial conditions, and steady state attributes need to be stated explicitly for each model. The time-interrupt approach is a shorthand approach where, only values which differ between the conditions need to be specified.

The calcium example will now be expanded to show how data obtained following the administration of two tracers were fitted. The study consisted of administration of an oral tracer (^{44}Ca) at time zero and the administration of an intravenous tracer (^{42}Ca) 1 hr later. In addition to the intravenous model already described (Fig 10.9) we now set up an analogous oral model using compartments 11-13, 15 and 16 (Fig 10.10). In addition, we add intestinal compartments 18 and 19 and a pathway to represent absorption from the intestine into serum (from compartment 18 to compartment 11). We use the parameter values obtained from fitting the intravenous data to fit the oral data, by equating all the parameters in the oral model to the values in the intravenous model. The intravenous tracer was administered one hour after the oral tracer. To model the later input of intravenous

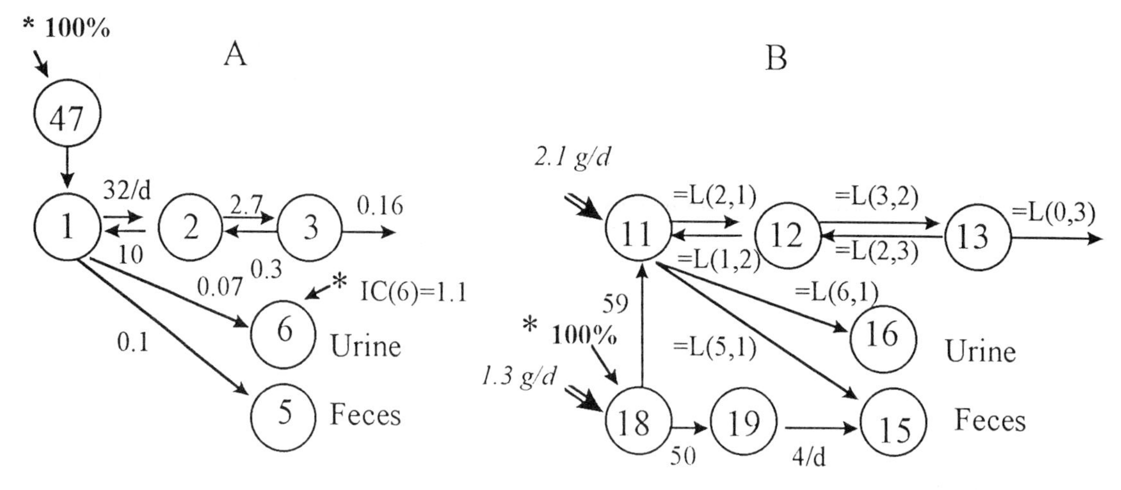

Fig 10.10 Model for calcium kinetics to fit data after A) intravenous tracer administration and B) oral tracer administration.

tracer, we add a 'dummy' compartment (compartment 47) to the model. We add a 'QO' to this compartment. (See Chapter 8 for examples) to reset the value of the compartment 47 equal to the value we specify at a given time. Here at t=0, the value for compartment 47 is 0. At 1 hr (0.042 d) we make the value 100 to represent 100% of the intravenous dose, but subtract any tracer initially in compartment 6. No absorption pathway is required for the intravenous model, although we add one, as described below in Section IX for a special purpose. Steady state values are calculated for each model.

VIII. Calculating functions

Once the model has been setup it is possible to enter functions to calculate parameters of interest, for example, for whole body studies: clearance from serum, absorption from the intestine, endogenous excretion. For enzyme studies one may calculate free enzyme, total bound enzyme etc. This is performed in WinSAAM by entering equations, designated G-functions. Equations can contain the four basic arithmetic operators (+, -, *, /), exponents and trigonometric functions and can be used to calculate for example, functions of parameters, as well as compartment masses or transport rates. In the calcium example, a function, G(3), is included to calculate absorption as the ratio of the parameter from the intestine into serum versus the total turnover of the intestinal compartment:

G(3)=L(11,18)/(L(11,18)+L(19,18))

IX. Solving inconsistencies

A large part of a modeling project involves resolving inconsistencies between observed and predicted values. Sometimes the model is incorrect, and the data are fitted by changing the model. In other cases, the model may fit data from some subjects, but not from others, even when parameter values are adjusted for each individual. In this case, it is important to check that observed data have been calculated and entered into the Input file correctly. In yet another situation, the model may fit the tracer data well, but may not fit the steady state data. In this case it is important to carefully examine the data and the experimental conditions to try and identify where the problem may lie. At best, it is useful to fit the data so that the difference can be explained kinetically while awaiting a biological explanation. In a model for selenium, to fit plasma data it was necessary to postulate the existence of several plasma pools (12). The biological existence of some of these pools has been confirmed subsequently.

In the calcium example, where the tracer study was combined with a balance study, absorption was calculated from the rate of appearance of oral tracer in serum. However, this value did not fit the tracee data. It was an overestimate, and the result was that tracee excreted in feces was underestimated. To begin to resolve this, we allowed absorption for the oral tracer to differ from absorption of dietary calcium. This was done by letting absorption in the iv tracer model represent absorption of dietary calcium. Absorption of the oral tracer was 53% versus 39% for dietary calcium. The same pattern was observed in all subjects in the study. The explanation proposed was that oral tracer was administered with breakfast, with 250 mg calcium, whereas lunch and dinner contained over 500 mg calcium each. As fractional calcium absorption is decreases with increasing calcium load, absorption from the later two meals was lower due to the higher calcium load.

X. Comparing data from two steady states

In the example described in Section IX, data were obtained under two conditions (oral and intravenous tracer administration) to provide additional information about a system. Often data are obtained in two conditions to identify the differences between the conditions. This ability to determine differences between sets of kinetic data is a powerful feature of modeling. Some examples are to compare: a normal with an abnormal condition, metabolism before and after treatment, two physiological conditions. As described previously, the models can be set up as duplicate models or using time-interrupts (see Tools, Chapter 8). With both methods, data from one condition are fitted, and then data from the second condition are fitted by systematically introducing differences in parameter values between the first and second condition. This is called the 'minimal principle'. It assumes that there are a minimal number of differences in parameter values between two conditions that will explain the kinetics. The aim is to identify the changes that are necessary and sufficient to explain the differences.

The logic behind this approach is that regulation of a system normally occurs at specific points. For example, installing traffic lights at a few key intersections can regulate traffic over whole region. The problem is to identify the key intersections, or key pathways. This approach was used to identify five sites of regulation of zinc metabolism *in vivo* (16). Radioisotope tracer data were obtained from human volunteers over 9 months while they consumed their regular diets with 10 mg Zn/d and for a further 9 months while they consumed an additional 100 mg Zn/d. Tracer was measured at seven sites: plasma, red blood cells, urine, feces, and over liver, thigh, and whole body. By systematically changing parameter values, differences between the normal zinc intake period and high intake period were explained by changes in five parameters, representing five sites of zinc regulation (16).

XI. Fitting a nonlinear model

As discussed in Chapter 9, most biological systems are nonlinear but can be treated as linear systems by using tracers. Examples of nonlinear systems may include receptor-binding, chemical reactions, hormone-metabolite, and drug metabolism. If tracers are not available, these systems are more difficult to model than linear systems because the effect of parameter changes on the calculated fit is not always intuitive. Models are, however, powerful tools, and modeling may be the only way to determine whether a hypothesis is consistent with observed data.

An example of a receptor binding model is that of Gex-Fabry and Delisi (6). This model was developed based on the known processes that occur during endocytosis and is therefore an *a priori* model. It was fitted to experimental data on epidermal growth factor binding and uptake by cells and was used to predict receptor down regulation and the rates of clustering of receptors prior to endocytosis.

Chemical reactions are often second or third order as substrates and enzymes bind to form products. For some reactions the chemistry has been determined and modeling the reactions involves setting up equations. About 30 commonly used chemical reactions are available in WinSAAM format in the Mathematical Models of Biological Systems Library (see Chapter 22). For other chemical reactions, the sequence may not be known but can be determined from the data. An example is the development of the model for BAM HI restriction endonuclease (7), an enzyme used to cut DNA. The reaction is slow, occurring over a period of hours. A model was developed based on the rate of disappearance of superhelical DNA and the appearance of nicked and linear DNA (8). Blood coagulation is another example of a complex dynamic system (10). A model for this system shows how a complex series of enzyme reactions results in the production of fibrin which causes blood clotting. An example of a hormone-metabolite model is the glucose-insulin system (see Chapter 7). This model shows how insulin released in response to an increase in blood glucose acts to increase glucose uptake by tissues and inhibit glucose production in the liver. It is the insulin level in an extravascular compartment that regulates these changes. An example of a pharmacodynamic system is the model for lithium kinetics in rats (4).

XII. Fitting tracer and tracee data simultaneously (non-steady state)

When systems are not in steady state it is necessary to fit both the tracer and tracee data simultaneously. This approach was used to model zinc kinetics in preterm infants (17). Because the infants increased their zinc intake and almost doubled in weight during the study, both tracee and tracer kinetics were modeled. This was necessary because the apparent loss of tracer from a pool (such as plasma) is affected by the expansion of the pool size as well as the processes of distribution. Two models were used: one for the tracer (stable zinc isotope) and one for the tracee (total zinc). Parameter values were equated between the models. Initial conditions for the tracer model were the sites of tracer administration (the gut for oral tracer and blood for intravenous tracer). The initial conditions for the tracee model were the zinc mass in different compartments, based on literature values for zinc concentration and tissue size for infants of varying gestational age. Zinc intake in the diet was modeled as an input function into the tracee model.

By modeling both tracer and tracee data, clinical interventions that perturbed zinc metabolism could be incorporated in the modeling. For example, preterm infants often require transfusions of blood cells until their hemopoeitic system develops. The transfused blood cells are from adults, who have ten-fold higher zinc concentration that infants. Therefore, each transfusion represents an infusion of zinc into the infants. This was modeled as an input into the preterm tracee model, and successfully predicted the rise in preterm erythrocyte zinc (17).

XIII. Summary: Developing models

This chapter has described the process of developing a model by using several examples of published models. Other papers where the model development is described carefully are the models of selenium kinetics (12, 14, 15) and the dynamic model for copper in cows (2, 3). Model development cannot be described in terms of a set of rules. Rather a set of guidelines is proposed for developing models. The aim is to develop the simplest model that will fit all the data and be consistent with known physiology. The approach described here is to develop the model based on data obtained by perturbing the system. Decisions on whether to add a pathway or where to locate an additional compartment are based on the shape of the predicted curves compared to the data, and on knowledge of the physiology and biochemistry of the system. Sometimes the model will contain constructs which are not consistent with known physiology. These may represent new insights into the system which will be confirmed through further experimentation, or they may represent experimental artifacts which will be disproved through further experimentation. In either case the model has represented a useful step in extending knowledge of the system and experimentation upon it.

Many aspects of developing models are learned only by trial and error. Another approach to understanding model development, rather than by model synthesis as described in this chapter, is by model decomposition. With this approach, models are examined by removing pathways or compartments to see the effect on the calculated fit to the data. Readers are encouraged to access models available with this book or through the model library (see Chapter 22) and to determine what role various compartments, pathways, and other model constructs play, by removing them from the model and seeing the effect on the model fit to the data. Every system has some unique features about the model that has been developed to describe it. By analyzing and testing a variety of models, investigators will be able to develop the intuition necessary to translate biological systems into mathematical models.

REFERENCES

1. Berman, M. 1969. Kinetic modeling in physiology. *FEBS Letters*. 2:S56-S57.
2. Buckley, W. T. 1991. A kinetic model of copper metabolism in lactating dairy cows. *Can. J. Anim. Sci.* 71:155-166.
3. Buckley, W. T. 1995. Copper metabolism in dairy cows: development of a model based on a stable isotope tracer. *In* Kinetic Models of Trace Element and Mineral Metabolism During Development. K. N. Siva Subramanian and M. E. Wastney, editors. CRC Press, Boca Raton, FL. 37-52.

4. Everts, H. B., H.-Y. Jang, R. C. Boston, and N. L. Canolty. 1996. A compartmental model predicts that dietary potassium affects lithium dynamics in rats. *J. Nutr.* 126:1445-1454.
5. Finklestein, L., and E. R. Carson. 1985. Mathematical Modelling of Dynamic Biological Systems. John Wiley and Sons, UK.
6. Gex-Faby, M., and C. DeLisi. 1984. Receptor-mediated endocytosis: a model and its implications for experimental analysis. *Am J Physiol.* 247:R768-R779.
7. Hensley, P., G. Nardone, J. G. Chirikjian, and M. E. Wastney. 1990. The time-resolved kinetics of superhelical DNA cleavage by BamHI restriction endonuclease. *J. Biol. Chem.* 265:15300-15307.
8. Hensley, P., G. Nardone, and M. E. Wastney. 1992. Compartmental analysis of enzyme-catalyzed reactions. *Meth Enzymol.* 210:391-405.
9. Jackson, A. J., and L. A. Zech. 1991. Easy and practical utilization of CONSAM for simulation, analysis, and optimization of complex dosing regimens. *J Pharm Sci.* 80:317-320.
10. Lawson, J. H., M. Kalafatis, S. Stram, and K. G. Mann. 1994. A model for the tissue factor pathway to thrombin. *J Biol Chem.* 269:23357--23366.
11. Neer, R., M. Berman, L. Fisher, and L. E. Rosenberg. 1967. Multicompartmental analysis of calcium kinetics in normal adult males. *J Clin Invest.* 46:1364-1379.
12. Patterson, B. H., O. A. Levander, K. Helzlsouer, P. A. McAdam, S. A. Lewis, P. R. Taylor, C. Veillon, and L. A. Zech. 1989. Human selenite metabolism: A kinetic model. *Am. J. Physiol.* 257 (Reg. Integr. Comp. Physiol. 26):R556-R567.
13. Patterson, B. H., and L. A. Zech. 1992. Development of a model for selenite metabolism in humans. *J. Nutr.* 122:709-714.
14. Patterson, B. H., L. A. Zech, C. A. Swanson, and O. A. Levander. 1995. An overview of selenium kinetics in humans. *In* Kinetic Models of Trace Element and Mineral Metabolism During Development. K. N. Siva Subramanian and M. E. Wastney, editors. CRC Press, Boca Raton.
15. Swanson, C. A., B. H. Patterson, O. A. Levander, C. Veillon, P. R. Taylor, K. Helzslouer, P. A. McAdam, and L. A. Zech. 1991. Human [74]Selenomethionine metabolism: a kinetic model. *Am. J. Clin. Nutr.* 54:917-926.
16. Wastney, M. E., R. L. Aamodt , W. F. Rumble, and R. I. Henkin. 1986. Kinetic analysis of zinc metabolism and its regulation in normal humans. *Am J Physiol.* 251:R398-R408.
17. Wastney, M. E., P. Angelus, R. M. Barnes, and K. N. Siva Subramanian. 1996. Zinc kinetics in preterm infants: a compartmental model based on stable isotope data. *Am J Physiol.* 271 (*Regulatory Integrative Comp. Physiol.* 40):R1452-R1459.
18. Wastney, M. E., J. Ng, D. Smith, B. R. Martin, M. Peacock, and C. M. Weaver. 1996. Differences in calcium kinetics between adolescent girls and young women. *Am. J. Physiol.* 271:R208-R216.

11

REJECTING HYPOTHESES AND ACCEPTING A MODEL

This chapter discusses methods for evaluating models. Observed data are compared to the calculated values obtained using models. The chapter begins by discussing criteria that are available in WinSAAM to evaluate fits through a series of ten examples. It then shows how to modify a model, based on the models lack of fit to data. Finally, criteria are discussed for rejecting or accepting a model.

I. Comparing calculated and observed data

Graphical, mathematical, and statistical methods are several ways to compare observed with model calculated values. It is important for the final model to fit all data simultaneously. With this approach the maximum amount of information is obtained from the data. By arbitrarily ignoring data (e.g., early points on a curve, or late points on the tail of a curve) information on the system is forfeited.

Graphical comparison is the most useful during initial model development and is one reason why graphics have been made an integral part of the WinSAAM program. Much of model development relies on power of observation to detect subtle differences between the predicted and observed data. Kinetic curves from biological systems are often complex due to the nature of the system or the experimental technique employed to uncover the nature of the system. Furthermore, the observed data may contain a mixture of compounds. For example, to fit plasma data on selenium kinetics, four plasma compartments that turned over at distinct rates were required (3). This resulted in a succession of peaks in the plasma curve and which were only detected by careful observation of the data. At the time the data were fit, it was not known that plasma contained multiple Se compounds.

Mathematical comparison includes the ratio of the calculated values, QC, to observed values, QO, which are available in WinSAAM. The residuals, or difference between the calculated and observed values, QO-QC, provide a second measure of model fit. The ratios, R(I), and residuals, RES(I), can be listed or plotted for any compartment that contains data. These measures provide quantitative values as a more rigorous assessment of the model fit. However, the most rigorous assessment of model fit to observed data is provided by statistical measures. These include the sum of squared residuals, the errors on the adjustable parameters and parameter correlations.

The sum of squared residuals refers to the total error between the model and all observed data. Sums of squared residuals can be used to compare the fit of different models to the same data set if the same weighting scheme is used. The sums of squared residuals are accessed by the command SS. The total sum of squared residuals can be listed by compartment using the command SS(I). Note that the sum of squared residuals for individual compartments will be related to the weight assigned to that compartment. It is useful to list SS(I) after iterating to see how the error is distributed and whether one compartment is contributing more error than the others. The error determined for each adjustable parameter is accessed by the command FSD(I) for fractional standard deviation (standard deviation/mean) or SE(I) for the standard error of each parameter. Parameter correlation is accessed by COR(I,J) to list the correlation between all adjustable parameters. The modeling aim is to fit all the experimental data with a model that has a minimum sum of squared residuals and whose parameter values have low FSD and low correlation.

II. Modifying your model

A model can under- or over-predict a curve in a number of ways. Deciding how to modify your model to improve the fit is the heart of developing models of biological systems. The decisions are based on knowledge of kinetics and the biology of the system. The following examples demonstrate some situations that are commonly seen when fitting data. In each case there are a number of ways the model could be altered to fit the data, but one solution is given in each case.

The first case shows a systematic deviation between the model-calculated fit and the observed data (Fig 11.1a). The curve fits the later data, but not the early data. (If the parameter is adjusted, the early data will be fitted, but not the later data). The data show that there is a delay before the tracer/material of interest reaches the compartment being sampled. Therefore, by adding a compartment to slow transfer of the model, the model fits the data (Fig 11.1b).

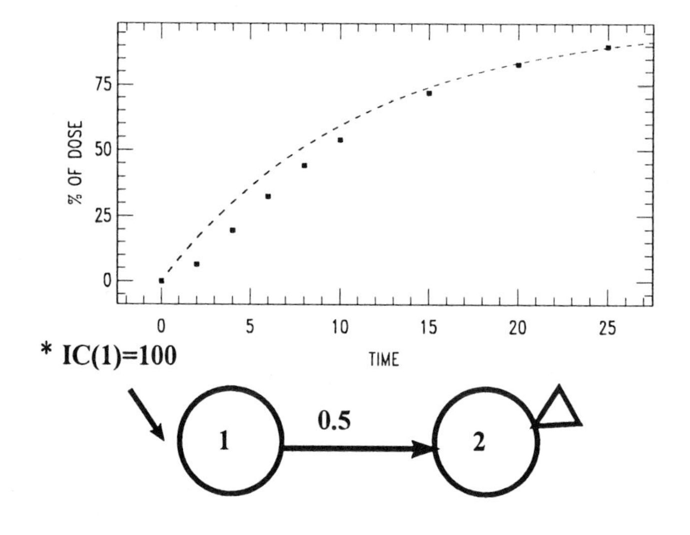

Fig 11.1a Data are not fitted by a 2-compartment model.

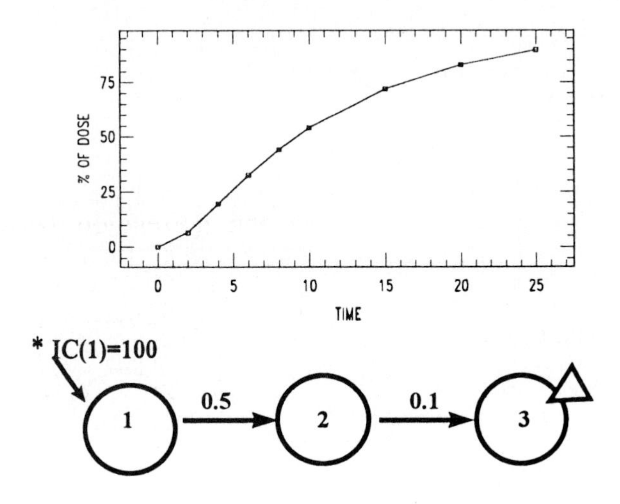

Fig 11.1b Data are fitted by adding a delay compartment.

In the second case, data are also not fitted early in the curve (Fig 11.2a) but notice that the data pattern differs to that in Fig 11.1a. These data indicate that there is a 'real' delay, i.e., material is actually held for a period of time and then released rapidly. It can be fitted by the model in Fig 11.2.b. These type of data are often observed when transport is involved, for example movement of material through the intestine before excretion in feces.

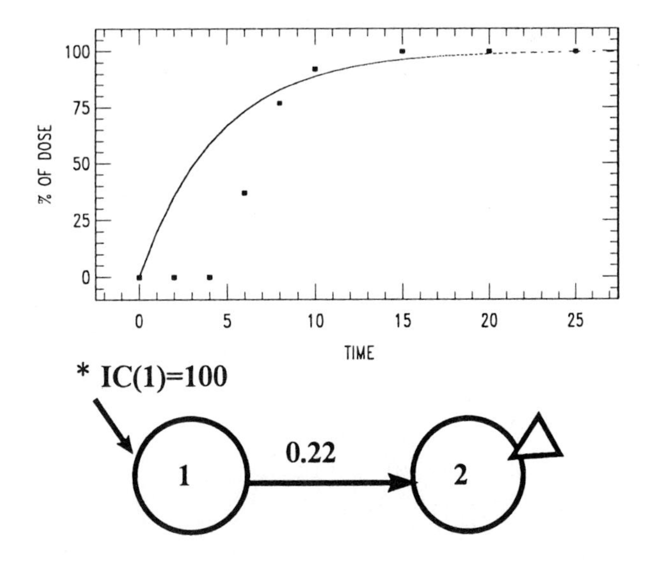

Fig 11.2a Data are not fitted by a 2-compartment model

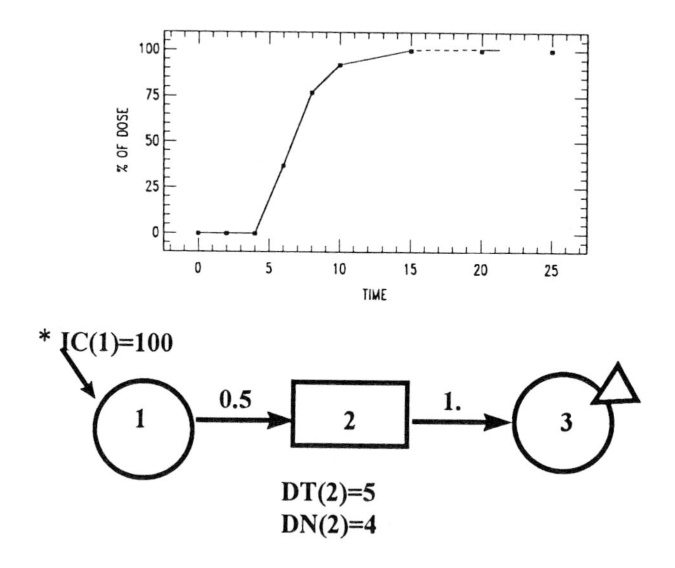

Fig 11.2b A 2-compartment model with a delay fits the data.

In the third case, the model does not predict the correct shape of the curve - the calculated peak occurs too early (Fig 11.3a). To fit these data, it is necessary to slow down the turnover of the compartment (compartment 1) that feeds into the sampled compartment (compartment 2), i.e., reduce the value of L(1,1), which is L(0,1) and L(2,1), (Fig 11.3b).

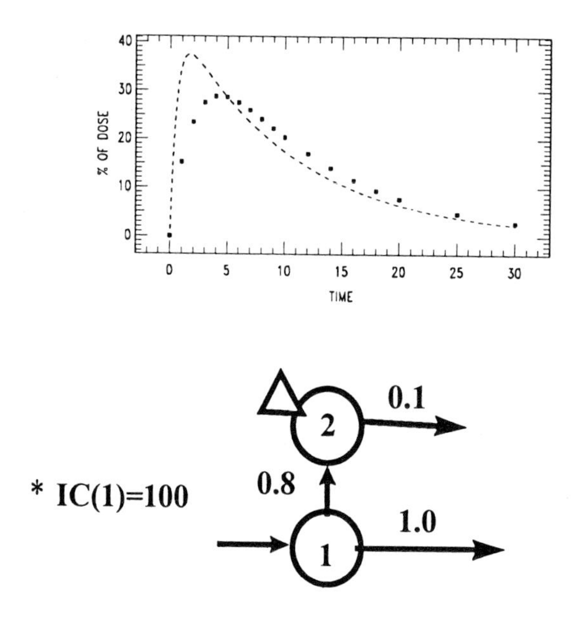

Fig 11.3a The model does not fit the peak data.

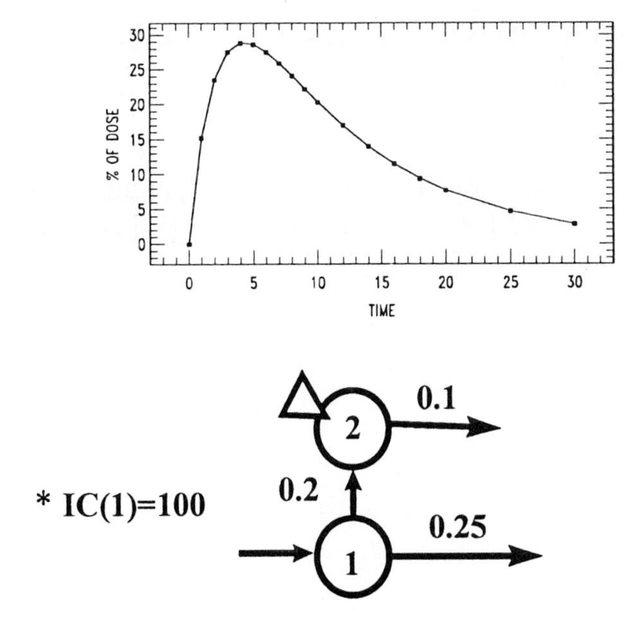

Fig 11.3b The model fits the data after slowing down the turnover of compartment 1.

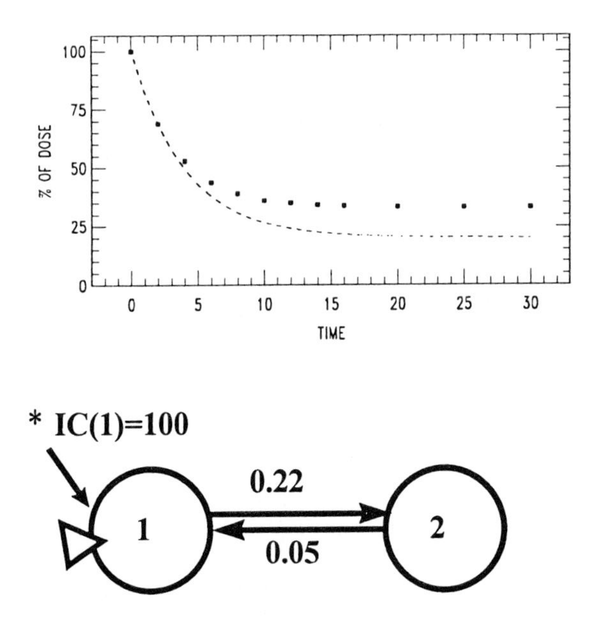

Fig 11.4.a Data are not fitted at the tail of the curve.

In the fourth case, data are underestimated at the end of the curve (Fig 11.4a). The data are fitted by increasing return of material from compartment 2 to compartment 1 (Fig 11.4b).

In the fifth case, data are not fitted at the tail of the curve (Fig 11.5a). These data, however, cannot be fitted by increasing the turnover of compartment 2 as this would raise the curve at the 2-3 minute time. The data indicate that a third compartment is required (Fig 11.5.b).

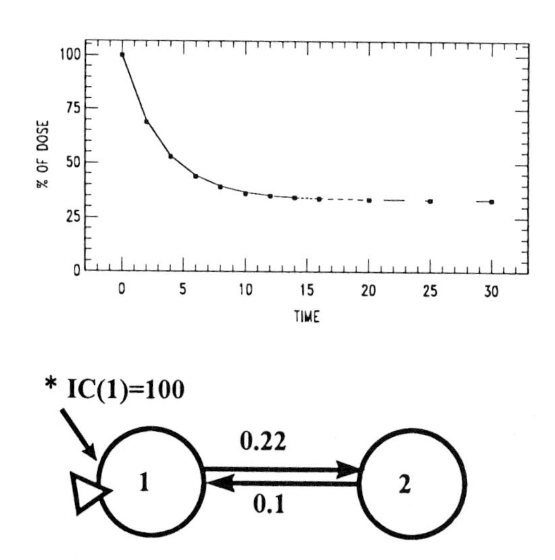

Fig 11.4.b Data are fitted by increasing turnover of compartment 2.

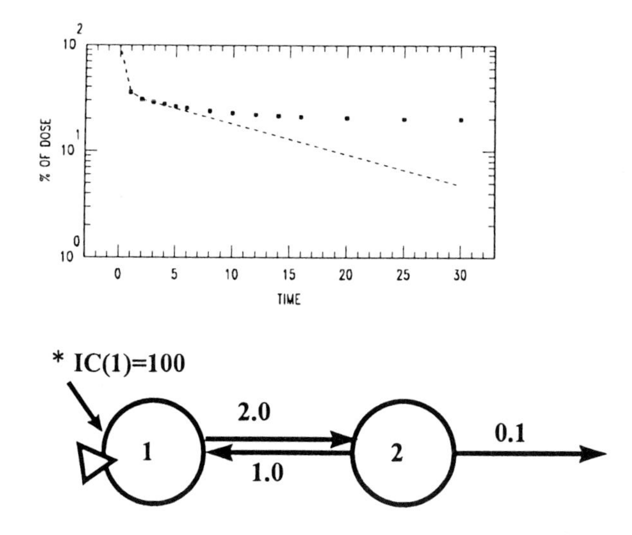

Fig 11.5.a Data are not fitted at the tail of the curve.

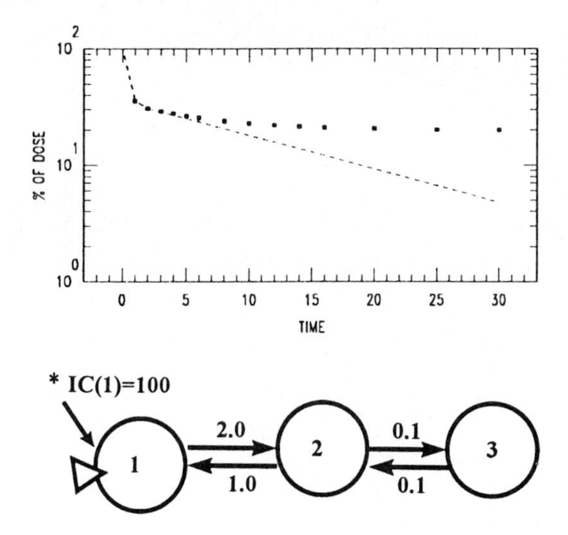

Fig 11.5b Adding a third compartment fits the data.

In the sixth case, data are not fitted at the tail (Fig 11.6a) but in this case, the data are fitted by 'summing' two compartments (Fig 11.6b). This situation is observed, for instance, when the tracer is attached to a compound that may be metabolized into an inactive form.

Notice that the data can also be fitted by assuming that the tracer was on the compound of interest (compartment 1) but also on a second compound that does not break down (Fig 11.6c).

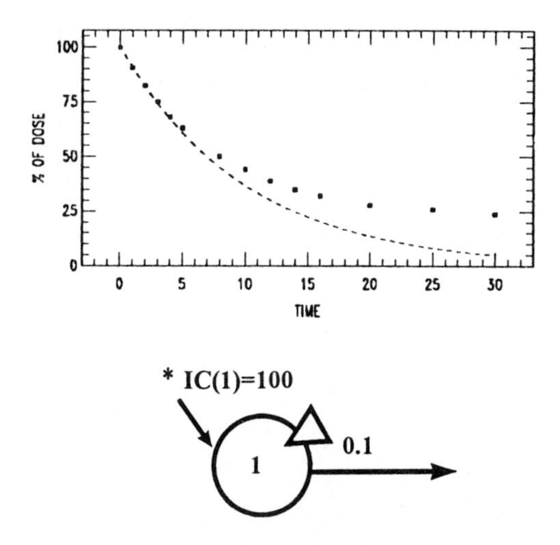

Fig 11.6a Data are not fitted at the tail by a one-compartment model.

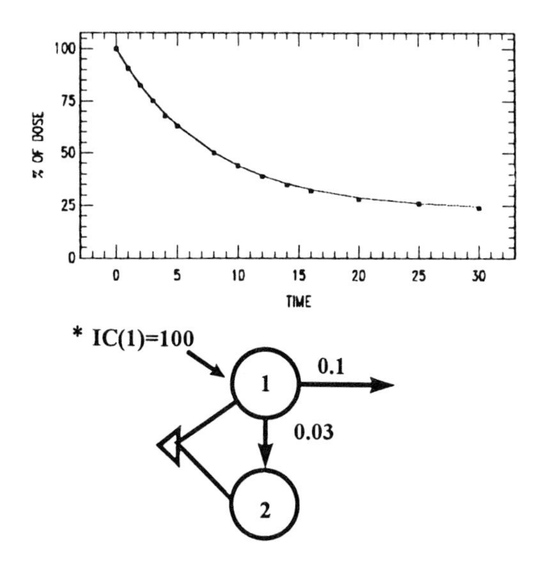

Fig 11.6b Data are fitted at the tail by adding a second compartment.

In the seventh case, the calculated curve decays, while the observed data decay slowly at first, and then rapidly (Fig 11.7a). Such data are often observed in lipoprotein studies. The data suggest that material is not being metabolized straight-away, but is probably moved between different species before being catabolized (Fig 11.7b).

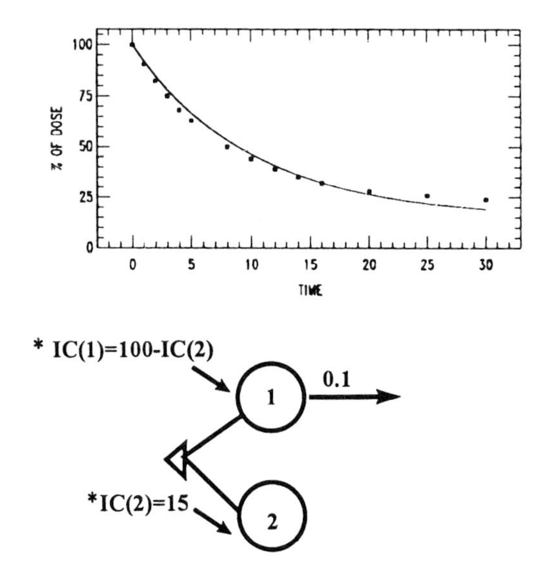

Fig 11.6c Fitting data by measurement of two species.

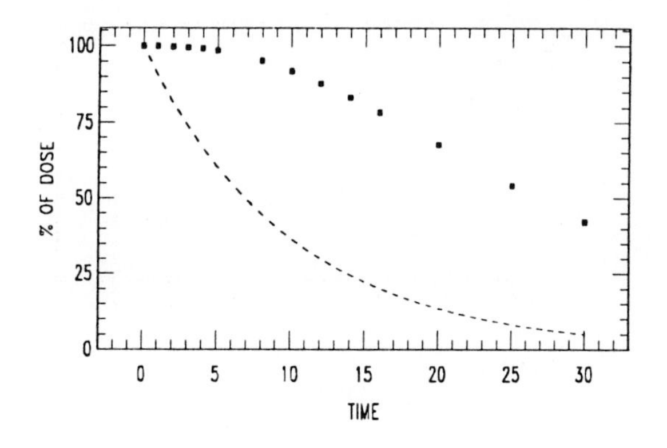

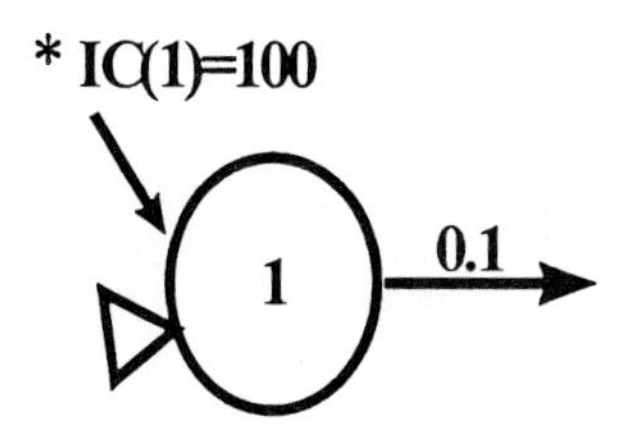

Fig 11.7a Data decay slowly initially, then more rapidly.

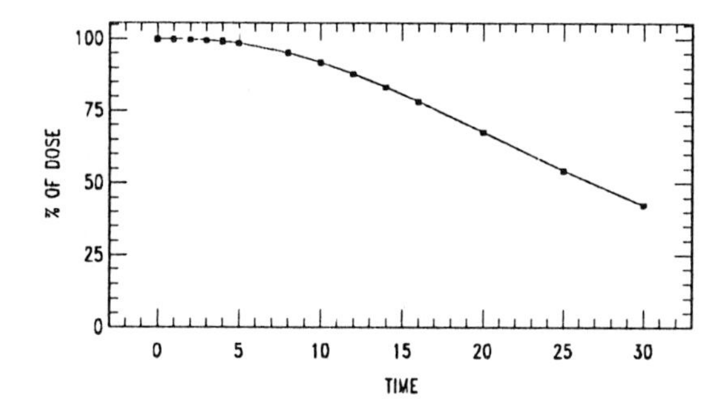

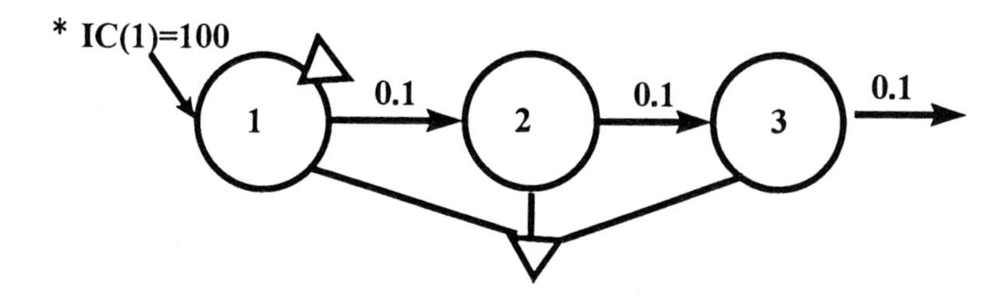

Fig 11.7b Data are fitted by a 'cascade' or series of compartments.

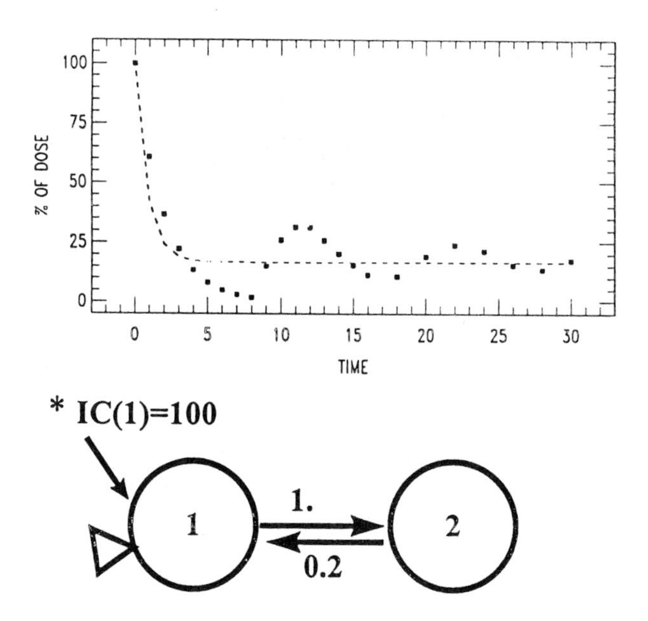

Fig 11.8a Data decay with oscillations.

In the eighth case, data show oscillations (Fig 11.8a). While they could be fitted using a sine function, mechanistically, the data imply that material recycles through a delay compartment (Fig 11.8b).

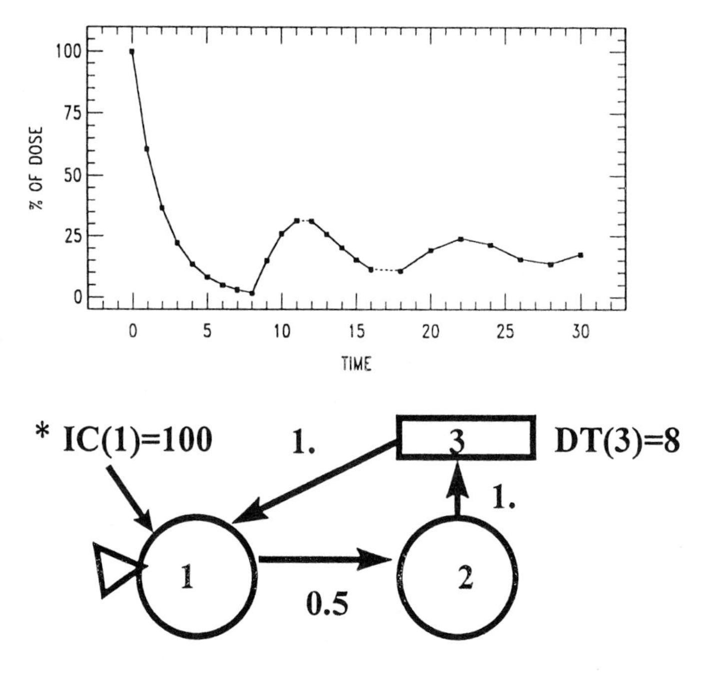

Fig 11.8b Oscillations are fitted by a model with recycling through a delay.

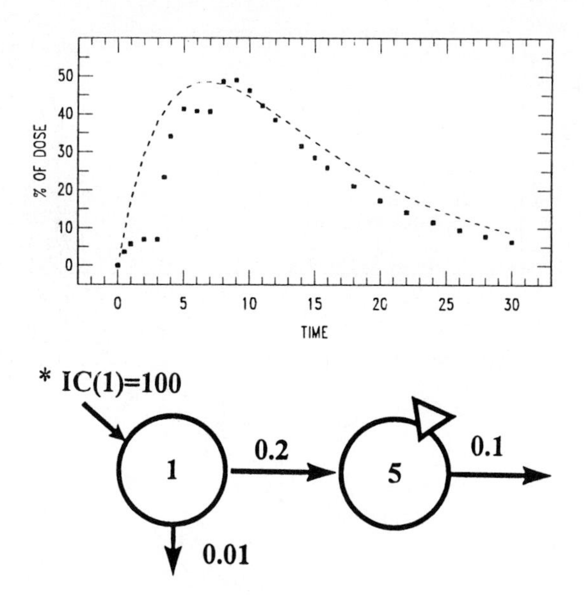

Fig 11.9a Data appear to have several peaks.

In the ninth case, data have several peaks (Fig 11.9a). These suggest that tracer is moving into the sampled compartment by several routes with delays (Fig 11.9b).

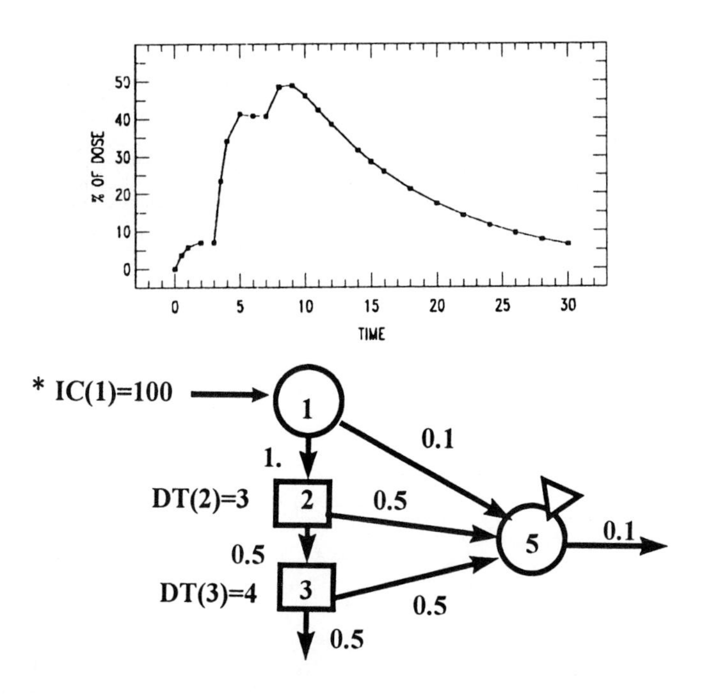

Fig 11.9b Multiple inputs.

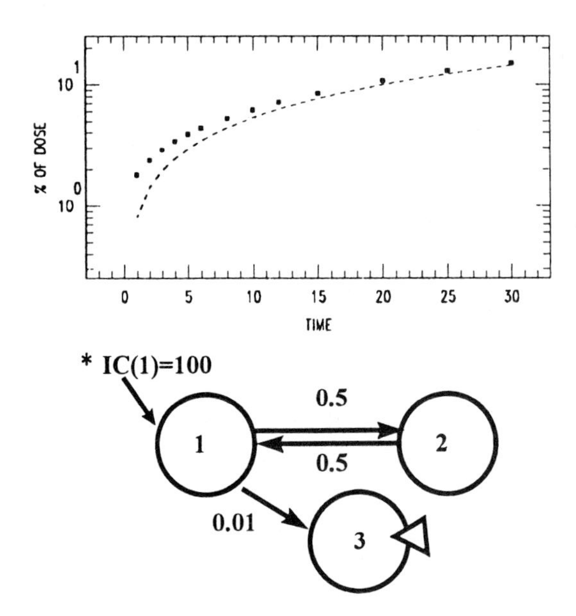

Fig 11.10a Model underestimates the initial appearance of tracer.

In the tenth case, the model under-calculates appearance of tracer in the sampled compartment (Fig 11.10a). This may imply that some tracer enters the sampled compartment directly (Fig 11.10b).

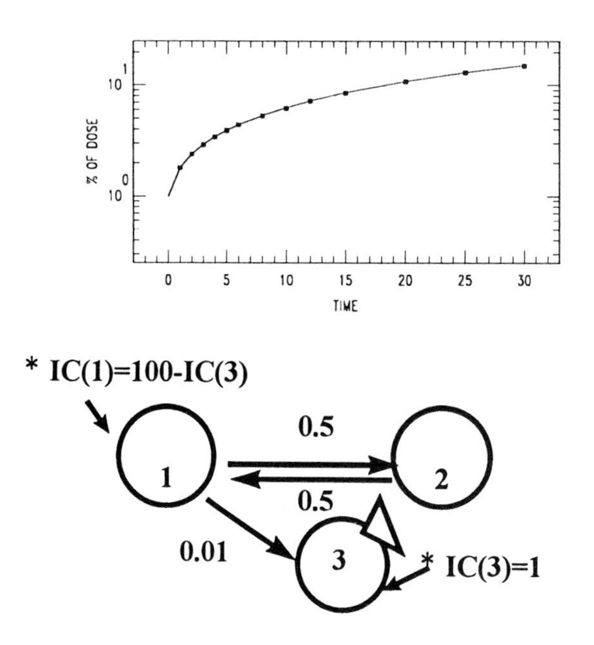

Fig 11.10b Data are fitted by adding a small initial condition on the sampled compartment.

III. Rejecting your model

A model may be nonunique or inconsistent (1). A nonunique solution occurs when the information contained in the data is inadequate to define the model, i.e., the model is too complex. The parameters are ill-determined (have large associated errors) and the changes in the parameter values do not improve the fit. The model can be simplified by reducing the number of parameters or by introducing constraints. Constraints may include fixing a parameter to a particular value (perhaps based on information obtained by another study) or by determining the ratio of a pair of parameters instead of the individual parameter values.

An inconsistent solution occurs when the model does not have sufficient freedom to fit the data (1), i.e., the model is too simple. It is indicated by consistent deviations between the calculated values and observed data. Inconsistencies can be solved by increasing the degrees of freedom in the model by adding new parameters, or by allowing more parameters to adjust during the fitting procedure.

IV. Deciding when a model is acceptable

There are two criteria for deciding whether a model is acceptable. The first relates to the biological mechanisms and the second relates to mathematical criteria. The first relates to whether the parameter values and pathways are consistent with biological properties of a system. It is perhaps the more important criteria for gauging the acceptability of a model. For instance, are parameters relating to uptake by tissues consistent with known rates of blood flow? Do the masses of the compartments approximate known values for the system? The model must be consistent with known information about the system. However, in addition, the model may contain new features that may be unknown and require experimental confirmation. If a new pathway or compartment is required by the data, it should be shown in the model with justification on why it is required, what it may represent, and should propose new studies for testing whether it does exist. Aspects of a model may be only operational in that they are required to fit features of the data but do not identify the physiological processes which generate them (2).

Mathematically, an acceptable model is one that is unique and consistent, i.e., where there is random scatter of data about the predicted values and the parameter errors are low. As a general guide, FSD of <0.5, parameter correlation -0.8 to 0.8, random sum of squared residuals and small sum of squared residuals between compartments are considered an acceptable fit. Berman (2) considered that a reasonable way to test the validity, or acceptability of a model, is to test its power for making predictions beyond the database used in its development.

A model can be considered to be acceptable when it fits all available data, is consistent with known biology, and is well-determined statistically.

REFERENCES

1. Berman, M. 1963. The formulation and testing of models. *Annal N Y Acad Sci.* 108:182-194.
2. Berman, M. 1982. Kinetic analysis and modeling: Theory and applications to lipoproteins. *In* Lipoprotein Kinetics and Modeling. M. Berman, S. M. Grundy, and B. V. Howard, editors. Academic Press, NY. 3-36.
3. Patterson, B. H., and L. A. Zech. 1992. Development of a model for selenite metabolism in humans. *J. Nutr.* 122:709-714.

12

MODEL SUMMARIZATION

Due to developments over the last two decades in computing and modeling software, the tools of modeling are now available to every biological investigator for probing and understanding biological systems. However, to be adopted and used widely by the biological community, models need to be explained in a language familiar to biologists, and published descriptions of models should clearly describe the background and functioning of the model (4). Models must be communicated clearly in terms of their physiological relevance and mathematical rigor. Model summarization is the final step of a modeling project and involves completing the steps necessary to prepare a scientific manuscript on the project. This chapter explores the area of publishing models by examining the communication of modeling projects.

I. Summarizing a Model for Publication

Historically the rules for publishing models have varied depending on whether the journal is in the theoretical (mathematical or engineering) or experimental (biological) domain, with each field stressing different aspects of the modeling and often using different definitions. Some criteria for summarizing models for publication are listed under Table 12.1 under the headings of model development, model description, model evaluation, discussion and others which include criteria external to the report (8).

1. Model Development

Model development includes discussion on the reason for developing the model and how the model was developed. This is important for establishing the biological importance of the study as it may have dictated the modeling approach applied (1). It is important to describe, in addition to the purpose, the biological relevance of the model in lay terms. Assumptions of the model need to be stated and the basis and validity of the assumptions provided. The software used to solve the model should be referenced. Finally the steps in developing the model should be carefully described in terms of models that were developed that *did not* fit the data. This approach was used to describe development of a whole body model for selenium kinetics in humans (7), a pharmacodynamic model for lithium in rats (3) and model for endonuclease kinetics (5). The final model is more easily understood by showing how simpler, or alternative models, were unable to fit the data.

2. Model Description

Model description includes listing the sources of any parameter values that were not derived through the modeling. It is helpful to include a model diagram. Diagrams can summarize a lot of description and show the relationship of sub-

Table 12.1. Some criteria for summarizing models (modified from (8))

1. Model Development:	**3. Model Evaluation**
Purpose	Model results vs data
Physiological relevance	Parameter uncertainties
Assumptions stated	Sensitivity analysis
Software specified	Model results vs expected trend
Steps described	Cross-validation
2. Model Description:	**4. Discussion:**
Sources for parameter values	Physiological evaluation
Model diagram	Comparison to previous models
Equations given	Limitations
Symbols defined	Suggestions for new studies
Parameter values	**5. Other:**
Glossary	Can model be reconstructed
Units specified	Is electronic form available

models. For compartmental models, diagrams are an alternative way of showing the model equations. The equations of the model should be listed implicitly (for example, as a diagram) or stated explicitly. All symbols need to be defined. For large models that contain many parameters, a glossary can make the manuscript more readable. It is important that units be specified for model parameters and all results. Brownell et al. (2) proposed a list of standard definitions for modeling studies.

3. Model evaluation

Model evaluation involves showing that the model fits observed data by direct comparison to the data or to expected trends. Once the model has been shown to be valid with respect to observations, parameter uncertainties need to be stated and sensitivity analysis should be reported for important parameters (See Chapters 15 and 16). As a further test (or validation) model predictions should be tested against data that were not used in model development.

4. Discussion

Discussion of the model should indicate its physiologic significance and limitations. The model should also be compared and contrasted to previous models for the system. This is important for indicating the new contributions of the study or highlighting differences between the conclusions of the current and previous studies. If previous models are ignored, a number of different models may exist for a system, but no new knowledge is obtained. Limitations of the study or model should be discussed and used to suggest further studies. The modeling results should also be used to predict new studies to expand or refine the current model or hypothesis.

5. Other

As a test of the completeness of the model description, the model should be able to be reconstructed from the manuscript (See Chapters 17-22). That is, the results and plots should be able to be replicated by the reader. An electronic form of the model should be made available to other users and the manuscript should mention how it can be obtained. One option is to use a model library on the Internet (See Chapter 22). Papers containing models can reference the URL for the library, (for example see (6)). As more journals are published electronically, the published article can contain a link to the model directly.

II. Summarizing Models for the Internet

The Internet is a useful medium for distributing working versions of published models (as described further in Chapter 22). In addition to the mathematical equations and a version of the model compatible with one or more modeling software packages, the model needs to be summarized as described above for publication, but in compact form. An example is shown in Fig 12.1. The summary is designed to provide an overview of the study. It lists the reference, a statement on the relevance of the study, and then summarizes the model development. Model development states the experimental species used, and information such as the tracer used, site of tracer administration, the sites of data collection and the length of the study. Under Model heading, the type of model (i.e., type of mathematical specification), information on the type of parameter, and the software used to develop the model are listed. The results are summarized and then, under Discussion, conclusions of the study and speculation based on the modeling results are stated. Included in this section are some limitations of the study, and some unanswered questions. More detailed information on the study such as parameter values, a model graphic, plots of the model calculated fits to data, and other model calculated values can be obtained from the reference, and some of this information is included in the Model Library (see Chapter 22). The Library, via the Internet allows other information that can add to the understanding of the final model, to be presented. For example, models that were rejected, and the sequence of model development can be shown. This background material is often excluded from publications due to page limitations. The Internet also allows for models to be presented in a more interactive way, and this may aid in the understanding, distribution, and use of models by other investigators.

Model 34: Zinc Kinetics in Rats

REFERENCE:

"Compartmental analysis of zinc kinetics in mature, male rats"
House WA and Wastney ME
Am. J. Physiol 273 (Reg Integ Comp Physiol 42):R1117-1125, 1997

Relevance : The model fits data from16 tissues in rats and extends earlier models of rat zinc kinetics. It can be used to evaluate pathophysiological conditions and dietary extremes on the metabolism of zinc.

DEVELOPMENT:

Experiment: Rats (n=84, male Sprague-Dawley, weight 370 g) on 34 mg Zn/kg diet were administered ^{65}Zn iv. Groups of two-six rats were studied 16 times over 4 days. At each time point blood, liver, spleen, kidneys, testes, right tibia, skeletal muscle, small intestine, cecum, and colon were sampled and whole body was counted.

Data: Radioactivity (expressed as % dose) and zinc in tissues. Data from all tissues were fitted simultaneously using the program SAAM/CONSAM.

Assumptions: IV tracer behaved in the same way as dietary zinc.

MODEL:

Graphic: Compartmental, based on the model for zinc in humans (Wastney et al., 1986). Human model was extended by adding compartments for skin, slow compartments in muscle and bone, and for segments of the intestinal tract.

Initial Conditions: For i.v. dose, IC(1)=100.

Parameters: Parameters are all constant.

Software: SAAM/CONSAM

RESULTS:

Compartments 10, 11 and 22 in the human model were identified as kidney, part of spleen and part of testes respectively in the rat.

Plots: Tissues, gastrointestinal segments.

DISCUSSION:

Conclusion: Zinc kinetics can be explained by 13 compartments in the body and 5 compartments in the GI tract. Loss of zinc occurs via urine excretion, endogenous excretion and loss of hair.

Speculation: Zinc is endogenously excreted into two segments of the gastrointestinal tract and absorption of endogenously excreted zinc occurs from four sites in the intestinal tract.

Limitations: Rats were studied for only 4 days following iv tracer administration. Longer studies are required to label slower turning over pools in muscle, skin and bone.

Unanswered questions:

Does zinc in the diet follow the same kinetics as zinc administered intravenously? How do zinc kinetics change during pathophysiological conditions and during extremes of dietary intake?

Fig 12.1 Description of a model for zinc kinetics in rats, summarized for the Internet.

REFERENCES

1. Berman, M. 1982. Kinetic analysis and modeling: Theory and applications to lipoproteins. *In* Lipoprotein Kinetics and Modeling. M. Berman, S. M. Grundy, and B. V. Howard, editors. Academic Press, NY. 3-36.
2. Brownell, G. L., M. Berman, and J. S. Robertson. 1968. Nomenclature for Tracer Kinetics. *Int J Appl Radiat Isotop.* 19:249-262.
3. Everts, H. B., H.-Y. Jang, R. C. Boston, and N. L. Canolty. 1996. A compartmental model predicts that dietary potassium affects lithium dynamics in rats. *J. Nutr.* 126:1445-1454.
4. Garfinkel, D. 1984. Modeling of inherently complex biological systems: Problems, strategies, and methods. *Math Biosci.* 72:131-139.
5. Hensley, P., G. Nardone, J. G. Chirikjian, and M. E. Wastney. 1990. The time-resolved kinetics of superhelical DNA cleavage by BamHI restriction endonuclease. *J. Biol. Chem.* 265:15300-15307.
6. House, W. A., and M. E. Wastney. 1997. Compartmental analysis of zinc kinetics in mature male rats. *Am J Physiol.* 273:R1117-R1125.
7. Patterson, B. H., O. A. Levander, K. Helzlsouer, P. A. McAdam, S. A. Lewis, P. R. Taylor, C. Veillon, and L. A. Zech. 1989. Human selenite metabolism: A kinetic model. *Am. J. Physiol.* 257 (Reg. Integr. Comp. Physiol. 26):R556-R567.
8. Wastney, M. E., X. Q. Wang, and R. C. Boston. 1998. Publishing, interpreting and accessing models. *J. Franklin Instit.* 335B:281-301.

13

MULTIPLE STUDIES

In many modeling situations, a set of values for the model parameters is regarded as characterizing an individual. The modeler, however, may be interested in estimating the distribution of parameter values in the population from which the individuals are drawn e.g., noninsulin-dependent diabetic subjects. The purpose of this chapter is to survey some approaches for the analysis of data from multiple studies and show how to implement population kinetic models using SAAM. The facility in SAAM for population data analysis is referred to as EMSA (Extended Multiple Studies Analysis). We compare results derived using EMSA (1) with those using the NONMEM (Nonlinear Mixed Effects Modeling) program (2). However, because EMSA assumes a multivariate Normal distribution of the kinetic parameters and NONMEM a log Normal distribution, we also compare results obtained when no assumptions are made about the parametric distribution of the kinetic parameters using the non-parametric maximum likelihood method (NPML) by Mallet (3).

Mathematical models are used for a variety of purposes including the summarization of large amounts of data, generating and testing hypotheses, characterization of the system being modeled, and prediction and control. Although population data analysis is new to the field of biological systems modeling, a move towards adoption of population pharmacokinetics as a routine procedure is now standard during drug development (4). The reason for adopting a population approach to pharmacokinetic studies is that it has become increasingly obvious that the drug should be studied in the target population, which may be different from the normal population (5). Clearly, from a pathophysiologic perspective, this same line of reasoning applies to the study of biological systems.

I. The Estimation Problem

1.1 Individual-based Parameter Estimation

Let t_i be the ith value of the independent variable time, $i = 1, 2, ..., n$. Although the choice of t is motivated by the example to be used later, in which time is the independent variable, t can be multidimensional. Let u represent system inputs, such as dose, dosage regimen, or tracer infusion rate as a function of time, and θ, the model parameter vector for the individual being modeled. Then, $Y(t_i) = Y_i$ is the value of the dependent variable at $\hat{\mu}$. The mathematical model is then given by

$$Y(t_i) = Y_i = F(t_i, u, \theta) + \varepsilon_i \qquad (13.1)$$

where F is a known function or structural model and ε_i is a random variable representing the error process. Assumptions regarding the distribution of ε_i's are required using both EMSA (multivariate Normal) and NONMEM (log Normal). Thus, given the model F and an appropriate input u, data of the form $\{(t_i, Y_i), i = 1, 2, ..., n\}$ are collected and we are interested in obtaining an estimate $\hat{\theta}$, of θ, together with some indication of the accuracy of this estimate. Standard approaches to the of solution Eq. 13.1 are ordinary, generalized, or weighted nonlinear least squares estimation of θ. These approaches are available in WinSAAM.

1.2. Population-based Parameter Estimation

For population data analysis, the model to be used must be a known model for the system under investigation. To begin to define a population model, Eq. 1 is recognized as applying to the observations on a single individual among many in a population (6,7). If similar data are available from each of *N* individuals, and the individual from whom each datum came can be explicitly noted, then a model treating the set of individual observations as a single unit of response can be stated.

In particular, the model F is regarded as being the same for all individual members of the population but each individual is viewed as being characterized by a specific constant parameter vector. The goal is to estimate the distribution of parameter values in the population. Application of such estimates include: a) evaluation of the structural model by examining model fit in the population, b) the characterization and comparison of populations, c) prediction at the population level and d) optimal sampling design. In addition, population distributions can be used as prior distributions in applying Bayes' Theorem to estimation, prediction, and control at the individual level (8).

Modification of the notation in Eq. 13.1 is required to handle the population problem. In particular, we introduce the subscript, j, to indicate individuals. Thus, $Y_j(t_{ij}) = Y_{ij}$ indicates the measurement at time t_{ij} on the jth individual or, equivalently, the ith measurement on the jth individual, $i = 1,2,...,n_j$, $j = 1,2,...,N$. Eq. 1 can now be rewritten as:

$$Y_j(t_{ij}) = Y_{ij} = F(t_{ij}, u_j, \theta_j) + \varepsilon_{ij} \qquad (13.2)$$

2.1. Naive Pooled Data Approach

Here, all data are combined and analyzed as though they came from a single individual. Thus, the subscript j is ignored in Eq. 13.2; so it is equivalent to individual-based kinetic analysis. It is appropriate to use an individual parameter estimation technique such as nonlinear least squares to obtain point estimates of the parameters. The mean residual (possibly weighted) is used to estimate the variance. This method is unattractive for a variety of reasons. Most fundamentally, it obscures the underlying notion of a population distribution. That is, it is not clear what the concept of a population distribution is and any appreciable amount of intraindividual variability will be undetected because it will be masked by the regression function F.

Formally, analysis by this method combines all sources of error, including measurement error and intraindividual variation. Consequently, there are no estimates available for the variance of the parameters at the population level. In computer simulation studies, this approach has been found to perform poorly (10,11,13).

2.2. Two-Stage Methods

The most commonly used methods for estimating the distribution of θ are probably those referred to as "two-stage" methods. As their name implies, these methods proceed in two-stages where in the first stage, each individual's set of data is separately analyzed using an individual-based parameter estimation technique. Nonlinear least squares and maximum likelihood methods are commonly used. The results are point estimates, $\hat{\theta}_j$, for the parameter vector θ_j, and their associated covariance matrices. The use of these methods requires sufficient experimental data on each individual to obtain reasonable point estimates for the $\hat{\theta}_j$.

Once the set of $\hat{\theta}_j$'s have been obtained, they may be used in various ways to obtain the desired population estimate. For example, it is often assumed that the population distribution of θ is Normal or Lognormal. In these cases, among others, it is only necessary to estimate the mean and covariance of the population density. The sample mean and the sample covariance of the $\hat{\theta}_j$'s may be used for this purpose.

All of the two-stage methods treat the $\hat{\theta}_j$'s as though they were observations rather than estimates of individual parameter vectors. Thus, one can expect these approaches to yield good estimates when each $\hat{\theta}_j$ is estimated with good precision and accuracy. In more typical applications, in which the $\hat{\theta}_j$ have moderate to large variances, one would at least expect the two-stage methods to

As before, F is a given structural model and u_j is the known input for the j
individual. The independent variable values t_{ij} are also known but the paramet
vector θ_j associated with the jth individual is unknown. The goal is estimatio
of the distribution of the population parameter vector Θ of which the unknow
$\theta_j = [\theta_1, \theta_2, ..., \theta_N]$ is a sample (observations).

Some approaches to this problem involve estimation of the individual θ_j'
while others do not. In any case, estimation of individual parameters is not th
goal of population analysis (8). However, individual components of variance may
be important contributors to both kinetic and pharmacologic variability in living
systems. Our philosophy is to be suspect of methods for population data analysis
that do not follow individuals in the population as a first step.

II. Methods

Two issues play a central role in considering the various approaches to
population data analysis. Firstly, is the amount and nature of the available data.
For example, in terms of drug levels, in many pharmacologic situations there is a
very large amount of data available but it consists of relatively small number of
measurements on each of a large number of individuals (10,11). In addition, the
quality of data may vary considerably across individuals. This situation is not
conducive to the estimation of individual parameter values but may be suitable fo
the population problem in pharmacokinetics. We emphasize that there is no plac
for data of poor quality in the analysis of biological systems either at t
individual level or the population level. Messy data should be discarded unless
is of the type which fits the paradigms proposed by Milliken and Johnston (12).

Secondly, an important consideration involves the extent and nature of
assumptions that one is willing to make about the form of the distribution of
population θ. We may assume, for example, that θ is multivariate Normal
thus, we need only estimate its mean vector and covariance matrix. On the
hand, we may wish to adopt a nonparametric approach in which no func
form is assumed for the density of θ. In all cases, assumptions will be nec
concerning the distributions of the ε_{ij}'s. Techniques for evaluati
appropriateness of these assumptions include analysis of a) identifiab
analysis of residuals, c) sensitivity analysis, and d) the use of variance
The ε_{ij}'s may involve measurement error, model misspecification error
in the measurement of Y_i and/or t_i.

underestimate the variances of the population parameters as a result of ignoring variability in the individual point estimates. We point out that the EMSA approach takes into consideration the variability in the individual point estimates, unlike other two-stage methods. Sheiner and Beal (10,11,13) have observed that a two-stage method underestimated the variances of the population parameters as a result of ignoring variability in the individual point estimates in several pharmacokinetic applications.

2.3. Nonlinear Mixed Effects Model

A third approach to the population problem has been proposed by Sheiner and Beal (10,11,13) and a version of it implemented by them in the computer program NONMEM (2). This method is based on an assumed multivariate Lognormal density for θ. Now, suppose that θ_j, the parameter vector for individual j, has components $\theta_{j1}, \theta_{j2}, ..., \theta_{jP}$, where p is the dimension of the parameter vector for the structural model. Under the Lognormal assumption, $\ln(\theta_{jk})$ are jointly distributed with a multivariate Lognormal density in the population for $k = 1,2,..., p$.

Denoting the components of the mean parameter vector for the population by $\ln(\theta_1)$, $\ln(\theta_2)$, ..., $\ln(\theta_p)$, each individual parameter vector component can be written as the population mean plus a random variable representing the deviation of the individual from the population mean. Thus,

$$\ln(\theta_{jk}) = \ln(\theta_k) + \xi_{jk}, \quad k = 1, 2, ..., p. \tag{13.3}$$

for each j, where ξ_{jk} have Lognormal densities. If the model were linear in the parameters and all densities were Normal, we would have an analysis of covariance problem with mixed effects. Hence, the acronym NONMEM, for Nonlinear Mixed Effects Model.

The basic idea is to expand the model about the population mean parameter, using the linear portion of the Taylor series expansion and to estimate θ by minimizing a criterion that corresponds to maximum likelihood estimation in the Normal case. It has been shown that this estimator has desirable large sample properties even when distributions are not Normal and its behavior has been studied in a variety of computer simulations (10,11,13). In those studies, the performance of the estimators, and the corresponding coefficients of variation of the parameter estimates in the population have been good. To date, however, no rigorous testing of NONMEM has been made comparing its results to standard sets of population parameter values obtained by entirely different techniques such as, the jackknife, the bootstrap, the jackknife-after-bootstrap, or Monte Carlo (14).

2.4. Nonparametric Maximum Likelihood Method

The nonparametric maximum likelihood method (NPML) advanced by Mallet (3) requires knowledge of the distributions of the ε_i, but makes no assumption regarding the form of the density for θ. NPML software is now available (Bestfit s.a.,Luxembourg). It has been shown that the maximum likelihood estimate of the density for θ is a discrete density with positive density at a number of points that is less than or equal to the number of individual samples (*N*). One can in principle find the location of these support points as well as their corresponding densities and thus calculate the maximum likelihood estimate of the desired function.

2.5. Bayesian Methods

For the individual case, Eq. 13.1, and letting $D = (Y_1, Y_2, ..., Y_N)$ represent the data. If h(θ)is the posterior density of θ, i.e., the density function for θ prior to consideration of D, h($D;\theta$) is the probability of the data given θ, which may be regarded as the likelihood of the data as a function of θ. h($\theta;D$) is the posterior density of θ given D, then Bayes' Theorem states:

$$h(\theta; D) = Kh(D;\theta)h(\theta) \tag{13.4}$$

where the constant K is the reciprocal of the integral of $h(D;\theta)h(\theta)$ integrated over all possible values of θ.

In the absence of sufficient prior information concerning the distribution of θ, one may regard $h(\theta)$ as locally uniform and constant as a statement of prior ignorance (15,16). In this case, Eq. 4 becomes:

$$h(\theta; D) = Kh(D;\theta)\,. \tag{13.5}$$

Now, calculation of $h(D;\theta)$ is the same as calculation of the likelihood function in earlier methods (7).

Two-stage methods are possible in which Bayes' Theorem (Eqs. 4 or 5) is used in the first stage. Then, after stage-one, we have an estimated posterior density for each individual. These could be combined, for example, by averaging these posterior densities at each value of θ (8). This approach does not require large amounts of data on each individual and takes into account the relative information content in the different individual data sets.

Analogues of both EMSA and NONMEM are also possible based on Bayes' Theorem. That is, given a functional form for the population density and densities for the ε_i's, the likelihood $h(D;\theta)$ can be calculated. Here, D represents the complete data set for all of the individuals and θ is the complete set of population parameters.

As before, F is a given structural model and u_j is the known input for the jth individual. The independent variable values t_{ij} are also known but the parameter vector θ_j associated with the jth individual is unknown. The goal is estimation of the distribution of the population parameter vector Θ of which the unknown $\theta_j = [\theta_1, \theta_2, ..., \theta_N]$ is a sample (observations).

Some approaches to this problem involve estimation of the individual θ_j's while others do not. In any case, estimation of individual parameters is not the goal of population analysis (8). However, individual components of variance may be important contributors to both kinetic and pharmacologic variability in living systems. Our philosophy is to be suspect of methods for population data analysis that do not follow individuals in the population as a first step.

II. Methods

Two issues play a central role in considering the various approaches to population data analysis. Firstly, is the amount and nature of the available data. For example, in terms of drug levels, in many pharmacologic situations there is a very large amount of data available but it consists of relatively small number of measurements on each of a large number of individuals (10,11). In addition, the quality of data may vary considerably across individuals. This situation is not conducive to the estimation of individual parameter values but may be suitable for the population problem in pharmacokinetics. We emphasize that there is no plac(for data of poor quality in the analysis of biological systems either at th individual level or the population level. Messy data should be discarded unless is of the type which fits the paradigms proposed by Milliken and Johnston (12).

Secondly, an important consideration involves the extent and nature of the assumptions that one is willing to make about the form of the distribution of the population θ. We may assume, for example, that θ is multivariate Normal and thus, we need only estimate its mean vector and covariance matrix. On the other hand, we may wish to adopt a nonparametric approach in which no functional form is assumed for the density of θ. In all cases, assumptions will be necessary concerning the distributions of the ε_{ij}'s. Techniques for evaluating the appropriateness of these assumptions include analysis of a) identifiability, b) analysis of residuals, c) sensitivity analysis, and d) the use of variance models. The ε_{ij}'s may involve measurement error, model misspecification error and error in the measurement of Y_i and/or t_i.

2.1. Naive Pooled Data Approach

Here, all data are combined and analyzed as though they came from a single individual. Thus, the subscript j is ignored in Eq. 13.2; so it is equivalent to individual-based kinetic analysis. It is appropriate to use an individual parameter estimation technique such as nonlinear least squares to obtain point estimates of the parameters. The mean residual (possibly weighted) is used to estimate the variance. This method is unattractive for a variety of reasons. Most fundamentally, it obscures the underlying notion of a population distribution. That is, it is not clear what the concept of a population distribution is and any appreciable amount of intraindividual variability will be undetected because it will be masked by the regression function F.

Formally, analysis by this method combines all sources of error, including measurement error and intraindividual variation. Consequently, there are no estimates available for the variance of the parameters at the population level. In computer simulation studies, this approach has been found to perform poorly (10,11,13).

2.2. Two-Stage Methods

The most commonly used methods for estimating the distribution of θ are probably those referred to as "two-stage" methods. As their name implies, these methods proceed in two-stages where in the first stage, each individual's set of data is separately analyzed using an individual-based parameter estimation technique. Nonlinear least squares and maximum likelihood methods are commonly used. The results are point estimates, $\hat{\theta}_j$, for the parameter vector θ_j, and their associated covariance matrices. The use of these methods requires sufficient experimental data on each individual to obtain reasonable point estimates for the $\hat{\theta}_j$.

Once the set of $\hat{\theta}_j$'s have been obtained, they may be used in various ways to obtain the desired population estimate. For example, it is often assumed that the population distribution of θ is Normal or Lognormal. In these cases, among others, it is only necessary to estimate the mean and covariance of the population density. The sample mean and the sample covariance of the $\hat{\theta}_j$'s may be used for this purpose.

All of the two-stage methods treat the $\hat{\theta}_j$'s as though they were observations rather than estimates of individual parameter vectors. Thus, one can expect these approaches to yield good estimates when each $\hat{\theta}_j$ is estimated with good precision and accuracy. In more typical applications, in which the $\hat{\theta}_j$ have moderate to large variances, one would at least expect the two-stage methods to underestimate the variances of the population parameters as a result of ignoring variability in the individual point estimates. We point out that the EMSA approach takes into consideration the variability in the individual point estimates, unlike other two-stage methods. Sheiner and Beal (10,11,13) have observed that a two-stage method underestimated the variances of the population parameters as a result of ignoring variability in the individual point estimates in several pharmacokinetic applications.

2.3. Nonlinear Mixed Effects Model

A third approach to the population problem has been proposed by Sheiner and Beal (10,11,13) and a version of it implemented by them in the computer program NONMEM (2). This method is based on an assumed multivariate Lognormal density for θ. Now, suppose that θ_j, the parameter vector for individual j, has components $\theta_{j1}, \theta_{j2}, \ldots, \theta_{jP}$, where p is the dimension of the parameter vector for the structural model. Under the Lognormal assumption, $\ln(\theta_{jk})$ are jointly distributed with a multivariate Lognormal density in the population for $k = 1,2,\ldots, p$.

Denoting the components of the mean parameter vector for the population by $\ln(\theta_1)$, $\ln(\theta_2)$, ..., $\ln(\theta_p)$, each individual parameter vector component can be written as the population mean plus a random variable representing the deviation of the individual from the population mean. Thus,

$$\ln(\theta_{jk}) = \ln(\theta_k) + \xi_{jk}, \quad k = 1, 2, \ldots, p. \tag{13.3}$$

for each j, where ξ_{jk} have Lognormal densities. If the model were linear in the parameters and all densities were Normal, we would have an analysis of covariance problem with mixed effects. Hence, the acronym NONMEM, for Nonlinear Mixed Effects Model.

The basic idea is to expand the model about the population mean parameter, using the linear portion of the Taylor series expansion and to estimate θ by minimizing a criterion that corresponds to maximum likelihood estimation in the Normal case. It has been shown that this estimator has desirable large sample properties even when distributions are not Normal and its behavior has been studied in a variety of computer simulations (10,11,13). In those studies, the performance of the estimators, and the corresponding coefficients of variation of the parameter estimates in the population have been good. To date, however, no rigorous testing of NONMEM has been made comparing its results to standard sets of population parameter values obtained by entirely different techniques such as, the jackknife, the bootstrap, the jackknife-after-bootstrap, or Monte Carlo (14).

2.4. Nonparametric Maximum Likelihood Method

The nonparametric maximum likelihood method (NPML) advanced by Mallet (3) requires knowledge of the distributions of the ε_i, but makes no assumption regarding the form of the density for θ. NPML software is now available (Bestfit s.a.,Luxembourg). It has been shown that the maximum likelihood estimate of the density for θ is a discrete density with positive density at a number of points that is less than or equal to the number of individual samples (N). One can in principle find the location of these support points as well as their corresponding densities and thus calculate the maximum likelihood estimate of the desired function.

2.5. Bayesian Methods

For the individual case, Eq. 13.1, and letting $D = (Y_1, Y_2, ..., Y_N)$ represent the data. If h(θ)is the posterior density of θ, i.e., the density function for θ prior to consideration of D, h(D;θ) is the probability of the data given θ, which may be regarded as the likelihood of the data as a function of θ. h(θ;D) is the posterior density of θ given D, then Bayes' Theorem states:

$$h(\theta; D) = Kh(D;\theta)h(\theta) \qquad (13.4)$$

where the constant K is the reciprocal of the integral of $h(D;\theta)h(\theta)$ integrated over all possible values of θ.

In the absence of sufficient prior information concerning the distribution of θ, one may regard $h(\theta)$ as locally uniform and constant as a statement of prior ignorance (15,16). In this case, Eq. 4 becomes:

$$h(\theta; D) = Kh(D;\theta). \qquad (13.5)$$

Now, calculation of $h(D;\theta)$ is the same as calculation of the likelihood function in earlier methods (7).

Two-stage methods are possible in which Bayes' Theorem (Eqs. 4 or 5) is used in the first stage. Then, after stage-one, we have an estimated posterior density for each individual. These could be combined, for example, by averaging these posterior densities at each value of θ (8). This approach does not require large amounts of data on each individual and takes into account the relative information content in the different individual data sets.

Analogues of both EMSA and NONMEM are also possible based on Bayes' Theorem. That is, given a functional form for the population density and densities for the ε_i's, the likelihood $h(D;\theta)$ can be calculated. Here, D represents the complete data set for all of the individuals and θ is the complete set of population parameters.

2.6. Extended Multiple Studies Analysis

Recently, a new procedure for the analysis of kinetic studies to produce population parameter estimates has been made available as a service of the Simulation Analysis And Modeling (SAAM) environment for kinetic data analysis and graphics. The Extended Multiple Studies Analysis (EMSA) facility within NIH-SAAM31 is an efficient and accurate maximum likelihood estimator rooted in application of dynamic systems theory of population density estimation. EMSA is underpinned by a theoretical cohesion based on maximum likelihood estimation applied to the nonlinear mixed-effects problem subject to a set of mixed linear and nonlinear constraints, the types of problems encountered in compartmental modeling of radiolabelled compounds in living systems (1).

EMSA is not available in the current version of WinSAAM yet, but an aggressive programming effort is underway to make this facility available soon. EMSA currently runs only in batch mode using NIH-SAAM31.

EMSA falls within a new category of two-stage procedures, which we call global-iterative two-stage (GITS) methods. The basic idea underlying global two-stage (GTS) and iterative two-stage (ITS) methods is to take advantage of the population variance-covariance matrix using the estimates of the individual parameter means and variance-covariances as observations. On the other hand, the standard two-stage (STS) methods also involve a first stage fitting of the data for each individual, but in the second stage population parameters are estimated by combining all individual estimates using basic statistical procedures for calculation of mean and standard deviation Sheiner and Beal (10,11,13). In contrast, EMSA combines the strategies of the GTS method in a first stage and in a second stage, it implements the strategy of the ITS methods (17,18). Furthermore, it is a third order method. The first-order NONMEM method (10,11,13) only makes use of the first partial derivatives of the Taylor expansion of the observation function in terms of the parameters, and the second-order method of Lindstrom and Birke makes use of only the first two partial derivatives (19).

With EMSA the data for each individual is analyzed separately using individual-based analysis. These estimates provide the initial estimates for the parameter values for *stage 1*. Both the data and the updated results are then concatenated to form the EMSA input file. In *stage 1* of the solution procedure, the data from each individual are reanalyzed automatically and consecutively in block yielding improved initial individual parameter estimates along with improved initial estimates for each individual study's respective variance-covariance matrix. Iterative refinements of these improved initial estimates are obtained by iteration cycles until the process converges, yielding more refined estimates for the individual kinetic parameter values and each study's associated variance-covariance matrix. *Stage 1* of the analysis ceases with the automatic generation of the multiple studies file (*msf*). *Stage 2* of the solution procedure commences using the *msf* from *stage 1* as the active set of values for the batch of kinetic studies. These are analyzed simultaneously in-block as the iteration cycles in *stage 2* are continued until the sufficient conditions are satisfied for an optimum convergent maximum likelihood solution yielding final estimates for the population mean vector μ and variance-covariance matrix Ω.

Formally stated, the problem under investigation can be described as one where a random kinetic experiment is performed N times under identical conditions in a known population of individuals and a set of p variates are measured in each experiment. When we perform the random kinetic experiment, one and only one outcome occurs. Thus, outcomes are mutually exclusive in that they cannot occur simultaneously. The information obtained at experiment j is a random p-component vector of observations $\theta_j = [\theta_{1j}, \theta_{2j}, ..., \theta_{pj}]^T$ and a symmetric p x p positive definite covariance matrix V_j of the observations. We assume that the random variable θ_j is distributed according to $N(\mu_j, V_j)$ where $\mu_j = [\mu_{1j}, \mu_{2j}, ..., \mu_{pj}]^T$ and that μ is a random variable distributed according to $N(\mu, \Omega)$ where $\mu = [\mu_1, \mu_2, ..., \mu_p]$ and Ω is the symmetric p x p positive definite covariance matrix of μ. Hence, the joint probability density function for θ_j and μ_j is

$$f(\theta_j, \mu_j) = f(\theta_j / \mu_j) f(\mu_j)$$

$$= \frac{1}{(2\pi)^{\frac{p}{2}} |V_j|^{\frac{1}{2}}} \exp\left\{-\frac{1}{2}(\theta_j - \mu_j)^T V_j^{-1} (\theta_j - \mu_j)\right\} \otimes \tag{13.6}$$

$$\frac{1}{(2\pi)^{\frac{p}{2}} |\Omega|^{\frac{1}{2}}} \exp\left\{-\frac{1}{2}(\mu_j - \mu)^T \Omega_j^{-1} (\mu_j - \mu)\right\}$$

the marginal probability density function of θ_j is

$$f(\theta_j) = \int_{-\infty}^{\infty} \Lambda \int_{-\infty}^{\infty} f(\theta_j, \mu_j) \prod_{i=1}^{p} d\mu_i$$

$$= \int_{-\infty}^{\infty} \Lambda \int_{-\infty}^{\infty} f(\theta_j / \mu_j) f(\mu_j) \prod_{i=1}^{p} d\mu_i \tag{13.7}$$

$$= \int_{-\infty}^{\infty} f(\theta_j / \mu_j) f(\mu_j) d\mu_j$$

and thus,

$$f(\theta_j) = \frac{1}{(2\pi)^{\frac{p}{2}} |V_j + \Omega|^{\frac{1}{2}}} \exp\left\{-\frac{1}{2}(\theta_j - \mu_j)^T (V_j + \Omega)^{-1} (\theta_j - \mu)\right\} \tag{13.8}$$

is the population distribution function. Correspondingly, the marginal distribution of the random variable θ_j is multivariate normal with meanvector μ and covariance matrix $(V_j + \Omega)$. The distribution is independent of μ_j, the mean vector of the individual study.

The natural logarithm of the likelihood is

$$Log\ L = -\frac{Np}{2}\log(2\pi) - \frac{1}{2}\sum_{j=1}^{N}\log\left|V_j + \Omega\right| - \frac{1}{2}\sum_{j=1}^{N}(\theta_j - \mu)^T (V_j + \Omega)^{-1}(\theta_j - \mu) \tag{13.9}$$

Maximizing log L with respect to μ and Ω we find that

$$\frac{\partial \log L}{\partial \mu} = 0 \Rightarrow \hat{\mu} = \left[\sum_{j=1}^{N}\left(V_j + \hat{\Omega}\right)^{-1}\right]^{-1}\sum_{j=1}^{N}\left(V_j + \hat{\Omega}\right)^{-1}\theta_j \tag{13.10}$$

and

$$\frac{\partial \log L}{\partial \Omega} = 0 \Rightarrow -\frac{1}{2}\sum_{j=1}^{N}\left(V_j + \hat{\Omega}\right)^{-1} + \frac{1}{2}\sum_{j=1}^{N}\left(V_j + \hat{\Omega}\right)^{-1}\left(\theta_j + \hat{\mu}\right)\left(\theta_j + \hat{\mu}\right)'\left(V_j + \hat{\Omega}\right)^{-1} = 0 \tag{13.11}$$

Features of Eqs. 13.10 and 13.11 are: (1) given values of V_j and $\hat{\Omega}$, $\hat{\mu}$ is directly estimable from Eq. 13.10 and (2) no closed form solutions to Eq 11 exist. To bootstrap the estimation procedure for $\hat{\Omega}$ we need to derive starting values for its estimates. The essence of bootstrapping is the idea that, in the absence of any other knowledge about a population, the distribution of values found in a random sample of size N from the population is the best guide to the distribution in the population. Thus, for the estimation of $\hat{\mu}$ and $\hat{\Omega}$ from Eqs. 10 and 11, initial estimates for the elements for $\hat{\mu}$ are obtained from the average of the separate study estimates because it makes sense to use the actual data from the given experimental situation. Although, different notions can be employed in choosing the particular values to bc used (14).

Then, using a 3-term Taylor series expansion of $\Gamma_{ii}\left(V_j + \hat{\Omega}\right)^{-1}$ about $\Gamma_{ii}V_j = \Gamma_{ii}\hat{\Omega}$ and substituting into Eq 13.11, where $\Gamma_{ii}(\mathrm{B}) = b_{ij}$, i.e., extracting the ij-th element of the matrix yields initial estimates for $\Gamma_{ii}\hat{\Omega}$, and with these, improved estimates of $\hat{\mu}$ are obtained using Eq. 13.10. Combining the newly improved bootstrapped estimate of $\hat{\mu}$ with the estimate of $\hat{\Omega}$, a Newton-Raphson iterative procedure is applied using Eq 13.11 to iteratively improve the estimate of $\hat{\Omega}$. The best estimate of $\hat{\Omega}$ is substituted into Eq 13.10 and an improved estimate of $\hat{\mu}$ is obtained. To obtain $\Gamma_{ij}(\hat{\Omega})$, and this is the second stage of the scheme, another Newton-Raphson iterative procedure is used which applies a Lagrangian function to the objective function to ensure that realistic covariance matrix elements emerge.

The most important ingredient of the new EMSA algorithm is exemplified by the population distribution function, Eq 13.8. Note that it preserves the error in the individual parameter estimates (V_j) which allows more precise estimates for the population distribution to be obtained. Thus, although Sheiner and Beal (10,11,13) have observed that a two-stage method underestimated the variances of the population parameters in their computer simulations, their two-stage approach ignored this variability in the individual point estimates.

III. Results

3.1. EMSA vs. NONMEM

We have compared results obtained using EMSA and NONMEM to a standard reference set of population parameter values obtained with the jackknife. We found that EMSA provides estimates for the dispersion of NE kinetic parameters which are less biased than estimates obtained using NONMEM. EMSA estimates of population standard deviation closely approximated the "true" jackknife population standard deviation (9).

Fig 13.1 below illustrates a SAAM31 deck ready for processing by EMSA. Data for only 2 individuals is shown to conserve space. Nevertheless, this is a runable deck and serves as a template.

```
A SAAM31 NEC001.001              NEPC003:PROTO-1:OAL:950630
2        25                                            19
P(3)      2.11
CCCC ENTER 3H[NE] INFUSION RATE (DPM/ML) - P(20)
    P(20)     1494786
    L(1,20)   1.
    L(0,20)=-L(1,20)
    IC(20)=P(20)
    L(2,1)      9.583758E-01   9.999998E-04    1.000000E+01
    L(1,2)      8.841792E-03   9.999998E-06    2.000000E+00
    L(0,2)      6.055759E-02   9.999998E-06    1.000000E+00
    P(90)       3.776262E+03   1.000000E+02    1.000000E+06
CCCC G(1) DEFINES THE RADIOACTIVITY OBSERVATIONS
XG(1)=F(1)/P(90)
CCCC ENTER THE MEAN STEADY-STATE PLASMA NE LEVEL - P(50)
    P(50)    119
    L(11,12)=L(1,2)
    L(12,11)=L(2,1)
```

```
   L(0,12)=L(0,2)
   IC(11)=P(50)*P(90)
   IC(12)=IC(11)*L(2,1)/L(1,2)+L(02)
CCCC L(12,21) IS THE STEADY-STATE (SS) EXTRAVASCULAR NE RELEASE
   L(12,21)  1.518508E+06  1.000000E+04   1.000000E+07
   L(0,21)=-L(12,21)
   IC(21)   1.
CCCC G(11) DEFINES THE NE MASS OBSERVATIONS
XG(11)=F(11)/P(90)
C
H DAT
101G(1)                                      FSD=.05
C            TIME (MIN)     3H-NE(DPM/ML)
             40             749.91
             50             692.73
             60             634.36
101G(1)                                      WT=0
             0
2            1                               50
111G(11)                                     FSD=.05
C            TIME (MIN)     NE(PG/ML)
             40             181
             50             177
             60             160
             70             165
             80             140
CCCC    STOP INFUSION
H PCC                       TC(1)
   L(1,20)=0.
CCCC    RESET TIME TO ZERO
H DAT                       TC(1)
   T         0.
102G(1)                                      FSD=.05
C            TIME (MIN)     3H-NE(DPM/ML)
             1              429.20
             2              183.94
             4              77.23
             6              66.62
             8              37.14
             10             36.55
             12             28.30
             14             24.76
             16             26.53
             18             23.58
             20             14.74
A SAAM31 NEC004.001            NEPC004:OAL:950630
2       25                                     20
H PAR
CCCC ENTER SUBJECT'S BODY SURFACE AREA (METERS SQUARED) - P(3)
   P(3)      1.76
CCCC ENTER 3H[NE] INFUSION RATE (DPM/ML) - P(20)
   P(20)     1184898
   L(1,20)  1.
   L(0,20)=-L(1,20)
   IC(20)=P(20)
   L(2,1)     5.924109E-01  9.999998E-04   1.000000E+01
   L(1,2)     9.780818E-03  9.999998E-06   2.000000E+00
```

```
  L(0,2)      4.756985E-02  9.999998E-06   1.000000E+00
  P(90)       7.527194E+03  1.000000E+03   1.000000E+04
CCCC G(1) DEFINES THE RADIOACTIVITY OBSERVATIONS
XG(1)=F(1)/P(90)
CCCC ENTER THE MEAN STEADY-STATE PLASMA NE LEVEL - P(50)
  P(50)   146
  L(11,12)=L(1,2)
  L(12,11)=L(2,1)
  L(0,12)=L(0,2)
  IC(11)=P(50)*P(90)
  IC(12)=IC(11)*L(2,1)/L(1,2)+L(02)
CCCC L(12,21) IS THE STEADY-STATE (SS) EXTRAVASCULAR NE RELEASE
  L(12,21)  3.253814E+06  1.000000E+04   1.000000E+07
  L(0,21)=-L(12,21)
  IC(21)   1.
CCCC G(11) DEFINES THE NE MASS OBSERVATIONS
XG(11)=F(11)/P(90)
H DAT
101G(1)                                          FSD=.05
C            TIME (MIN)     3H-NE(DPM/ML)
             40             413.45
             50             347.64
             60             242.55
101G(1)                                          WT=0
             0
2            1                                   50
111G(11)                                         FSD=.05
C            TIME (MIN)     NE(PG/ML)
             40              159
             50              151
             60              129
             70              160
             80              131
CCCC    STOP INFUSION
H PCC                       TC(1)
  L(1,20)=0.
CCCC    RESET TIME TO ZERO
H DAT                       TC(1)
  T          0.
102G(1)                                          FSD=.05
C            TIME (MIN)     3H-NE(DPM/ML)
             1              274.92
             2              131.09
             4              67.40
             6              57.85
             8              42.46
             10             45.64
             12             33.44
             14             31.84
             16             30.25
             18             28.66
             20             24.94
102G(1)                                          WT=0
             0
2            1                                   20
```

Fig 13.1 Runable template SAAM31 deck for norepinephrine population kinetics modeling.

To perform a multiple studies analysis using SAAM31, fit the data pertaining to each study separately with the appropriate model. Update the parameters of each study to the (best) estimated values. Join the updated study files to form a single file of all studies (a multiple studies file). Make two alterations to each study within this file as follows: In columns 10 through 19 of line 1 (the A SAAM31 line) insert the project and study id's. The project id is a 3 alphabetic character descriptor viz: NEC, the study id is a 7 numeric character descriptor viz: 004.001. The full project/study descriptor could be NEC004.001 and a second study in this project set would be NEC005.001. To all but the last study, add a line 2 indicating that model code 19 is to be used in processing the study viz: 2 ..... 47 spaces ....19. On the last study add a 2 line which refers to model code 20 viz: 2 ..... 47 spaces ....20. Submit this multiple studies file to the batch version of SAAM31 for processing viz: SAAM31 < MultStud.INP > MultStud.OUT. Here MultStud.INP is the multiple studies input file prepared as per step above and MultStud.OUT is one of the output files (the standard batch file) created in conjunction with the use of SAAM31.

In conjunction with performing a multiple studies analysis with SAAM, several output files in addition to the 'standard' output file are generated viz: A) A file called STUDIES (also referred to as the Population Values File). This file is in fact a normal model specification section of a SAAM input file and it contains the following for your convenience 1) the appropriate project-identified on the A SAAM line, 2) the population parameter values, their upper and lower limits, and their uncertainties on parameter lines under H PAR, 3) the normal equations under H MAT. This file must be modified to include dependencies and initial conditions as appropriate since neither of these classes of parameters can be directly adjustable. B) A file called FORT.1 (also referred to as the Covariance Matrix File). This file contains the covariance matrices derived from the individual studies and in fact is a complete SAAM input file. It can under certain circumstances be directly submitted to SAAM for processing viz: SAAM31 < FORT.1 > FORT_1.OUT and will produce the same studies file but an attenuated standard output file (FORT_1.OUT for the above command). This is a 'permanent' file and will not be overwritten by simply running another SAAM job. Indeed subsequent runs calling for model code 19/20 processing will simply append covariance matrix data to this file. To avoid this you will need to either delete or rename this file. C) A file called LABELS (also referred to as the Parameter Names File). This file simply contains the names of all adjustable parameters in the multiple studies decks. It is only important to you if you wish to resubmit the covariance matrix file FORT.1 for analysis. This is a temporary file and will be replaced with a new version if another multiple studies analysis is conducted. D) A file called CD16FILE. This is a scratch file and can be removed at any time. To locate salient output you should refer to either the standard output file or the STUDIES file.

Amongst other routine results, the standard output file contains the following: a) Population estimates of the parameters and their covariances based on model code 16 (the new procedure), b) A listing of the deviations of each study's parameter estimates from the population estimates, c) The original algorithm estimates of the population parameters and their uncertainties. These are usually less conservative than those derived using the new algorithm (model code 16, based on the full likelihood function).

3.2. EMSA vs. NONMEM vs. Standard Two-Stage Procedure

We have compared the results of EMSA using NONMEM and the standard two-stage procedure. The data for this study come from the test theophylline data provided with the distribution of the NONMEM package. The model is the simple one-compartment model with first-order absorption depicted in Fig 13.2.

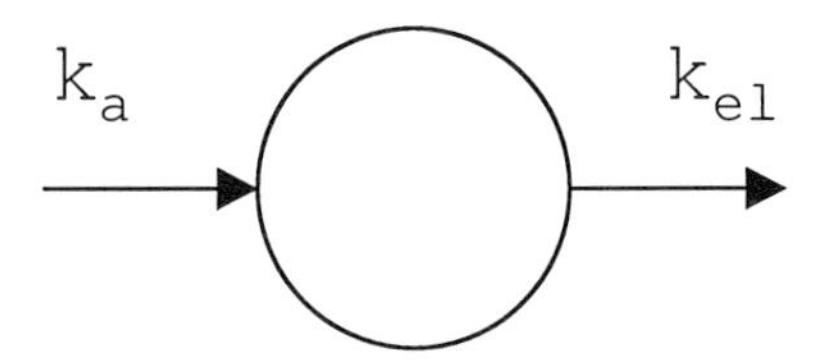

Fig 13.2 First-order absorption model for theophylline kinetics.

Table 13.1. Results of theophylline test problem with 12 individuals.

Method	k_a	k_{el}
Two-Stage	1.56±0.05	0.080±0.003
NONMEM	2.77±0.71	0.078±0.007
EMSA	2.05±1.63	0.083±0.010

As shown in Table 13.1, the two-stage method clearly underestimates the population standard deviation, as expected, for both k_a and k_{el}. However, the NONMEM estimates for population standard deviation are also downwards biased compared to the EMSA estimates.

3.3. EMSA vs. Nonparametric Maximum Likelihood Method

EMSA assumes a multivariate normal distribution of the kinetic model parameters. In the non-parametric maximum likelihood method (NPML), no assumptions are made about the parametric distribution of the kinetic parameters (3). The likelihood $l_j(\theta, x_j)$ of the observations x_j in individual j, for any value of the parameters θ, is readily derived from the two-compartment norepinephrine kinetic model and the measurement error model. We assume a homoscedastic error model with a coefficient of variation of 5% which is the measurement error for the NE assay in our laboratory (20).

Table 13.2. EMSA compared to nonparametric maximum likelihood method.

	EMSA N=19		Nonparametric Maximum Likelihood N=19	
	Mean	**SD**	**Mean**	**SD**
L(2,1)	0.742	0.173	0.838	0.376
L(1,2)	0.008	0.003	0.008	0.004
L(0,2)	0.053	0.020	0.052	0.030
Vd	4.41	1.63	4.48	2.20

The means were not significantly different. However, the standard deviation with NPML was significantly greater for L(2,1) and L(0,2) [F-test p values < 0.05] compared to EMSA. More basic work needs to be done in this area. However, based on jackknife estimates for standard deviation, EMSA estimates for standard deviation are closer to those of the jackknife standard. Thus, for this data set, NPML estimates of standard deviation for L(2,1) and L(0,2) were upwards biased.

IV. Summary and Discussion

We have defined the population estimation problem in several contexts and indicated some general application and verification. This is currently an active area of research and development in our laboratories. We conclude this chapter by briefly discussing major advantages and disadvantages of some of the methods. For the reader interested in perusing issues of interest, the book by Davidian and Giltinan (18) is particularly important. The references there and in this chapter should provide direction.

Several steps have been taken in the development of the GITS third-order EMSA method to exclude the sources for bias alluded to by Sheiner and Beal (10,11,13) and Mentre and Mallet (21). First, the global individual parameter estimates obtained in the first stage of EMSA serve only as initial estimates for the parameters of the population-based model. Then, in the second stage, the population-based model is solved simultaneously in-block by iterative refinement of those initial parameter estimates and their corresponding variance-covariance matrices, yielding individual parameter estimates adjusted for estimation error in the initial parameter estimates and their variance-covariance. The latter set of parameter estimates and their variance-covariance thus comprise the set of observations from which population values are converged upon after undergoing a final phase of iterative refinement.

Finally, if ε_i is the random error in the population kinetic parameter estimates, and δ_i is the random error induced by the ε_i in the estimated population kinetic parameter estimates, then the errors induced in the population parameter estimates, δ_i, as a result of the error $\varepsilon_{i,}$ have variance asymptotically equal to the second partial derivatives of the observation function at values equal to or close to the estimated kinetic parameter estimates. Hence, two random features need to be considered: the kinetic variability which is directly related to the population (Ω), and the δ_i error which is directly relatedto the experimental design. The GITS third-order EMSA method provides values for the δ_i errors in the population parameter estimates. This allows one to judge the goodness-of-fit of the population-based kinetic model to the population data and provides a quantitative estimate of the variability that is not physiologic in origin.

This is in sharp contrast to the standard two-stage method evaluated by Sheiner and Beal (10,11,13) and Mentre' and Mallet (21) which involves a first stage fitting to the data for each individual and then in the second stage, population parameters are estimated by combining all the corresponding individual estimates. That approach is analogous to individual-based kinetics modeling with assembling of individual kinetic parameter values from each individual and calculating means and standard deviations in the usual way. Our results presented in this chapter using this approach are in close agreement with the observations of Sheiner and Beal (10,11,13) and Mentre' and Mallet (21), i.e., that a two-stage method that does not take into consideration variability in the individual estimates will yield population estimates which are downwards biased.

The results presented in this chapter may also have important implications for population data analysis in general, and pharmacology in particular, because, in pharmacology, routine individual data are usually limited, i.e., sparse, noisy, obtained from studies with various experimental designs, or obtained during the course of routine therapeutic drug monitoring in hospitalized patients. Here the limited data on any one individual is toosparse to provide good estimates of the parameters in individuals but enough data may be available on a large number of individuals. NONMEM was developed to overcome the obstacles imposed by limited data and provide the first two moments of the population distribution of pharmacokinetic parameters. NONMEM is now a standard for population pharmacokinetic analysis. However, since NONMEM, applied in the context of messy data, does not follow individual kinetics and therefore does not take into account individual components of variance, it is possible that these unknown individual components of variance could influence population parameter estimates. For example, it is known that ignoring individual error variances leads to underestimation of population density using two-stage methods (10,11,13). Thus, individual components of variance may play a significant role in population data analysis since NONMEM underestimated the population standard deviations for k_a and k_{el} compared to EMSA (Table 1).

Based on the titles of published abstracts on the internet of the meeting on non-parameteric and semi-parameteric methods in population data analysis presented in Geneva (February 11, 1997), there appears to be much interest in non-parameteric and semi-parameteric methods in population data analysis. Nevertheless, for our norepinephrine kinetic data (Table 2), estimates obtained using NPML were poorer than those obtained using EMSA since we know the "true" population distribution of these parameters based on a set of standard kinetic values obtained using the jackknife (9).

The observations in a tracer kinetics experiment define a set of observational parameters that are functions of the basic kinetic parameters of the model of the system. The problem of identifiability is concerned with whether the observational parameters uniquely specify the basic kinetic parameters. As such, it depends only on the functional relation between the two levels of parameters and not on errors of observation and the estimation procedure. It should be checked before performing the experiment. A strong point of the population-based studies presented in this chapter is that they are based on a compartmental model structure which is uniquely identifiable (Linares et. al. (20)). The importance of identifiability in population density estimation cannot be overemphasized. In addition, the population-based norepinephrine kinetics model presented here is embedded within a framework (EMSA) which combines the application of dynamic systems theory and Bayesian probabilistic thinking to the population estimation problem. A major benefit of this approach is the ability of the population-based norepinephrine kinetics model to perform recursive estimation"updating" of the population density as new norepinephrine kinetic data becomes available. This applies to other sets of experimental data.

REFERENCES

1. Lyne, A., Boston, R., Pettigrew, K., L. Zech. EMSA: a SAAM service for the estimation of population parameters based on model fits to identically replicated experiments. *Comput. Methods Progr. Biomed.* 38:117-151, 1992.
2. Beal, S.L., L.B. Sheiner. *NONMEM Users Guide, Part I.* Technical Report, Division of Clinical Pharmacology, University of California, San Francisco, CA. 1980.
3. A., Mallet. A maximum likelihood estimation method for random coefficient regression models. *Biometrika* 73:645-656, 1986.
4. Whiting, B., Kellman, A.W., J. Grevel. Population pharmacokinetics: theory and clinical application. *Clin. Pharmacokinet.* 11:387-401, 1986.
5. L. Aarons. Population pharmacokinetics. *Intl. J. Clin. Pharmacol. Ther. Toxicol.* 30:520-522.
6. L.B., Sheiner. Analysis of pharmacokinetic data using parameteric models-I: regression models. *J. Pharmacokinet. Biopharm.* 12:93-117, 1984.

7. Katz, D., Population Density Estimation. *Prog. Food Nutrit. Sci.* 12:325-338, 1988.
8. Katz, D., D.Z., D'Argenio. Implementation and evaluation of control strategies for individualizing dosage regimens with application to aminoglycoside antibiotics. *J. Pharmacokinet. Biopharm.* 14:523-537, 1986.
9. Linares, O.A., Zech, L.A., Supiano, M.A., J.B. Halter. Evaluation of bayesian estimation in comparison to a first-order method for population kinetics modeling: application to norepinephrine metabolism in humans. *J. Invest. Med.* 43(suppl 3):478A, 1995.
10. Sheiner, L.B., S.L. Beal. Evaluation of methods for estimating population pharmacokinetic parameters. I. Michaelis-Menten model: routine clinical pharmacokinetic data. *J. Pharmacokinet. Biopharm.* 8:553-571, 1980.
11. Sheiner, L.B., S.L. Beal. Evaluation of methods for estimating population pharmacokinetic parameters. III. Monoexponential model:routine pharmacokinetic data. *J. Pharmacokinet. Biopharm.* 11:303-319, 1983.
12. Milliken, G.A. and D.E. Johnson. *Analysis of Messy Data. Volume I: Designed Experiments.* Chapman & Hall, New York, 1992.
13. Sheiner, L.B., S.L. Beal. Evaluation of methods for estimating population pharmacokinetic parameters. II. Biexponential model and experimental pharmacokinetic data. *J. Pharmacokinet. Biopharm.* 9:635-651, 1981.
14. B.F.J. Manly. *Randomization, Bootstrap, and Monte Carlo Methods in Biology.* Chapman & Hall, New York, 1997.
15. Box, G.E.P., G.C. Tiao. *Bayesian Inference in Statistical Analysis.* Addison-Wesley, New York, 1973.
16. West, M., J. Harrison. *Bayesian Forecasting and Dynamic Models.* Springer-Verlag, New York, 1989.
17. Steimer, J., A. Mallet, J. Golmard, and J. Boisvieux. Alternative approaches to estimation of population pharmacokinetic parameters: comparison with nonlinear mixed-effect model. *Drug. Metab. Rev.* 15: 265-292, 1984.
18. Davidian, M. and D. M. Giltinan. *Nonlinear Models for Repeated Measurement Data*, London: Chapmann & Hall, 1995.
19. Lindstrom, F. T. and D. S. Birke. Estimation of population pharmacokinetic parameters using destructively obtained experimental data: a simulation study of the one-compartment open model. *Drug Metab. Revs.* 15: 195-264, 1984.
20. Linares, O. A., J. A. Jacquez, L. A. Zech, M. J. Smith, J. A. Sanfield, L. A. Morrow, S. G. Rosen, and J. B. Halter. Norepinephrine metabolism in humans: kinetic analysis and model. *J. Clin. Invest.* 80: 1332-1341, 1987.
21. Mentre, P.L., A. Mallet. A maximum likelihood estimation method for random coefficient regression models. *Biometrika* 645-656, 1986.

14

INFORMATION IN THE MODEL

Once a model has been developed, the model can help determine many features about the system. These include not only the structure of the system, such as the number of compartments and their connectivity, but also model-based calculations. For example, metabolic systems are often studied kinetically to determine the rates at which compounds are synthesized and metabolized. Synthesis and catabolism however often occur at sites that differ from the usual sampling compartment (e.g., tissue vs. blood). Model parameters can be used to calculate production rate (PR) and fractional catabolic rate (FCR) in addition to other parameters; how long a particle from one compartment spends in another compartment, the fraction of any compartment that is transferred to another compartment, residence times, transit times, and the number of times a particle recycles through a compartment. This chapter will discuss the information that can be gained about a system from a model and then describe some model-based calculations using examples.

I. Structure

The structure of a model when drawn graphically or listed mathematically provides a visual picture of how compartments are linked. The fractional transfer coefficients, L(i,j), provide information on pathways in a system and the relative flow along each pathway. The total flow out of a compartment is the sum of all pathways out of a compartment (i.e.. the sum of L(j,i)) and is called L(i,i). The reciprocal of L(i,i) is the turnover time of the compartment. This information can be used to compare the turnover of compartments in various parts of a system such as a tissue compartment or a blood compartment. The steady state information on a model consists of tracee inputs, U(I), compartment masses, M(i), and transport rates, R(i,j). The tracee inputs define where tracee enters the system, as well as the rates. For some systems, such as calcium in humans, there are two tracee inputs representing dietary calcium intake and resorption from bone (4). The masses of the compartments are related to turnover rates in that compartments with the slower turnover times have the largest mass. Transport rates show the flow of tracee along various pathways and are the product of the fractional transfers, L(i,j), and mass of compartment j. Transport rates may be large even when the fractional transfer coefficient is small because a compartment mass is large.

II. Model-based Calculations

Model-based calculations are described in papers by Berman (1,3) and are summarized in Table 14.1 Residence time, $\underline{t}(i,j)$ is the average time that a particle from compartment j is expected to spend in compartment i. It is equivalent to the L-inverse, LI(i,j) and is also the mass in compartment I generated by a unit infusion into compartment j. FCR (i,i) is the probability per unit time for a particle in compartment I to leave that compartment and not return (1). It is also equal to the reciprocal of the residence time, LI(i,j), Production rate is the rate of entry of new material into a compartment and is the product of M(i) x FCR(i,i). It includes material from the outside (or newly synthesized material), and also, any material entering the compartment for the first time via other compartments. Production rate can be calculated by dividing the mass of compartment I, M(I), by the L-inverse, LI(i,j).

The transfer coefficient, TR(i,j), is the fraction of material in compartment j that will reach compartment i. It is calculated as the normalized L-inverse, LI(i,j)/LI(i,i), and in some references is referred to as LN(i,j). The number of times an average particle will recycle through a compartment is equal to the residence time, divided by the transit or turnover time, minus 1.

Most values can be determined from the fractional transfer coefficients and also from the steady state parameters (Table 14.1). These approaches are described for vitamin A by Lewis et al, (2) and for lipoproteins by Berman (1). The calculations are shown for a two-compartment model (Fig 14.1) where the WinSAAM text input file (Fig 14.2) was used to generate the values in Fig 14.3-5.

Table 14.1 Some model-based Calculations

NAME	SYMBOL	COMPARTMENTAL MODEL	TRACEE
Fractional Catabolic Rate	FCR(I)	1/LI(I,I)	
Production Rate	PR(I)	M(I)/LI(I,j)	M(I)*FCR(i,I)
Residence Time	t(I,J)	LI(I,j)	M(I)/PR(i)
Transfer fraction	TR(I,J)	LI(I,J)/LI(I,I)	
Turnover	L(I,I)		
Turnover time		1/L(I,I)	

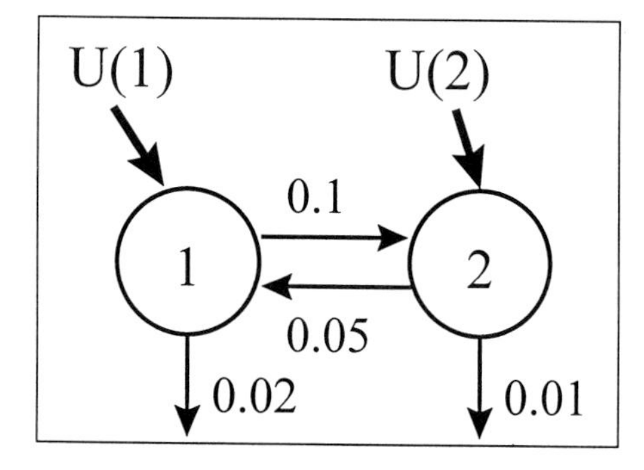

Fig 14.1 Two-compartment model

```
A SAAM31              CH14EG01.saam
C Example for calculating residence time,
C transfer coefficients, production rates and
C fractional catabolic rate
H PAR
   L(2,1)    0.1
   L(1,2)    0.05
   L(0,1)    0.02
   L(0,2)    0.01
H STE
   U(1)      10.
   U(2)
   M(1)      1              0.1             100
H PAR
C P(1) is the probability that a particle in compartment 1
C exits the system per unit time, FCR(1,1)
C P(2) is the probability that a particle exits
C compartment 2 per unit time, FCR(2,2)
   P(1)=L(0,1)+L(2,1)*L(0,2)/(L(0,2)+L(1,2))
   P(2)=L(0,2)+L(1,2)*L(0,1)/(L(0,1)+L(2,1))
C P(3) is the reciprocal area under F(1)
C P(4) is the reciprocal under F(2)
   P(3)=1/(0.695*5.994+0.30472*75.823)
   P(4)=1/(-0.650*5.994+0.6509*75.823)
C P(5) is the fraction of particles from compartment 1
C reaching compartment 2 per unit time
C and is equivalent to residence time
   P(5)=L(2,1)/(L(0,1)+L(2,1))
C P(6) is the production rate for compartment 1
C P(7) is the production rate for compartment 2
   P(6)=P(1)*M(1)
C    P(7)=P(2)*M(2)
H DAT
XUF(1)=10
XG(1)=0.5*EXP(-1.833*t)+0.5*EXP(-0.1667*T)
XG(2)=F(1)/UF(1)
100
   U(1)
   U(2)
   M(1)
   M(2)
   R(0,1)
   R(2,1)
   R(1,2)
   R(0,2)
101G(2)
            2
            4
            8
            INF
C ********* END OF MODEL
```

Fig 14.2 Text input file for model graphic in Fig 14.1, for calculating residence times, transfer coefficients, production rates and fractional catabolic rates.

Residence Times and Transfer Coefficients

Residence time can be calculated in WinSAAM by three methods (Fig 14.3); from the L-inverse matrix, the levels in a compartment resulting from a unit infusion, and the ratio of the mass of a compartment compared to the tracee input. The results show that the values are equivalent e.g., LI(1,2) is 22.7 and is equivalent to M(1)/U(2) or the average time a particle will spend in compartment 1 if it enters the system via compartment 2.Calculation of transfer coefficients is demonstrated in Fig 14.3 using the TR(i,j) command and ratio of the L(i,j), as P(5) (from text input file in Fig 14.2).

Calculating residence time from LI(i,j):
>calc LI
>li(i,j)

	1	*2*
1	*27.2*	*22.7*
2	*45.4*	*54.5*

Calculating residence time from assymptotic level with constant infusion
>prin qc(1)

Comp	*Category*	*T*	*QC*
1	*G(2)*	*2.0*	*1.86*
1	*G(2)*	*4.0*	*3.28*
1	*G(2)*	*8.0*	*5.43*
1	*G(2)*	*1E35*	*27.2*

Calculating residence time from steady state parameters, M(i,j)/U(j)
Residence time for compartment 1:
>U(1) *10*
>M(1) *27.2*
>M(2) *454.*
[M(1)/U(1) = 27.2, M(2)/U(1) = 45.4]
Residence Time for compartment 2:
>U(1)=0
>U(2)=100
>solv
>M(1) *2272*
> M(2) *5454*
[M(1)/U(2) = 22.7, M(2)/U(2) = 54.5]
Calculating transfer coefficients using LI:
>calc LI
>tr(i,j)

	1	*2*
1	*1.00*	*0.833*
2	*0.833*	*1.00*

Calculating transfer coefficient TR(2,1) using L's:
>P(5) *0.833*

Fig 14.3 Calculation of residence times and transfer coefficients for a model (Fig 14.1 and Fig 14.2). The symbol '>' precedes a WinSAAM command and values in italics are returned by the program in response.

Fractional Catabolic Rate

Fractional catabolic rate can be calculated five ways (Fig 14.4); from the ratio of the L(i,j), the reciprocal of the L-inverse, directly using the FCR(i,j) request, and from the area under the curve, and from the ratio of the tracee inputs and masses. In each case, for the example shown, the FCR(1,1) is 0.036 and FCR(2,2) is 0.0183. This means that 3.6% of particles in compartment 1 and 1.8% of compartment 2 leave irreversibly per unit time.

```
Calculating FCR from the ratio of the L(i,j):
    >p(1)          0.036
    >p(2)          0.0183

Calculating FCR from the reciprocal of LI:
    >calc LI
    >li(i,j)
                  1             2
    1           27.2          22.7
    2           45.4          54.5

    [1/LI(1.1)=0.036, 1/LI(2,2)=0.0183]

Calculating FCR in WinSAAM:
    >fcr(I,i)
                  1             2
    1           0.036         0.044
    2           0.022         0.0183
Calculating FCR from the area under the curve:
   >P(3)          0.036
   >P(4)          0.022
Calculating FCR from steady state parameters
   > U(1)=10
   >solv
   >M(1)        272
   >M(2)        454
  [U(1)/M(1)=0.036, U(1)/M(2)=0.022]
   >U(2)=1000
    >solv
   >M(1)           22,727
    >M(2)          54,545
  [U(2)/M(1)=0.044, U(2)/m(2)=0.0183]
```

Fig 14.4 Calculation of FCR in WinSAAM using the model in Fig 14.1 and listed in the text input file in Fig 14.2.The symbol '>' precedes a WinSAAM command, and values in parentheses were returned in response to the command. Five approaches are shown; using the ratio of the fractional transfer coefficients, the WinSAAM FCR command, the reciprocal of L-inverse, from the area under the curve, and using steady state parameters.

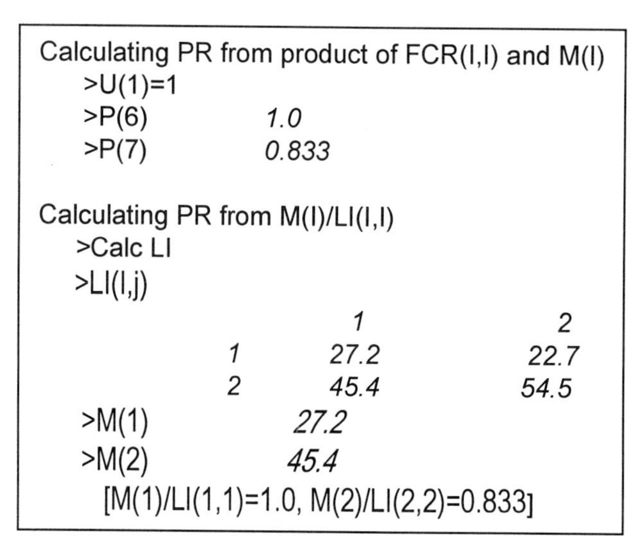
Calculating PR from product of FCR(I,I) and M(I)

>U(1)=1

>P(6) 1.0

>P(7) 0.833

Calculating PR from M(I)/LI(I,I)

>Calc LI

>LI(I,j)

	1	2
1	27.2	22.7
2	45.4	54.5

>M(1) 27.2

>M(2) 45.4

[M(1)/LI(1,1)=1.0, M(2)/LI(2,2)=0.833]

Fig 14.5 Two approaches in WinSAAM for calculating production rates, using the model in Fig 14.1 and listed in the text input file in Fig 14.2.The symbol '>' precedes a WinSAAM command, and values in parentheses were returned in response to the command.

Production rates

Production rate can be calculated by two approaches in WinSAAM: as the product of FCR(I,I) and M(I), and from M(I)/L-inverse (Fig 14.5).

Number of Cycles

The relationship between transit time, residence time, transfer coefficient, and number of cycles, is demonstrated using some simple models as shown in Fig 14.6. The models are simplified so that the values can be verified intuitively by the reader. Note that transit time is the same for compartment 1(and for compartment 2) across all models, as the sum of the pathways out of compartment 1and 2, and are 0.5 and 0.2 respectively in all models.

Residence time is the same as the turnover time for compartment 1 in models A and B, but is zero if the input is into compartment 2 as no material returns to compartment 1. Residence time is longer for compartment 1 in Models C and D, as some recycles back from compartment 2. For compartment 2, residence time is the reciprocal of turnover of compartment 2 for Model A, but for Model B, some material entering compartment 1 does not enter compartment 2, and so the residence time is lower (4 vs. 5). Residence time for compartment 2 increases in Models C and D when recycling occurs.

The transfer coefficient is unity when all material entering the system enters a compartment. It is less than unity when material is lost before entering a compartment (e.g., amount of material entering compartment 1 from compartment 2 when the input is into 2, in Model C) and is zero when none of the material is transferred (e.g., in models A and B none of compartment 2 enters compartment 1).

The number of cycles is the (residence time/transit time) minus one. No recycling occurred in Models A or Model B and recycling was higher in Model C than Model D, as material was lost from compartments 1 and 2 in this model.

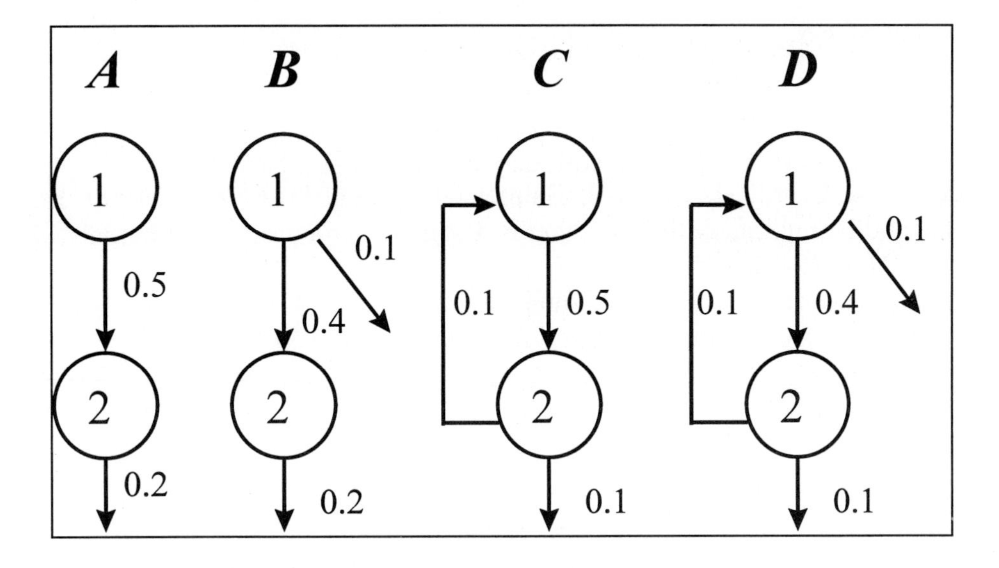

Fig 14.6 Four models used to demonstrate model-based calculations in Table 14.2

Table 14.2 Parameters calculated for the Models in Fig 14.6.

PARAMETER	*Model A*	*Model B*	*Model C*	*Model D*
TURNOVER TIME, 1/L(I,I)				
Compartment 1	2	2	2	2
Compartment 2	5	5	5	5
RESIDENCE TIME, LI(I,I)				
Compartment 1 (input 1)	2	2	4	3.3
Compartment 1 (input 2)	0	0	2	1.6
Compartment 2 (input 1)	5	4	10	6.6
Compartment 2 (input 2)	5	5	10	8.3
TRANSFER COEFFICIENT, TR(I,j)				
Compartment 1 (input 1)	1	1	1	1
Compartment 1 (input 2)	0	0	0.5	0.5
Compartment 2 (input 1)	1	0.8	1	0.8
Compartment 2 (input 2)	1	1	1	1
Number of cycles (Residence time/transit time)-1				
Compartment 1	0	0	1	0.65
Compartment 2	0	0	1	0.66

REFERENCES

1. Berman, M. 1982. Kinetic analysis and modeling: Theory and applications to lipoproteins. *In* Lipoprotein Kinetics and Modeling. M. Berman, S. M. Grundy, and B. V. Howard, editors. Academic Press, NY. 3-36.
2. Lewis, K. C., M. H. Green, J. Balmer Green, and L. A. Zech. 1990. Retinol metabolism in rats with low vitamin A status: a compartmental model. *J Lipid Res.* 31:1535-1548.
3. Wastney, M. E., W. E. H. Hall, and M. Berman. 1984. Ketone body kinetics in humans: a mathematical model. *J Lipid Res.* 25:160-173.
4. Wastney, M. E., J. Ng, D. Smith, B. R. Martin, M. Peacock, and C. M. Weaver. 1996. Differences in calcium kinetics between adolescent girls and young women. *Am. J. Physiol.* 271:R208-R216.

15

ERRORS IN COMPARTMENTAL MODELING

In this chapter we define error and discuss the types of error pertinent to compartmental modeling. The effects of many sources of error, or the propagation of errors, is discussed. We give methods provided in the compartmental modeling software package WinSAAM to incorporate error into the model and provide examples using these methods.

I. Introduction

Error can be defined as a deviation from some "true" value. In many experimental situations, the true value is unknown and can only be estimated with uncertainty, or error. In experimental science, errors fall into two categories, as defined by Mandel, 1964. The first includes mistakes resulting from departures from a prescribed experimental procedure and can encompass mistakes in the design as well as in the implementation of the experiment. The second category, which we will focus on, includes error that results from fluctuations in the conditions surrounding an experiment, for example, variability in temperature, measurement errors arising from the instruments and differences arising from measurements made by different technicians. Such errors may be due to random variation or to bias. Random errors may represent variability from subject to subject, within-subject variability, as well as other sources of error. Bias represents systematic error. Random errors affect the reproducibility, or precision, of an experiment. In this chapter we focus primarily on random, rather than systematic error.

The observed data may be viewed as representing deterministic or systematic components plus a stochastic component which has an element of randomness. The random components may represent variability from subject to subject, intra-individual variability, as well as other sources of error. (See Kendall and Buckland for a review of different types of error). Experimental error is any error in an experiment, whether due to random variation or to bias. This term typically refers to probabilistic variability over repeated experiments. A single experiment can be thought of as sample from an infinite number of such experiments; errors arise from sampling effects.

Consider an infinite number of measurements of the same quantity using the same measurement process. Let x be a single measurement and T be the "true" measurement. Let m be the mean of an infinite population of measurements of the same quantity using the same measurement process. Then $x - T$ represents total error and can be decomposed into two components: $x - T = (x - m) + (m - T)$,

The term *(x - m)* represents a random deviation of the measurement from its true mean, and the term *(m - T)* represents the deviation of the mean of the repeated measurements from the "true" measurement, or the bias. When there is no bias, *m* coincides with *T*. In practical terms, *T* represents a reference value; the existence of such a value is necessary to determine bias. A bias that varies day-to-day but averages to zero over a long period of time is called a "factor of precision."

Random errors affect the precision, or reproducibility of an experiment, while systematic errors affect the accuracy, or proximity to the true value. The term "repeatability" refers to within-run precision, while the term "reproducibility" refers to between-run precision of a method. In some cases, random errors are referred to as "indeterminate " errors, and systematic errors as "determinate" errors. A measurement process with a small standard deviation (e.g., small variability) has high precision, where "small" is relative and depends on the measurement process. Accuracy refers to the proximity of the true value and is a measure of the size of the total error. Accuracy typically refers to a measurement process rather than to a single measurement. Systematic errors affect the validity of a measurement. Repeating the experiment can reduce the error due to lack of accuracy but it will not reduce bias. However, the process of calibration can reduce bias to the extent that it becomes negligible.

We begin by discussing the sources of error that the modeler needs to take into account. Then we present the concept of error in statistics as applicable to compartmental modeling. Topics here include reliability and validity; direct measures of error including the standard error and the coefficient of variation, the use of residuals, and finally, tests of significance. We then turn to how errors are handled in WinSAAM. WinSAAM allows the user to provide statistical weights for the data, giving indications of lack of fit or uncertainty for the various components, and, through the iterative process, provides visual as well as quantitative measures for the modeler to use in assessing error associated with various parts of the model.

II. Sources of error in compartmental modeling

Several types of error can be distinguished in the process of compartmental modeling. First, the experiment is a source of two types of error: mistakes in design and execution, and measurement error. Second, the model itself is almost surely in error, as no model can perfectly imitate nature. Third, the modeler must choose which points to fit; he may err, choosing some wrong points. Closely allied to this source of error is lack of fit. We distinguish here between error due to lack of fit and error arising from metabolic differences which gives rise to between- and within-subject variability. While the compartmental model typically consists of a set of differential equations where there is no explicit error term, error can be taken into account in the modeling effort as explained below.

1. The experiment

The experiment is a source of many of the errors that the modeler faces; some he can take into account in his modeling, others he cannot. If the method of measurement has known error for the various types of data used in the model, this error can be assigned to the appropriate data, and used in building the model; this topic is treated below. The modeler can "model" the experiment, including mistakes in execution in the code describing the experiment (e.g., the text input file in WinSAAM). The more the modeler knows about the execution of the experiment, the more accurately will his model reflect the underlying kinetics.

The modeler is frequently presented with data that have already been collected and has had no say in the design of the experiment. In the absence of knowledge about a system, sample collections may have been scheduled at fixed intervals, or the schedule may have been based on convenience. Samples which were scheduled to be collected hourly may actually have been collected at intervals of *approximately* an hour. For example, a sample listed as being collected 30 minutes after dosing with a tracer may have been collected 45 minutes after dosing. This leads to indeterminacy about what processes are occurring and when they are occurring. The collection time should always be recorded when it occurred, rather than when it should have occurred.

2. Decisions Made by the Modeler

The modeler, faced with the data, must decide which points to fit: should he try to simulate the peaks and valleys he sees in his data? He may worry about fitting random noise in the data, especially if he has few data sets with which to work. Trial and error and growing familiarity with the data will help him decide which patterns in the data likely represent features of metabolism and which represent random error. Knowledge of the physiology of the experimental subject(s) can provide insight into what processes might be happening.

When multiple studies are available, averaging data for samples across studies can lead to mistakes in describing the kinetics. Such averaging has been done in kinetic studies to eliminate "noise" in the data. However, it can obscure or eliminate features that characterize the kinetics seen in most subjects, but at different times. A pool that appears and decays rapidly may be captured on the increase, on the decline, or both; it may well not be captured at its peak. Thus the experimental data may reflect different phases of a process that occurs in all subjects. In addition, processes can occur at different times in different subjects; a pool that appears 1 hour after dosing in one subject may appear in another subject after 2 hours. The timing of data collection may obscure or miss the appearance and disappearance of pools in different subjects. Data from each study should be fitted separately.

3. The Model

A third area of uncertainty for the modeler is model misspecification. All models are wrong, some are just more useful than others; there are infinitely many models that fit the data, and infinitely many models that don't: It is hard to find one of the former. Finally, the modeler may find results in the literature that disagree with his model. These may suggest that another model structure should be considered. On the other hand, those results may themselves be in error. In case of a disagreement, the assumptions underlying both the model and the findings from the literature should both be scrutinized in order to attempt to determine the source of the inconsistency.

A search of the literature should have been undertaken before beginning the construction of a compartmental model. The situation in which a model for the system under study already exists is discussed elsewhere in this book. Assuming that this is not the case, the model under development is compared to the findings of others with respect to consistency during the process of modeling. Lack of agreement with these findings may well indicate an error in the model. However, after checking the data and trying alternative models, an apparent inconsistency that persists may point to a new interpretation of existing data or even a new finding.

III. The concept of error in statistics

Error in statistics refers to measurement error and to sampling error. Measurement error is the extent to which repeated measurements on the same experimental subject disagree. Sampling error refers to variability in a parameter value due to calculation based on a single sample. If many such samples were drawn and the parameter recalculated many times, the standard error of the estimates would be an indication of the sampling error for that parameter. In an experiment, the error may be thought of as effects due to chance fluctuations in the subject and effects attributable to environmental conditions not controlled by the experiment, that is, all effects not attributable to the experiment.

1. Measures of error

Several measures of scatter are used in statistics. The variance can be estimated when there are two or more measurements of the same quantity. The variability of a statistic (e.g., a mean) is measured by the standard error (the square root of the variance), and is equal to the standard deviation divided by the square root of the number in the sample.

There are two components to the variance, denoted s^2, of a random sample: the sampling variance due to differences between members of the sample, and the measurement variance, which is due to random (or systematic) errors in measuring a sample. The concept of relative error, which facilitates comparisons of variability, is used in modeling. Coefficients of variation (CV) are common relative measures in which the unit of measurement is canceled by dividing by the mean. The element coefficient of variation is defined as the standard deviation divided by the mean. The coefficient of variation is sometimes called the relative standard deviation; in WinSAAM, it is called the fractional standard deviation (FSD).

2. Components of variance

Error is defined as variability remaining in the data after a model has been fitted. In compartmental modeling, error refers to the deviation from the common underlying model for each individual subject, after model parameters have been adjusted to fit his individual data. The error remaining after a model is fit to a set of data can reflect variability from many sources. Two major components of error are between-subject error and within-subject error. An experiment can reflect a variety of sources of error or components of error variance. For example, differences in assays from several laboratories can reflect differences in measurement technique among technicians, differences in instrument calibration, and differences in handling of samples.

When an experiment has been performed under the same conditions on each subject more than once, within-subject error can be estimated simply by taking the standard deviation of the (two or more) values of each parameter and testing to see if they are significantly different via a *t*-test. Error in this case is a type of sampling error, in which two samples are collected, and each contains some random noise. However, such repeated-measures study designs in which subjects are studied more than a single time are infrequent in kinetics studies because of the intensity of sample collection.

More frequently, when subjects are studied more than a single time, it is under different conditions, and a test of changes in parameter values due to the change in conditions is desired. An example of such a repeated measures design was a two-phase selenium tracer study in which subjects were studied once fasting and once non-fasting in order to determine the effects of fasting status on selenium kinetics (Patterson et al., 1989). The two phases were separated by a washout period and order of fasting was randomized. Because subjects were not studied under identical conditions, within-subject variability due to sampling could not be estimated. Instead, differences in parameter values between the two study phases were attributed to fasting status, although some of the difference was almost certainly due to sampling variability.

Because WinSAAM permits modeling an entire experiment, tests for parameter differences in the same subjects, but at different times, are possible. For each subject, both phases of this study were fitted in the same model file, thus simulating the entire experiment. Data from the first study were fitted first. Then data from the second study were fitted, initially using parameter values from the first phase. With a goal of an adequate fit to the data for both studies simultaneously, parameter values were left the same, adjusted to a new value adequate for both studies, or changed from the first study to the next. Attempts were made to change as few parameter values as possible. In order to test for significant differences, a "difference" parameter was created for those parameters that did change, (e.g., P(70) = P(5) - P(15), where the parameter was denoted P(5) in the first study, and P(15) in the second). The value of the "difference" parameter (e.g., P(70)) was asked for as a dependent parameter (e.g., limits were given) in time interrupt zero. After an adequate fit was achieved for both phases, the deck was solved iteratively to obtain FSD's for the adjustable parameters.

The significance of the "difference" parameters was determined by looking at their FSD's. A small FSD indicated the variability of the difference was small relative to the difference itself. For example, an FSD < 0.3 indicated that the variability of the difference was within one standard deviation of the value of the difference; in this study, this was taken to be not significantly different from zero. This method provided a test for parameters that changed significantly as a result of the change in fasting status.

3. Propagation of Errors

The propagation of errors refers to the way in which uncertainties are propagated, or carried over, from the data points to the parameters and how the uncertainties of the determination of parameters are carried over, or propagated, to the final results of the experiment (see Bevington, 1969, for a general discussion). The formulas that give the variance of a function of several random variables are called the law of propagation of errors. Both random and systematic errors can be propagated, and we discuss and give formulas for both. As before, capital letters represent true values of the random variables, while small letters represent their realizations, or measurements. We are assuming that the measured quantities are independent. If this is not the case, the covariance between them must be taken into account; formulas are given in Bevington.

Linear case: u is a linear combination of measured quantities x, y, and z, and a, b, and c are constants:

$$u = ax + by + cz + \cdots \tag{15.1}$$

For random error, the variance and standard deviation of u are, respectively,

$$V(u) = \sigma_u^2 = a^2\sigma_x^2 + b^2\sigma_y^2 + c^2\sigma_z^2 + \cdots \tag{15.2}$$

$$\sigma_u = \sqrt{a^2\sigma_x^2 + b^2\sigma_y^2 + c^2\sigma_z^2 + \ldots} \tag{15.3}$$

For systematic error, where the errors are,

$$\Delta u = a\Delta x + b\Delta y + c\Delta z + \cdots \tag{15.4}$$

Nonlinear case

A technique called linearization is frequently used to approximate the standard deviations of non-linear functions. This method, which is based on the Taylor expansion, makes the assumption that for a small change in the independent variable, the corresponding change in the function is approximately linear. Here only the first term which is linear in that expansion is used.

For a function $Y = f(X)$, consider a small change, or error d in X, with a corresponding change e in Y, where e is very small relative to X.

$$\delta = f(X+\varepsilon) - f(X) \tag{15.5}$$

$$\frac{f(X+\varepsilon) - f(X)}{\varepsilon} \approx \frac{df(X)}{dX} . \tag{15.6}$$

Substituting (15.5.) in the numerator of (15.6) and solving,

$$\delta = \frac{df(X)}{dX} \cdot \varepsilon = Y_X^{'} \cdot \varepsilon \tag{15.7}$$

$Y_X^{'}$ is the derivative of Y with respect to X, and is taken at or very close to x, the measured value of X.

Equation (15.5.) is in the form of a linear relationship between d and e, so that the variance is calculated as in (15.2.):

$$V(\delta) = \left(Y_X^{'}\right) \cdot \sigma_\varepsilon^2 \tag{15.8}$$

For the general case of several random variables, or measurements, let U be a function, or derived quantity, of several uncorrelated random variables,

$$U = f(X, Y, Z, \cdots) \tag{15.9}$$

$$\sigma_x^2 = \left(\frac{\partial f}{\partial X}\right)^2 \sigma_{e_1}^2 + \left(\frac{\partial f}{\partial Y}\right)^2 \sigma_{e_2}^2 + \left(\frac{\partial f}{\partial Z}\right)^2 \sigma_{\varepsilon_3}^2 + \cdots \tag{15.10}$$

In (15.10) the partial derivatives are taken at or close to $x, y, z, \ldots$, respectively.

We now consider some special cases:

When the measured variables are multiplicative,

$u = axy$.

In this case,

$$\sigma_u^2 = a^2 y^2 \sigma_x^2 + a^2 x^2 \sigma_y^2$$

Dividing through by $u^2 = a^2 x^2 y^2$ and rearranging terms, we obtain

$$\frac{\sigma_u^2}{u^2} = \frac{\sigma_x^2}{x^2} + \frac{\sigma_y^2}{y^2} \tag{15.11}$$

Note that in the case of a function that is purely multiplicative, the law of propagation of errors takes the simple form:

$$\left(CV_u\right)^2 = \left(CV_x\right)^2 + \left(CV_y\right)^2 . \tag{15.12}$$

This relationship also holds for functions of the form: $u = \frac{axy}{vz}$

$$\left(CV_u\right)^2 = \left(CV_x\right)^2 + \left(CV_y\right)^2 + \left(CV_v\right)^2 + \left(CV_z\right)^2 . \tag{15.13}$$

This is of particular interest in compartmental modeling as the CV can be entered as an error term for measured data. Note that this relationship implies the following: The relationship in (15.14) holds with the squares of the relative standard deviations:

$$\frac{\sigma_u}{u} = \sqrt{\left(\frac{\sigma_x}{x}\right)^2 + \left(\frac{\sigma_y}{y}\right)^2 + \left(\frac{\sigma_v}{v}\right)^2 + \left(\frac{\sigma_z}{z}\right)^2} . \tag{15.14}$$

If u is obtained by raising x to a power:

$$u = ax^{\pm m}, \quad \left(\frac{\partial u}{\partial x}\right) = \pm\, amx^{\pm m-1} = \pm \frac{mu}{x},$$

and the $CV_u = m\frac{\sigma_x}{x}$. (15.15)

For exponentials: $u = ae^{\pm mu}$,

$$\left(\frac{\partial u}{\partial x}\right) = \pm\, ame^{m} = \pm mu. \text{ Then } CV_u = m\sigma_x. \qquad (15.16)$$

For systematic error, relative standard deviations are again used:

$$\frac{\Delta u}{u} = \frac{\Delta x}{x} + \frac{\Delta y}{y} + \frac{\Delta v}{v} + \frac{\Delta z}{z} \qquad (15.17)$$

IV. How errors are handled in WinSAAM

In the following discussion, we make extensive use of the glucose-insulin model developed by Bergman et al.(1), and recoded for WinSAAM by Boston. The model is given first in section 1. In each of the following sections, we describe tools available in WinSAAM for assessing errors.

1. Glucose-insulin model

We now give the glucose-insulin model that will be used extensively below.

```
 1: A SAAM31 BERGMANS INSULIN SENSITIVITY MODEL USING SAAM
 2: 2          25
 3: H PAR
 4: C
 5: C P(1)=SG
 6: C P(3)/P(2)=SI
 7: C P(4)=GB
 8: C P(5)=IB
 9: C P(6)=GO
10: C P(7)=SI
11: C
12:     P(1)       2.748289E-02  0.000000E+00   1.000000E+02
13:     P(2)       4.869298E-02  0.000000E+00   1.000000E+02
14:     P(3)       2.667146E-05  0.000000E+00   1.000000E+02
15:     P(4)       9.000000E+01
16:     P(5)       3.500000E+00
17:     P(6)       2.820000E+02
18:  L(0,7)=P(2)
19:     UF(7)      1                                      8G 7
20:  P(7)=P(3)/P(2)
21: C
22: C COMPT. 6 = GLUCOSE
```

continued

```
23: C COMPT. 7 = REMOTE INSULIN
24: C COMPT. 8 = PLASMA INSULIN
25: C
26:  IC(6)=P(6)
27:  IC(7)=0
28: C
29: H DAT
30: X UF(6)=-(P(1)+F(7))*F(6)+P(1)*P(4)
31: X G(7) = P(3)*(FF(8)-P(5))
32: X FF(8)= F(8)
33:    P(7)
34: 106                                          WT=0.0
35:              2             178
36:              3             262
37: 106                                          FSD=.04
38:              4             270
39:              5             260
40:              6             255
41:              8             237
42:              10            220
43:              14            199
44:              16            199
45:              19            177
46:              22            165
47:              24            168
48:              25            168
49:              27            160
50:              30            139
51:              40            105
52:              50            84
53:              60            75
54:              70            75
55:              90            83
56:              110           86
57:              130           89
58:              155           89
59:              180           88
60: 108QL
61:              2             2.46
62:              3             43.33
63:              4             68.43
64:              5             59.17
65:              6             47.07
66:              8             30.
67:              10            24.27
68:              14            15.71
69:              16            18.71
70:              19            15.19
```

```
  71:              22            166.25
  72:              24            147.78
  73:              25            95.18
  74:              27            69.15
  75:              30            37.92
  76:              40            9.39
  77:              50            6.78
  78:              60            3.82
  79:              70            4.07
  80:              90            2.84
  81:              110           3.36
  82:              130           4.13
  83:              155           2.46
  84:              180           3.10

> deck
* DECK BEING PROCESSED
PRE-PROCESSING TIME :       4.510 SECS
> solv
*** MODEL CODE 10 SOLUTION
SOLUTION TIME :        .710 SEC
> plot p(6)
** NO PARTIALS FOUND IN COMMAND *  35
> swit conn on
* CONNECT THE FIGURES  (CONN ON )
> plot q(8)
```

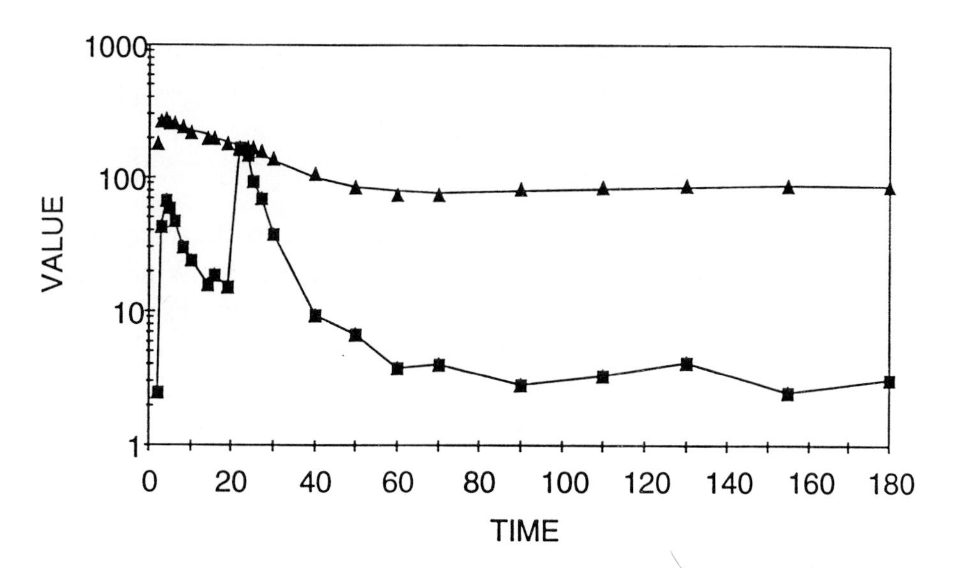

Fig 15.1 Glucose and insulin data: upper curve shows model passing through glucose data; lower curve shows insulin data connected by lines.

The upper curve shows the model fit to the glucose data, while the lower curve shows the appearance and disappearance of insulin in the plasma (Fig. 15.1).

2. Statistical Weights

The user can provide a measure of the variability of each data point in column 42 of the line where that data point is listed. This measure, called the "standard deviation" in WinSAAM can be the coefficient of variation, the standard deviation or some other measure of the confidence we have in a given data point. For example, we may have far greater confidence in plasma data than in data from the urine or feces, but have no other measure of the precision of that data. In this case, we assign a larger weight to the plasma data. For example, measures of selenium in the plasma will likely be more precise than measures of selenium in the urine or feces in part because incomplete collections of urine or feces would result in an underestimate of total Se excreted, and collections are likely to be at least marginally incomplete.

A measure of error can be assigned to each data point by the modeler before the analysis is undertaken. This measure of error, which is treated as a standard deviation, or "SD" in WinSAAM, is entered in field 5 under the heading "H DAT." Examples of such measures are:

SD = X where X is the standard deviation

FSD = X where X is the fractional standard deviation (coefficient of variation). The effect of this is to assign a weight of 1/X times the value of the datum. An example of assigning a weight (uncertainty) of 10% to each datum in a block would be: FSD = .1. When this field is left blank, Wt=1 is assumed by default.

RQO = X: a standard deviation proportional to the square root of the observation is assigned to the data. Note that the weight would be proportional to $\dfrac{1}{\left(x\sqrt{QO}\right)^2}$.

Using the SD, a weight *W(k)* that is proportional to the reciprocal of the variance (the standard deviation squared) is then calculated by WinSAAM. Each data point is then assigned a weight for that datum. When an observation =0, the weight for that observation is set to zero. An SD can be designated for each data point (observation) individually, or blocks of data can be weighted by entering the desired weight (in field 5) on the field modification card.

$W(k) \propto \dfrac{1}{SD(k)^2}$. In all cases, the weights are then normalized so that $\Sigma W(k) = N$.

When no SD's are provided for any of the data, all data points are given a weight of one. When the SD or FSD is known, it can be used; when it is unknown, the modeler can choose a hypothetical standard deviation that reflects his confidence in the data. For example, if he has far more confidence in plasma data than in urine data, he may give the first a very small estimated FSD, say

FSD=.001, and thus a large weight while he gives the urine data a larger FSD, say FSD=.1, and thus a smaller weight. The weights can be plotted: e.g., PLOT W(3) plots weights for data in component 3.

The types of weights used can have a profound effect on the fit, as shown in the following example:

```
 1: A SAAM31
 2: H PAR
 3:     L(0,1)      9.902604E-03  0.000000E+00    1.000000E+03
 4:     K(1)        9.958450E-01  0.000000E+00    1.000000E+02
 5:     IC(1)       1
 6: H DAT
 7: 101                                          FSD=.1
 8:                 1                .99
 9:                 2                .98
10:                 4                .96
11:                 5                .93
12:                 10               .91
13:                 14               .82
14:                 20               .71
15:                 25               0.2
```

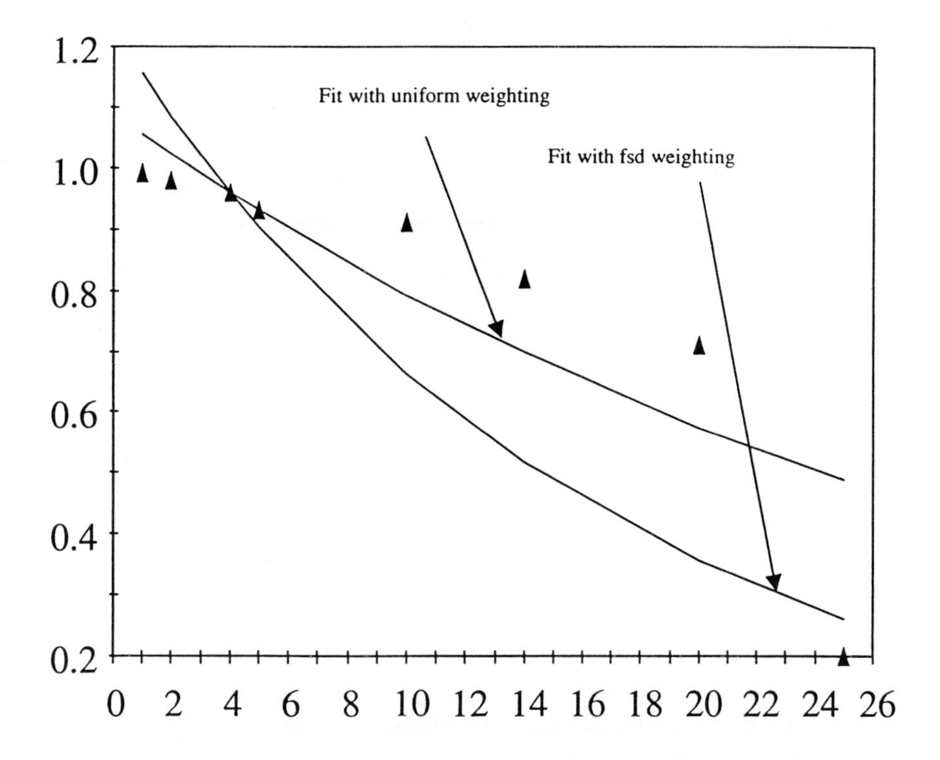

Fig 15.2 Plot illustrating the effects of using uniform vs. fsd weighting

This figure shows two lines, one giving the fit when fsd weighting is used. The other shows the fit when "FSD=.1" is changed to "Wt=1." Note the strong influence of the small value at time 25 when fsd weighting is used, compared to when all points are weighted equally.

In WinSAAM, the FSD of a parameter can be requested by typing: FSD(I); this can be accessed only after iteration has been performed.

```
> solv
*** MODEL CODE 10 SOLUTION
SOLUTION TIME :          .000 SECS
> iter
* CORRECTION VECTOR ESTIMATED

CONVERGENCE MEASURES
 IMPROVEMENT IN SUM OF SQUARES =       .00(%)
 FINAL VALUE OF CONAB =    6.515E+00
 LARGEST CHANGE (      .07 %) WAS IN PAR( 3, 0)

* CORRECTION VECTOR ESTIMATED

CONVERGENCE MEASURES
 IMPROVEMENT IN SUM OF SQUARES =       .00(%)
 FINAL VALUE OF CONAB =    0.000E+00
 LARGEST CHANGE (      .00 %) WAS IN PAR( 0, 0)

ITERATION TIME :          .270 SECS

DISTRIBUTION OF SQUARES
 COMP   SUM OF SQUARES
   6      4.7652E+02
> fsd(i)
* VALUES MAY NOT RELATE TO CURRENT PARAMETERS
* P ( 1)        2.749E-02      FSD( 1)    6.212E-02
* P ( 3)        2.665E-05      FSD( 2)    1.483E-01
* P ( 2)        4.869E-02      FSD( 3)    1.581E-01
```

3. Residuals

A residual, $e_i = Y_i - \hat{Y}_i$, which is a measure of unexplained variation, is the difference between the ith fitted and observed values. In least squares analysis, the technique used by WinSAAM, errors are assumed to be independent, with a mean of zero and a constant variance σ^2, and to follow a normal distribution. A listing and plot of the residuals for each component can reveal when these assumptions are violated. The distribution of residuals should appear to be a relatively uniform band. If residuals increase with time, creating a "funnel" shape, the variance is not constant, but increases with time. A weighted least squares analysis can be used to correct this problem when the variances of the individual points (σ_i^2 are known), or are of the form σ^2 / W_i, where the W_i are known. Residual plots exhibiting a trend suggest that the model may be missing some important feature. In WinSAAM, a listing and plots of the residuals against time are obtained by the commands: PRINT RES(.) and PLOT RES(.), where "." designates the component of interest. Returning to the glucose-insulin model, we have:

```
➢ print res(6)

*** NAME :      6
CURRENT KOMN
 #   COMP  TC   CATEGORY         T            RESIDS.
  2    6    0   F ( 6)      2.000E+00   -9.373E+01
  4    6    0   F ( 6)      3.000E+00   -4.769E+00
  6    6    0   F ( 6)      4.000E+00    8.459E+00
  8    6    0   F ( 6)      5.000E+00    3.899E+00
 10    6    0   F ( 6)      6.000E+00    4.498E+00
 12    6    0   F ( 6)      8.000E+00   -2.225E+00
 14    6    0   F ( 6)      1.000E+01   -8.258E+00
 16    6    0   F ( 6)      1.400E+01   -8.994E+00
 18    6    0   F ( 6)      1.600E+01    1.738E-01
 20    6    0   F ( 6)      1.900E+01   -9.182E+00
 22    6    0   F ( 6)      2.200E+01   -8.826E+00
 24    6    0   F ( 6)      2.400E+01    4.025E+00
 26    6    0   F ( 6)      2.500E+01    9.323E+00
 28    6    0   F ( 6)      2.700E+01    1.179E+01
 30    6    0   F ( 6)      3.000E+01    5.154E+00
 32    6    0   F ( 6)      4.000E+01    3.724E+00
 34    6    0   F ( 6)      5.000E+01   -2.106E+00
 36    6    0   F ( 6)      6.000E+01   -4.774E+00
 38    6    0   F ( 6)      7.000E+01   -2.866E+00
 40    6    0   F ( 6)      9.000E+01    3.766E+00
 42    6    0   F ( 6)      1.100E+02    3.692E+00
 44    6    0   F ( 6)      1.300E+02    4.062E+00
 46    6    0   F ( 6)      1.550E+02    1.933E+00
 48    6    0   F ( 6)      1.800E+02   -4.089E-01
> plot res(6)
> plot q(6)
```

We scan the residuals to look for exceptionally large values. Here, none stand out, with the possible exception of the 2 minute time point; this point has been noted and is unweighted (i.e., for this point, "wt=0" has been placed in field 4).

Here the model fit was plotted along with the residuals as an aid linking the residuals visually with the corresponding points; they can be plotted alone, according to personal preference. The distribution of residuals should appear to be a relatively uniform band . Residuals increasing with time creating a "funnel" shape, suggest that the variance is not constant, but increases with time. A

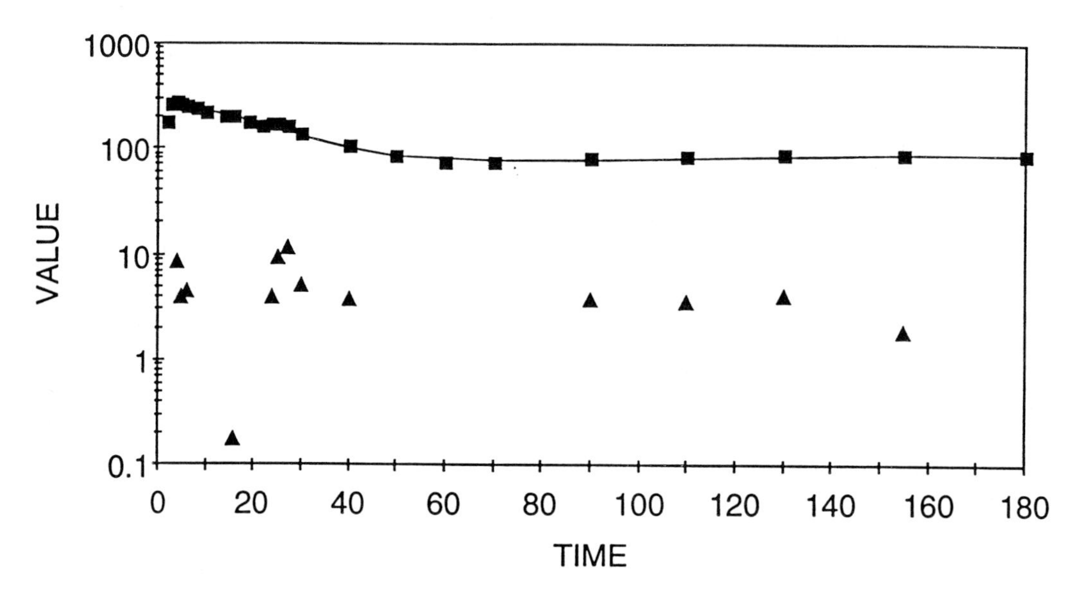

Fig 15.3 Plot of glucose values, model fit, and residuals.

weighted least squares analysis can be used to correct this problem when the variances of the individual points (σ_i^2 are known), or are of the form σ^2 / W_i, where the W_i are known.

4. SS(I):

SS(I), or the sums of squares for error (i.e., the sums of the squared residuals) for the I'th component can be used to assess error (fit) in different components. Typing "SS(I)" (with I not specified) will produce a list of the sums of squares for all components in the deck.

```
> ss(i)
SS( 6)   4.7654E+02
```

In problems where there are several components, the size of the sums of squares can be compared. Large values or some sums of squares relative to others suggest that the fit of that component is not as good as that for the others. Sometimes the fit can be improved, sometimes not.

5. Critical points

Some data points may be regarded as being particularly accurately measured, or of such import that the calculated values must coincide with these values. These can be specified as "critical." Individual values or blocks of data can be so designated. If only these points are to be used as "weighted data," in calculating the normal equations, this option can be specified on line 4 of the WinSAAM problem input file. In the output, estimates of the standard deviations and fractional standard deviations are given for all adjustable parameters, along with the matrix of correlation coefficients.

If the user has an *a priori* measure of the uncertainty of a parameter, he can use this estimate to "hold" the parameter near the initial estimate if it cannot be resolved from the data. This measure of uncertainty is entered in field 7 (character space 62) of the line on which the parameter is defined.

6. Partial derivatives

Partials derivatives indicate the instantaneous rate of change of a calculated value with respect to a parameter vs.time. To obtain partials, we enter the command "part."

```
> part
* PARTIALS ESTIMATED
PARTIALS TIME :     2.090 SECS
> resi
CORRECTION VECTOR TIME :     .110 SECS
> plot p(6) p(1) p(2) p(3)
> adju
PARAMETER   VALUE   LOW-LIMIT  HI-LIMIT
P ( 1)   2.7483E-02 0.0000E+00 1.0000E+02
P ( 3)   2.6670E-05 0.0000E+00 1.0000E+02
P ( 2)   4.8693E-02 0.0000E+00 1.0000E+02
> deck
UPDATE?
* DECK BEING PROCESSED
PRE-PROCESSING TIME :     9.950 SECS
> solv
*** MODEL CODE 10 SOLUTION
SOLUTION TIME :     .710 SECS
> iter
* PARTIALS ESTIMATED
* CORRECTION VECTOR ESTIMATED

CONVERGENCE MEASURES
 IMPROVEMENT IN SUM OF SQUARES =     .00(%)
 FINAL VALUE OF CONAB =  0.000E+00
 LARGEST CHANGE (     .00 %) WAS IN PAR( 0, 0)

** 2ND ITERATION ABORTED BECAUSE YOU ARE EITHER TOO CLOSE OR
   TOO FAR FROM A FIT **
ITERATION TIME :     1.200 SECS

DISTRIBUTION OF SQUARES
 COMP  SUM OF SQUARES
   6    4.7654E+02
```

We print the partials for p(1), p(2), and p(3) with respect to component (6), and compare their relative magnitudes. (Note that the "p" associated with "6" indicates partials, whereas the "p" associated with (2) and (3) refers to the parameters p(2) and p(3).

```
> print p(6) p(1),p(2),p(3)
*** NAME :    6
CURRENT KOMN
WITH RESPECT TO: P ( 1)
 #   COMP TC   CATEGORY          T          PARTIAL
  2    6   0   F ( 6)      2.000E+00   -3.633E+02
  4    6   0   F ( 6)      3.000E+00   -5.301E+02
  6    6   0   F ( 6)      4.000E+00   -6.865E+02
  8    6   0   F ( 6)      5.000E+00   -8.317E+02
 10    6   0   F ( 6)      6.000E+00   -9.660E+02
 12    6   0   F ( 6)      8.000E+00   -1.203E+03
 14    6   0   F ( 6)      1.000E+01   -1.401E+03
 16    6   0   F ( 6)      1.400E+01   -1.695E+03
 18    6   0   F ( 6)      1.600E+01   -1.799E+03
 20    6   0   F ( 6)      1.900E+01   -1.911E+03
 22    6   0   F ( 6)      2.200E+01   -1.966E+03
 24    6   0   F ( 6)      2.400E+01   -1.951E+03
 26    6   0   F ( 6)      2.500E+01   -1.929E+03
 28    6   0   F ( 6)      2.700E+01   -1.864E+03
 30    6   0   F ( 6)      3.000E+01   -1.736E+03
 32    6   0   F ( 6)      4.000E+01   -1.251E+03
 34    6   0   F ( 6)      5.000E+01   -8.296E+02
 36    6   0   F ( 6)      6.000E+01   -5.031E+02
 38    6   0   F ( 6)      7.000E+01   -2.601E+02
 40    6   0   F ( 6)      9.000E+01    3.512E+01
 42    6   0   F ( 6)      1.100E+02    1.578E+02
 44    6   0   F ( 6)      1.300E+02    1.842E+02
 46    6   0   F ( 6)      1.550E+02    1.599E+02
 48    6   0   F ( 6)      1.800E+02    1.185E+02
CURRENT KOMN WITH RESPECT TO: P ( 2)
 #   COMP TC   CATEGORY          T          PARTIAL
  2    6   0   F ( 6)      2.000E+00    0.000E+00
  4    6   0   F ( 6)      3.000E+00    0.000E+00
  6    6   0   F ( 6)      4.000E+00    1.003E+00
  8    6   0   F ( 6)      5.000E+00    3.447E+00
 10    6   0   F ( 6)      6.000E+00    6.487E+00
 12    6   0   F ( 6)      8.000E+00    1.507E+01
 14    6   0   F ( 6)      1.000E+01    2.720E+01
 16    6   0   F ( 6)      1.400E+01    5.970E+01
 18    6   0   F ( 6)      1.600E+01    7.862E+01
 20    6   0   F ( 6)      1.900E+01    1.084E+02
 22    6   0   F ( 6)      2.200E+01    1.385E+02
 24    6   0   F ( 6)      2.400E+01    1.593E+02
 26    6   0   F ( 6)      2.500E+01    1.707E+02
 28    6   0   F ( 6)      2.700E+01    1.961E+02
 30    6   0   F ( 6)      3.000E+01    2.392E+02
```

```
    32    6    0   F ( 6)     4.000E+01    3.821E+02
    34    6    0   F ( 6)     5.000E+01    4.785E+02
    36    6    0   F ( 6)     6.000E+01    5.240E+02
    38    6    0   F ( 6)     7.000E+01    5.313E+02
    40    6    0   F ( 6)     9.000E+01    4.718E+02
    42    6    0   F ( 6)     1.100E+02    3.673E+02
    44    6    0   F ( 6)     1.300E+02    2.611E+02
    46    6    0   F ( 6)     1.550E+02    1.582E+02
    48    6    0   F ( 6)     1.800E+02    9.059E+01
  CURRENT KOMN WITH RESPECT TO: P ( 3)
   #   COMP TC   CATEGORY        T          PARTIAL
     2    6    0   F ( 6)     2.000E+00    0.000E+00
     4    6    0   F ( 6)     3.000E+00   -1.259E+03
     6    6    0   F ( 6)     4.000E+00   -1.774E+04
     8    6    0   F ( 6)     5.000E+00   -4.680E+04
    10    6    0   F ( 6)     6.000E+00   -8.627E+04
    12    6    0   F ( 6)     8.000E+00   -1.811E+05
    14    6    0   F ( 6)     1.000E+01   -2.804E+05
    16    6    0   F ( 6)     1.400E+01   -4.609E+05
    18    6    0   F ( 6)     1.600E+01   -5.364E+05
    20    6    0   F ( 6)     1.900E+01   -6.327E+05
    22    6    0   F ( 6)     2.200E+01   -7.422E+05
    24    6    0   F ( 6)     2.400E+01   -8.775E+05
    26    6    0   F ( 6)     2.500E+01   -9.583E+05
    28    6    0   F ( 6)     2.700E+01   -1.115E+06
    30    6    0   F ( 6)     3.000E+01   -1.307E+06
    32    6    0   F ( 6)     4.000E+01   -1.538E+06
    34    6    0   F ( 6)     5.000E+01   -1.433E+06
    36    6    0   F ( 6)     6.000E+01   -1.241E+06
    38    6    0   F ( 6)     7.000E+01   -1.043E+06
    40    6    0   F ( 6)     9.000E+01   -7.000E+05
    42    6    0   F ( 6)     1.100E+02   -4.482E+05
    44    6    0   F ( 6)     1.300E+02   -2.792E+05
    46    6    0   F ( 6)     1.550E+02   -1.540E+05
    48    6    0   F ( 6)     1.800E+02   -8.144E+04
```

The partials are given with respect to the time points available for the given component. It is a good idea to print the partials before plotting them. If the partials are negative (i.e., the rate of change is decreasing), they will not be plotted on the default plot, whose y axis is in logarithmic units and thus accommodates only positive values. To deal with this situation, plot the partials, (you will see nothing on the plot for the negative values), then choose "Plot options" on the "View" menu. Choose "axes" then "Y". You will see that "logarithmic" is selected; click that box to unselect it. Upon choosing "okay" you will see a plot showing your partials.

We notice that the partials for p(1) and p(2) range from -363 to +118, so that they can be plotted together, while those for p(3) are very small, so that they must be plotted separately. Those for p(1) and p(2) take on both negative and positive values, so the y scale must be linear, not logarithmic. Similarly, those for p(3) are all negative so that they too must be plotted on a linear scale. The 2 plots are shown below.

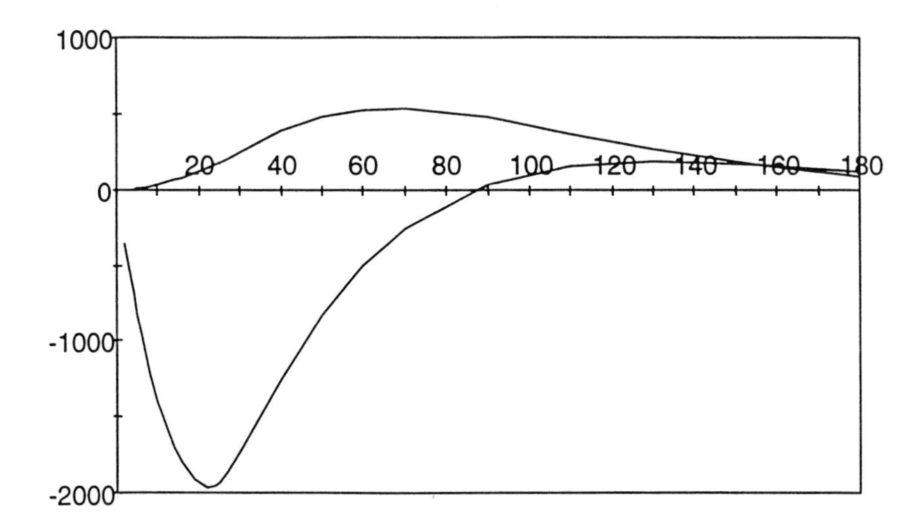

Fig 15.4. Graph of partial derivatives of component (6) relative to parameters P(1) and P(2) from issuing the > plot p(6) p(1) p(2) command.

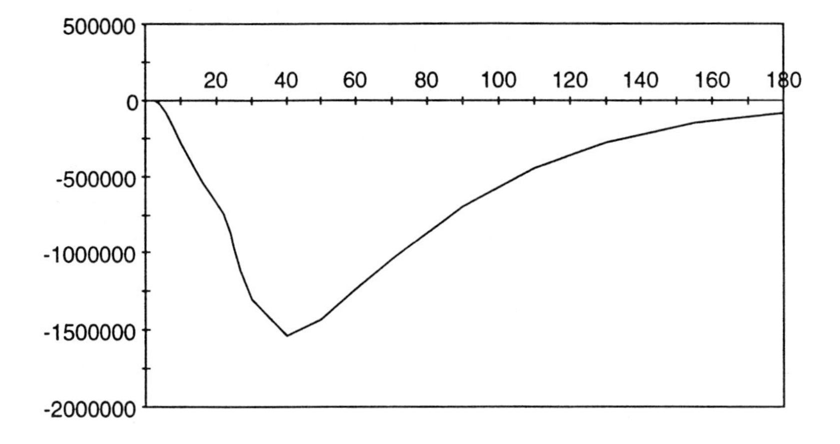

Fig 15.5 Graph of Partial Derivative of Component (6) Relative to P(3) from issuing the > plot p(6) p(3) command

7. Correlation Analysis

The covariance matrix of the adjustable parameters can be printed typing a "4" in column 1 and a "3" in column 2 on line 4, the print options line. Extremely large variances indicate problems in the solution, such as non-convergence, or estimates hitting boundaries or general identifiability problems. In some cases, checking the starting values will help. Here we give the correlation matrix, which is obtained from batch WinSAAM (e.g., type "SAAM" after typing "deck" and "solv").

```
ADJUSTABLE PARAMETERS
 1  P( 1, 0)=  2.7482890E-02 ...
 2  P( 3, 0)=  2.6671460E-05 ...
 3  P( 2, 0)=  4.8692980E-02 ...

CORRELATION MATRIX
 COLUMN  1    2    3
ROW 1  1.00  -.87  -.78
ROW 2  -.87  1.00  .96
ROW 3  -.78  .96  1.00
```

The correlation matrix is indexed by (row/column) numbers corresponding to parameters; the numbers and their corresponding parameters are given under "adjustable parameters (we have abbreviated the output in the interest of readability)." We notice that the 3 adjustable parameters in this example are highly correlated. As a rule of thumb, the modeler should examine correlations > 0.90 to decide if both of these highly correlated parameters are necessary to provide an adequate fit to the data. He might try fitting the data fixing one of these parameters and allowing the other to adjust.

V. Tests of significance and inference

Tests of significance are only available for parameter values; there is no formal test of the model correctness. When multiple studies are combined using model code 11 (see Chapter 13) for a population analysis, standard errors are available for testing the statistical significance of the model parameters. In most cases, the number of studies is small, so that tests should refer to the t distribution, with the degrees of freedom equal to one fewer than the number of studies. Tests for equality of parameter values between groups, (e.g., $m_1 = m_2$ for two groups), tests that a parameter is equal to a given value (e.g., $m = m_0$), tests that a parameter equals zero (e.g., $m_0 = 0$) can be performed.

WinSAAM also provides an indication of the inherent variability for each adjustable parameter for each subject. The standard deviation and the FSD are provided in the WinSAAM output when iterations are requested. These measures are based on the perturbations of the parameters in the "correction vector" calculated repeatedly during the iterative process, and provide a measure of the variability of that parameter as well as a test of whether it is significantly different from zero.

Inference in compartmental modeling is related to the model structure (e.g., does a pool exist?) and to its parameters (e.g., is a parameter value significantly different from zero? Does its value differ from that of the same parameter in the same system observed under other conditions?). The calculation of a 95% confidence interval about a parameter value can answer the latter questions. If the confidence interval about a parameter value excludes zero, by inference, the parameter is established. If the 95% confidence intervals for two parameter values do not overlap, this suggests that the values are different.

Feldman (1983) suggests the extra sum-of-squares principle as tool for making inferences on model structure (see Draper and Smith, 1981 for a more complete discussion). Using this method, the sums of squares for the more complex model are compared with those for the simpler model:

$$F = \frac{\Delta SS \,/\, \Delta df}{SS_{\min} \,/\, df_{\min}}$$

where $SS = \Sigma(\text{observed - predicted})^2$ and the degrees of freedom, df = data points - fitted parameters, and the F value is referred to the F tables. The larger model has a tighter fit, and therefore a smaller SS than does the simpler model. A small value of F would indicate that the simpler model would be adequate. Note that the SS_{min} are those from the more complex model, which requires A note of caution, however, is that this test may be invalid when the simpler model is obtained by setting some parameter(s) in the more complex model to zero as in this case, where a parameter is set to a boundary value, the numerator of the F ratio may not be distributed as a chi-square, as required for a valid F test.

The modeler should be aware of problems that arise when multiple tests are performed on data from a given model. When multiple parameters are tested say, each at a significance level of p = 0.05, the overall significance level increases as the number of tests increases. Thus the probability of one or more type I errors (rejecting the null hypothesis when it is false) is $1 - (1\text{-}p)^C$, where C is the number of tests, a result due to Bonferroni. When two tests are conducted, the probability of a type one error increases to 10%; for ten tests, this probability rises to 40%. To control the error rate, the overall significance level can be divided by the number of tests, i.e., each parameter is tested at the α /C level. If the overall significance level was 0.05 and the number of parameters to be tested was two, each would be tested at 0.025; if ten parameters were to be tested, the significance level for each test would be 0.005. In practice, the multiple testing problem is often disregarded; it should however be acknowledged.

References

1. Bergman, R. N. and Bowden, C.R. (1981) the minimal model approach to quantification of factors controlling glucose disposal in man. In: Carbohydrate Metabolism. C. Cobelli and R. N. Bergman (editors), Wiley Press, Pp. 269-296.
2. Bevington, Philip R. (1969). *Data Reduction and Error Analysis for the Physical Sciences.* McGraw-Hill, New York.
3. Crocker, L. and Algina, J. (1986). Introduction to Classical & Modern Test Theory. Holt, Rinehart and Winston, Inc, Fort Worth.
4. Draper, N. R., and Smith, H. (1981). Applied Regression Analysis. John Wiley & Sons, New York.
5. Feldman, H. A. (1983). Compartmental Models and Direct Estimation in (G. Forti and D. Rodbard, eds.) Computers in Endocrinology. Raven Press, New York.
6. Mandel, J. (1964). The Statistical Analysis of Experimental Data. Interscience Publishers (a division of John Wiley & Sons) New York.
7. Kendall, M. G., and Buckland, W. R. (1957) A Dictionary of Statistical Terms. Hafner, London.
8. Patterson, B. H., Levander, O. A., Helzlsouer, K., McAdam, P. A., Lewis, S. A., Taylor, P. R., Veillon, C. and Zech, L. A. (1989). Human selenite metabolism: a kinetic model. Am. J. Physiol. 257 (Regulatory Integrative Comp. Physiol. 26): R556-R567.

16

TESTING ROBUSTNESS: SENSITIVITY, IDENTIFIABILITY, AND STABILITY

This chapter discusses four tools to test the robustness of a model. Sensitivity refers to the relative influence of parameters on a solution. Identifiability refers to the uniqueness of the model and of its parameters. Estimability is the property of a data fitting process with respect to returning the same estimates of parameters more or less independently of starting points for estimation and noise in the data. Stability is the behavior of a system with respect to a perturbation. We begin with a mathematical description of these terms, then describe how they are obtained in WinSAAM and discuss their use. We illustrate the concepts described using examples.

I. Sensitivity

Sensitivity refers to calculations and analyses performed to describe the relative dependency of the model state variables on the model parameters. From a practical perspective, it represents calculations which reveal the impact of assumptions made in conjunction with model development. We discuss three types of sensitivity: point sensitivity (also called the partial derivative), relative sensitivity, and overall sensitivity.

Consider a model:

$$y = y(t,\theta) \tag{16.1}$$

where θ is a vector of parameters, $\theta_1,\ldots,\theta_l$. In WinSAAM, the vector would consist, for example, of all the L(j,i), k(j) and P(i), DT(i), and other adjustable parameters used to define the model.

The relationship between the state variables and point sensitivity can be shown using the first two terms in the Taylor expansion. The solution to (16.1) for any value of θ close to some value θ_0, is approximately given by:

$$y = y(t,\theta_0) + \sum \frac{\partial y}{\partial \theta}\partial\theta + \mathrm{K} \tag{16.2}$$

or

$$y = y(t,\theta_0) + G^T \cdot \partial\theta + \mathrm{K} \tag{16.3}$$

G is the partial derivative matrix of y with respect to θ. An element of G (the partial derivative of the solution taken with respect to some θ) is also referred to as the sensitivity of y with respect to θ, or the point sensitivity. While sensitivity is with respect to a particular parameter, θ_i , affecting a specific response, say y_j , in the discussion we suppress the subscripts "i and j" in the interest of simplicity. Note that the values of the elements of G are influenced by the 'actual' parameter values (i.e. the sensitivity of y with respect to θ, as defined here, is not just dependent on t and the structure of the model, but also on the actual values of θ). An alternate scale-free sensitivity, relative sensitivity S, is:

$$S = \frac{\theta}{y} G \tag{16.4}$$

or

$$S = \frac{\partial \ln(y)}{\partial \ln(\theta)} \tag{16.5}$$

To represent the *overall* sensitivity of a state variable (as opposed to point sensitivity) to a parameter over some time interval we use IG, where overall sensitivity with respect to some θ_i is across an interval 0 to T.

$$IG = \int_0^T \frac{\partial y}{\partial \theta} dt \tag{16.6}$$

If weights are used in the data fitting process,

$$IG = \int_0^T w(t) \left(\frac{\partial y}{\partial \theta} \right) dt \tag{16.7}$$

Example 1: One Compartment Model
Consider a one compartment model with a loss pathway, L, then:

$$\begin{aligned} y(t,\theta) &= e^{-Lt} \\ G(t) &= -t \cdot e^{-Lt} \\ S(t) &= -L \cdot t \\ IG_L &= \left(e^{-Lt}\left(L \cdot T + 1\right) - 1\right) / L^2 \\ &= \frac{y(T)\left(1 - S(T)\right) - 1}{L^2} \end{aligned} \tag{16.8}$$

WinSAAM provides both the point sensitivity and the relative sensitivity; these can be used in the model fitting process. The partials matrix is available either in conjunction with the first stage of iterative model fitting (ITER) or in association with the PARTials command. If the partials were computed in the process of refining parameters, they may no longer reflect the *current* value of the parameter. Recall that the point sensitivity (indeed all sensitivity measures) is a function of the parameter value. WinSAAM blocks access to the relative sensitivity if it detects this 'misalignment' with the following message, which appears after the "PART" command is issued:

```
** CURRENT SENSITIVITIES ARE UNAVAILABLE.
 TO GENERATE SENSITIVITIES EITHER USE THE PARTIALS COMMAND
OR RESTORE THE SOLUTION PRIOR TO THE LAST ITERATION *  37  ".
```

In WinSAAM, the partials matrix is calculated with the command "PART." The partials can be requested using the command for example, for component 4 and parameter L(2,3): "p(4) L(2,3)." The element S(I,J) is the fractional change in the ith calculated datum QC(I) given a fractional change in the Jth adjustable parameter x(J):

S(I,J) = (dQC(I)/QC(I))/(dx(J)x(J))

There is additional information on sensitivities in the Win SAAM "Help" file.

The overall sensitivity can be calculated in WinSAAM as demonstrated in the following example.

Example 2: Chemical Reaction

Consider the bi-directional chemical reaction (Daniels and Alberty, 1960):

$$CH_3COOC_2H_5 + OH \leftrightarrow CH_3COO + C_2H_5OH$$

Data were collected on hydroxide removal following the introduction of hydroxide into an ethyl acetate solution. A question regarding the reverse process, the reaction of acetate and ethanol to reform ethyl acetate and hydroxide is considered. In particular, we want to find out if setting the reverse reaction rate constant to some arbitrary very small number influences our estimation of the forward rate constant from the data. The WinSAAM input file is given in Table 16.1.

Table 16.1 WinSAAM input file for sensitivity calculation

```
   1: A SAAM31
   2: C
   3: C ANALYSIS OF A BI-DIRECTIONAL BIMOLECULAR REACTION
   4: C
   5: C CH3COOC2H5 + OH <==> CH3COO + C2H5OH
   6: C
   7: C F(1) = CH3COOC2H5
   8: C F(2) = OH
   9: C F(3) = CH3COO
  10: C F(4) = C2H5OH
  11: C
  12: H PAR
  13:     P(1)       1.055168E-01  0.000000E+00   1.000000E+02
  14: C
  15: C THE REVERSE REACTION IS TOO SLOW TO OBSERVE HERE
  16: C THUS THE RATE CONSTANT IS SIMPLY SET
  17: C
  18: C P(2)=0.0001
  19: C
  20: C FREE UP P(2) TO ALLOW DETERMINATION OF ITS PARTIAL
COMPONENT
  21: C
  22:     P(2)      .0001                            1000
  23: C
  24:   IC(1)=.00486
  25:   IC(2)=.00980
  26:   IC(3)=0
  27:   IC(4)=0
  28: H DAT
  29: X UF(1)=P(2)*F(3)*F(4)-P(1)*F(1)*F(2)
  30: X UF(2)=P(2)*F(3)*F(4)-P(1)*F(1)*F(2)
  31: X UF(3)=P(1)*F(1)*F(2)-P(2)*F(3)*F(4)
  32: X UF(4)=P(1)*F(1)*F(2)-P(2)*F(3)*F(4)
  33: C
  37: C NOTE THAT: 1) F(3) = F(4) AND
  38: C            2) IC(1)-F(1) = IC(2)-F(2)
  39: C
  48: 103                                          SD=.0001
  49:             0             0
  50:             178           .00088
  51:             273           .00116
  52:             531           .00188
  53:             866           .00256
  54:             1510          .00335
  55:             1918          .00377
  56:             2401          .00406
```

In order to estimate the partials, we issue the following commands:

```
> deck
* EDIT BUFFER BEING WRITTEN INTO DECK
* DECK BEING PROCESSED
*** L(0,1) ADDED AS A DUMMY.*3*
PRE-PROCESSING TIME :          1.533 SECS
* KOMN GENERATED BY DECK DIFFERS FROM PREVIOUS CORE KOMN
> solv
*** MODEL CODE 10 SOLUTION
SOLUTION TIME :          0.167 SECS
> part
* PARTIALS ESTIMATED
PARTIALS TIME :          0.283 SECS
```

and plot the partials matrix for parameters p(1), and p(2) with the result shown in Fig 16.1.

> Plot P(3)p(1),p(2) <== P(3) means Partials for component '3'

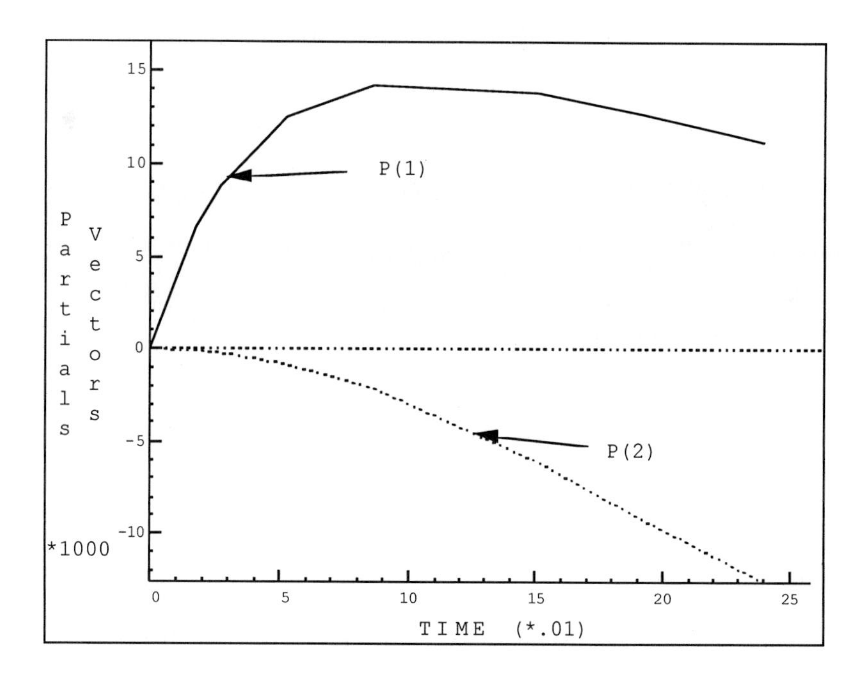

Fig 16.1 Plot of partials for component 3 with respect to parameters P(1) and P(2).

We can see that the partials of component 3 with respect to p(1) seem considerably larger than those with respect to p(2) Printing the values may be more useful.

```
> prin p(3)p(1),p(2)
----------------------------------------------
*** NAME :     3
CURRENT KOMN
WITH RESPECT TO: P ( 1)
  #  COMP TC  CATEGORY          T           PARTIAL
   3    3   0  F ( 3)       0.000E+00    0.000E+00
   6    3   0  F ( 3)       1.780E+02    6.531E-03
  10    3   0  F ( 3)       2.730E+02    8.810E-03
  18    3   0  F ( 3)       5.310E+02    1.248E-02
  26    3   0  F ( 3)       8.660E+02    1.419E-02
  42    3   0  F ( 3)       1.510E+03    1.379E-02
  52    3   0  F ( 3)       1.918E+03    1.268E-02
  64    3   0  F ( 3)       2.401E+03    1.119E-02
CURRENT KOMN
WITH RESPECT TO: P ( 2)
  #  COMP TC  CATEGORY          T           PARTIAL
   3    3   0  F ( 3)       0.000E+00    0.000E+00
   6    3   0  F ( 3)       1.780E+02   -5.821E-05
  10    3   0  F ( 3)       2.730E+02   -2.328E-04
  18    3   0  F ( 3)       5.310E+02   -8.149E-04
  26    3   0  F ( 3)       8.660E+02   -2.095E-03
  42    3   0  F ( 3)       1.510E+03   -6.054E-03
  52    3   0  F ( 3)       1.918E+03   -9.080E-03
  64    3   0  F ( 3)       2.401E+03   -1.257E-02
```

This confirms our suspicion that the influence of this assumption increases with time across the experiment.

The relative sensitivity values for p(2) are too small to plot on the same figure as for p(1). Further, they cannot be plotted on a logarithmic plot (at least directly) as they are negative. The values are as follows:

```
> prin s(3)p(1),p(2)     <== s denotes sensitivity
-------------------------------------------------------
*** NAME :     3
CURRENT KOMN
WITH RESPECT TO: P ( 1)
  #  COMP TC  CATEGORY          T           SENS.
   3    3   0  F ( 3)       0.000E+00    9.999E+30
   6    3   0  F ( 3)       1.780E+02    8.769E-01
  10    3   0  F ( 3)       2.730E+02    8.218E-01
  18    3   0  F ( 3)       5.310E+02    6.991E-01
  26    3   0  F ( 3)       8.660E+02    5.804E-01
  42    3   0  F ( 3)       1.510E+03    4.254E-01
  52    3   0  F ( 3)       1.918E+03    3.563E-01
  64    3   0  F ( 3)       2.401E+03    2.921E-01
CURRENT KOMN
WITH RESPECT TO: P ( 2)
```

```
 #   COMP  TC   CATEGORY          T           SENS.
  3    3    0   F ( 3)    0.000E+00    9.999E+30
  6    3    0   F ( 3)    1.780E+02   -7.407E-06
 10    3    0   F ( 3)    2.730E+02   -2.058E-05
 18    3    0   F ( 3)    5.310E+02   -4.327E-05
 26    3    0   F ( 3)    8.660E+02   -8.121E-05
 42    3    0   F ( 3)    1.510E+03   -1.770E-04
 52    3    0   F ( 3)    1.918E+03   -2.417E-04
 64    3    0   F ( 3)    2.401E+03   -3.110E-04
```

We see here that, in a relative sense, the dependency of our final estimates on p(2) is 3 to 4 orders of magnitude less than that for p(1). (Note that the sensitivity at time zero approaches infinity and should be ignored.)

Next we increase p(2) by several orders of magnitude to see if the picture changes:

```
> p(i)
P( 2, 0) 1.0000E-04 A
P( 1, 0) 1.0552E-01 A
> p(2)=.1
> solv
*** MODEL CODE 10 SOLUTION
SOLUTION TIME :     0.183 SECS
> part
* PARTIALS ESTIMATED
PARTIALS TIME :    0.267 SECS
```

Then we print and examine the partials

```
> prin p(3)p(1),p(2)
-----------------------------------------------------------------
*** NAME :      3
CURRENT KOMN
WITH RESPECT TO: P ( 1)
 #   COMP  TC   CATEGORY          T          PARTIAL
  3    3    0   F ( 3)    0.000E+00    0.000E+00
  6    3    0   F ( 3)    1.780E+02    6.470E-03
 10    3    0   F ( 3)    2.730E+02    8.632E-03
 18    3    0   F ( 3)    5.310E+02    1.170E-02
 26    3    0   F ( 3)    8.660E+02    1.240E-02
 42    3    0   F ( 3)    1.510E+03    1.068E-02
 52    3    0   F ( 3)    1.918E+03    9.462E-03
 64    3    0   F ( 3)    2.401E+03    8.385E-03
CURRENT KOMN
WITH RESPECT TO: P ( 2)
 #   COMP  TC   CATEGORY          T          PARTIAL
  3    3    0   F ( 3)    0.000E+00    0.000E+00
  6    3    0   F ( 3)    1.780E+02   -3.644E-05
 10    3    0   F ( 3)    2.730E+02   -1.136E-04
 18    3    0   F ( 3)    5.310E+02   -5.760E-04
 26    3    0   F ( 3)    8.660E+02   -1.581E-03
 42    3    0   F ( 3)    1.510E+03   -3.748E-03
 52    3    0   F ( 3)    1.918E+03   -4.841E-03
 64    3    0   F ( 3)    2.401E+03   -5.757E-03
```

Next, we print and examine the sensitivities of component 3 with respect to parameters p(1) and p(2).

```
> prin s(3)p(1),p(2)
---------------------------------------------------------------------
------*** NAME :     3
CURRENT KOMN
WITH RESPECT TO: P ( 1)
 #   COMP TC   CATEGORY          T            SENS.
  3    3    0   F ( 3)      0.000E+00     9.999E+30
  6    3    0   F ( 3)      1.780E+02     8.728E-01
 10    3    0   F ( 3)      2.730E+02     8.134E-01
 18    3    0   F ( 3)      5.310E+02     6.772E-01
 26    3    0   F ( 3)      8.660E+02     5.437E-01
 42    3    0   F ( 3)      1.510E+03     3.815E-01
 52    3    0   F ( 3)      1.918E+03     3.212E-01
 64    3    0   F ( 3)      2.401E+03     2.764E-01
CURRENT KOMN
WITH RESPECT TO: P ( 2)
 #   COMP TC   CATEGORY          T            SENS.
  3    3    0   F ( 3)      0.000E+00     9.999E+30
  6    3    0   F ( 3)      1.780E+02    -4.658E-03
 10    3    0   F ( 3)      2.730E+02    -1.015E-02
 18    3    0   F ( 3)      5.310E+02    -3.159E-02
 26    3    0   F ( 3)      8.660E+02    -6.569E-02
 42    3    0   F ( 3)      1.510E+03    -1.268E-01
 52    3    0   F ( 3)      1.918E+03    -1.558E-01
 64    3    0   F ( 3)      2.401E+03    -1.798E-01
```

While the partials present much the same picture as before, the relative sensitivities for p(2) have increased noticeably. However, p(2) still does not influence the predicted values. We conclude that p(2) is not identifiable from this experiment.

Example 3: Sensitivity in the Development of a Calcium Model

For a variation of the calcium model discussed earlier (Chapter 3), we determine the relative sensitivity of the solution to the current parameter values in order to isolate those parameters likely to influence the solution to move toward a 'better' fit.

The salient part of the text input file and solution commands are as follows:

```
  1: A SAAM31
  2: H PAR
  3:     IC(1)=100
  4:     P(1)       6.35                          100
  5:     P(21)      11.74                         100                 .1
  6:     P(22)      10.4                          100                 .1
  7:     L(1,2)     117.5                         1000
  8:     L(2,1)     180                           1000
  9:     L(2,3)     2.84                          100
 10:     L(3,2)     13.18                         100
 11:     L(3,4)     .27
```

```
  12:     L(4,3)   1.32
  13:     L(0,4)   .33
  14:     L(7,1)   5.4
  15:  L(8,1)=L(9,8)
  16:  L(9,8)=L(6,9)
  17:     L(6,9)   2.24                              100
  18:     K(1)     1.26                              100
  19:     L(5,1)   .6                                100


   compt. 5 - urine
   compt. 6 - feces
   compt. 7 - milk

> free all
   4:     P(1)     6.35         2.116667E+00   1.905000E+01
   5:     P(21)    11.74        3.913333E+00   3.522000E+01      .1
   6:     P(22)    10.4         3.466666E+00   3.120000E+01      .1
   7:     L(1,2)   117.5        3.916667E+01   3.525000E+02
   8:     L(2,1)   180          6.000000E+01   5.400000E+02
   9:     L(2,3)   2.84         9.466667E-01   8.520000E+00
  10:     L(3,2)   13.18        4.393333E+00   3.954000E+01
  11:     L(3,4)   .27          9.000000E-02   8.100001E-01
  12:     L(4,3)   1.32         4.400000E-01   3.960000E+00
  13:     L(0,4)   .33          1.100000E-01   9.899999E-01
  14:     L(7,1)   5.4          1.800000E+00   1.620000E+01
  17:     L(6,9)   2.24         7.466667E-01   6.720000E+00
  18:     K(1)     1.26         4.200000E-01   3.780000E+00
  19:     L(5,1)   .6           2.000000E-01   1.800000E+00
> deck
* EDIT BUFFER BEING WRITTEN INTO DECK
* DECK BEING PROCESSED
PRE-PROCESSING TIME :      3.783 SECS
> solv
*** MODEL CODE 10 SOLUTION

1
***COMPS ISOLATED IN STEADY STATE *8*
***  TC( 0)-  5  6  7 10 11 12 13 14 15 16 17 18 19
SOLUTION TIME :      2.417 SECS
> q(i)    <== identify the components for which partials exist
     5
     6
     7
    10
    19
> swit log of    <== sensitivity can go negative
* LINEAR PLOT MODE (LOG OFF)
> part    <== determine the partials
* PARTIALS ESTIMATED
PARTIALS TIME :      29.667 SECS
> plot s(1)l(2,1),l(1,2),l(3,2),l(2,3)
> plot s(1)l(4,3),l(3,4),l(0,4)
```

The plots resulting from the two plot commands in the Text input file are shown in Figs 16.2 and 16.3. Plots of the sensitivities of component 1 to additional parameters to see the effects on component 1 of the movement of calcium into the urine, feces and milk, are created by the following command (Fig 16.4):
> plot s(1)L(5,1),L(6,9),L(7,1)

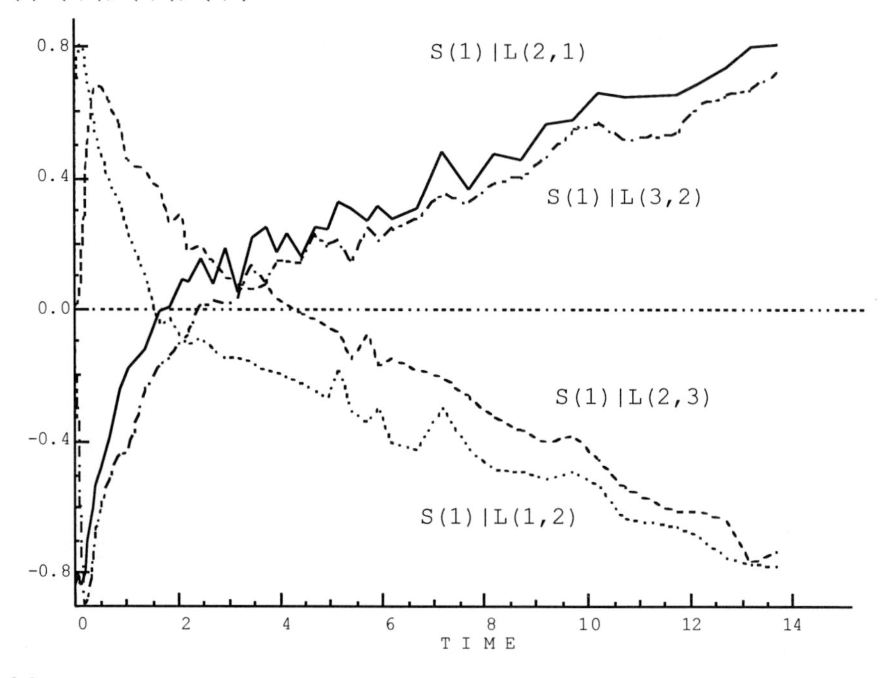

Fig 16.2 Plot of sensitivities of component 1 to parameters L(2,1), L(1,2), L(3,2) and L(2,3).

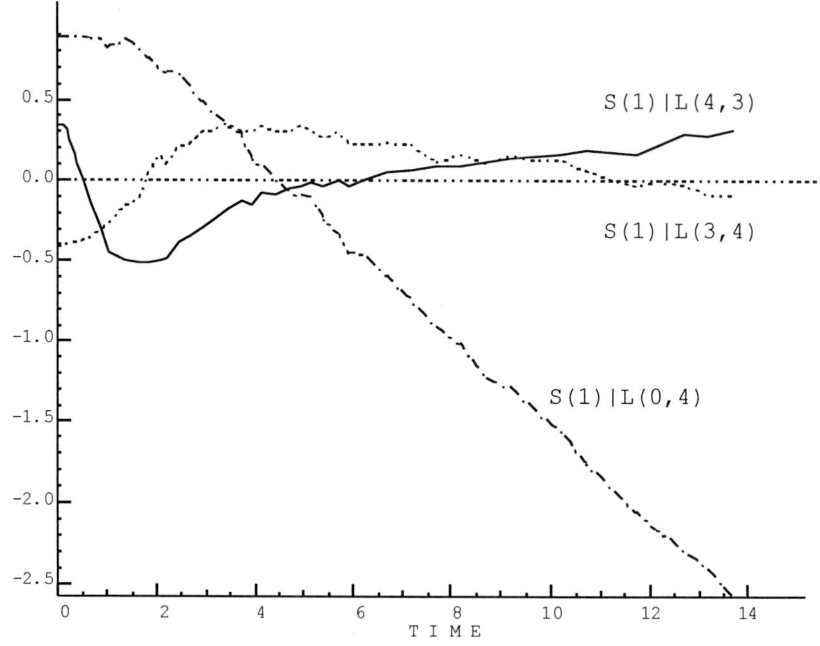

Fig 16.3 Plot of sensitivities of component 1 to parameters L(3,4), L(4,3), and L(0,4).

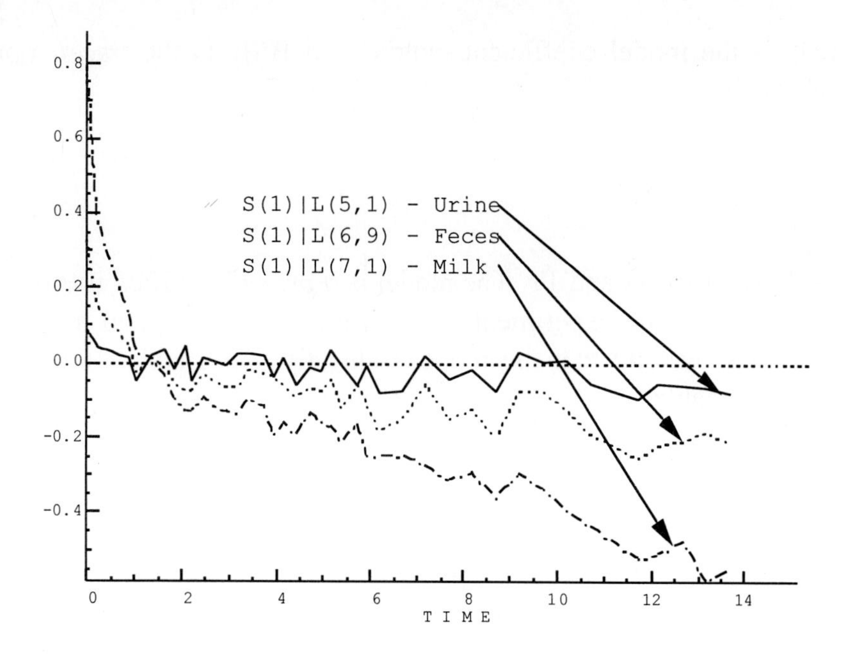

Fig 16.4 Plot of sensitivities of component 1 to parameters L(5,1), L(6,9) and L(7,1).

A number of observations regarding the temporal dependency of the model fit can be drawn in regard to these sensitivity plots.

II. Identifiability

As applied to modeling, identifiability is a mathematical concept which attempts to steer model development largely on the basis of an analysis reflecting whether features of a model for a presumed system can be extracted from a proposed experiment. Selecting a protocol for an experiment directed at elucidating a 'compartmental' model and its features involves selecting time points for observations, observation sites, sites for tracer input, as well as forms of tracer application. It is seldom possible to sample all compartments, and the number of compartments to which tracer can be applied is limited. Each choice we make here stands a chance of mitigating against identifying aspects of the model, even in an error-free situation. If the lack of identifiability makes it impossible to estimate parameters or test relevant hypotheses, we should like to know this before conducting the experiment. If the experiment is adequate to identify the model, we are still left with the estimability problem, that of resolving the parameters amidst noisy data (i.e,. data containing error).
For a model of the form (Jacquez, 1985):

$$f' = A(\theta)\cdot f + B(\theta)\cdot U \tag{16.9}$$

where $A(\theta)$ is the model coefficient matrix, and $B(\theta)$ is the tracer input with measurements

$$\eta = C(\theta) \cdot f \tag{16.10}$$

Often C is a matrix of 0's and 1's. The model is *a priori* identified if θ is uniquely determined by η. If the experimental observations were to yield a finite set of parameter values, we say that the model is locally identifiable. Otherwise the model is unidentifiable.

From the above,

$$\eta = \int_0^T c e^{A(t-r)} B \cdot U(r) \cdot dr \tag{16.11}$$

Taking the Laplace transform of (16.11) yields

$$L_\eta(s) = C\left(sI_m - A\right)^{-1} B \cdot L_u(s) \tag{16.12}$$

or

$$L_\eta(s) = \Phi(s) L_u(s) \tag{16.13}$$

$\Phi(s)$ is called the transfer function of the experiment. Inverting $\Phi(s)$ leads to the familiar R(t):

$$R(t) = C \cdot e^{At} \cdot B \tag{16.14}$$

Thus, just as identifying θ in $\Phi(s)$ implies model identification, so does identifying θ in R(t).

Example 4: Identifiability of a 2-compartment Model

The experimental response of the model (Fig 16.5), (if it indeed represents the system) to a perturbation in just one compartment (compartment 1) yields only 3 pieces of information. Hence a model of no more than 3 parameters can be fitted to the response, based purely on that response.

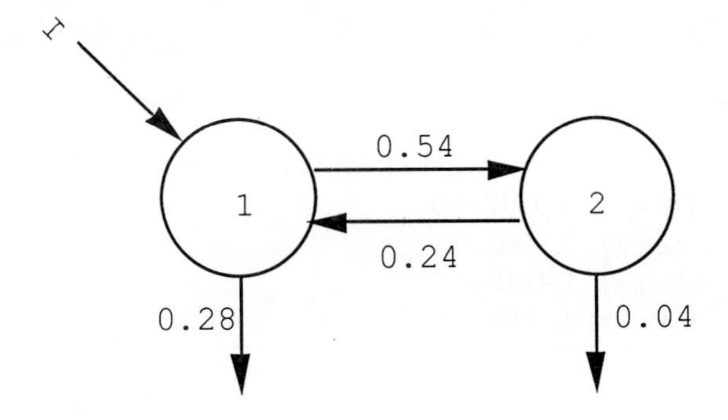

Fig 16.5 Two compartment model

The model is written as follows:

```
3:  L(2,1)  .54
4:  L(1,2)  .24
5:  L(0,2)  .04
6:  L(0,1)  .28
```

The experiment is represented as follows:

7: IC(1) 100

To illustrate the futility of attempting to identify this model based on the experiment we generate experimental data and attempt to estimate values for the parameters

```
> deck
* DECK BEING PROCESSED

***ALL WEIGHTS=1.*2*
PRE-PROCESSING TIME :       1.050 SECS
> solve
*** MODEL CODE 10 SOLUTION
SOLUTION TIME :       0.500 SECS
> rand/fsd=.4 q(1)    <== generate random data with fsd = .4
> prin q(1)        <== display synthetic data
---------------------------------------------------------------------
*** NAME :     1
CURRENT KOMN
 #  COMP TC  CATEGORY         T            QC          QO          QC/QO
  1    1   0  F ( 1)    0.000E+00   1.000E+02   3.878E+01     2.5787
  2    1   0  F ( 1)    1.000E-01   9.219E+01   1.269E+02     0.7263
  3    1   0  F ( 1)    2.000E-01   8.510E+01   1.224E+02     0.6953
  4    1   0  F ( 1)    3.000E-01   7.867E+01   2.918E+01     2.6966
  5    1   0  F ( 1)    4.000E-01   7.284E+01   5.219E+01     1.3957
  6    1   0  F ( 1)    5.000E-01   6.755E+01   7.206E+01     0.9373
  7    1   0  F ( 1)    6.000E-01   6.274E+01   2.909E+01     2.1571
  8    1   0  F ( 1)    7.000E-01   5.837E+01   6.141E+01     0.9506
  9    1   0  F ( 1)    8.000E-01   5.441E+01   8.276E+01     0.6574
```

the fit yielded:

```
> fsd(i)
* VALUES MAY NOT RELATE TO CURRENT PARAMETERS
* L ( 2, 1)   1.027E+00    FSD( 1)  7.563E+00
* L ( 1, 2)   3.169E-01    FSD( 2)  7.344E+00
* L ( 0, 2)   1.200E-01    FSD( 3)  1.985E+01
* L ( 0, 1)   9.333E-02    FSD( 4)  8.215E+01
```

Subsequent fits based on progressively smaller fsd's yielded similarly unidentified values:

```
> deck                    <== rebuild the input
UPDATE?
* DECK BEING PROCESSED

***ALL WEIGHTS=1.*2*
PRE-PROCESSING TIME :     1.317 SECS
> solv
*** MODEL CODE 10 SOLUTION
SOLUTION TIME :     0.500 SECS
> rand/fsd=.1 q(1)              <== reduce the error to 10%
> iter                    <== start the fitting process
* PARTIALS ESTIMATED
* CORRECTION VECTOR ESTIMATED
 LARGEST CHANGE (   0.00 %) WAS IN PAR( 1, 2)

ITERATION TIME :    4.733 SECS
DISTRIBUTION OF SQUARES
 COMP  SUM OF SQUARES
   1   4.9779E+01
> fsd(i)                  <== examine the identifiability
* VALUES MAY NOT RELATE TO CURRENT PARAMETERS
* L ( 2, 1)   8.296E-01    FSD( 1)  4.408E+00
* L ( 1, 2)   2.505E-01    FSD( 2)  4.316E+00
* L ( 0, 2)   1.200E-01    FSD( 3)  9.164E+00
  L ( 0, 1)   9.333E-02    FSD( 4)  3.878E+01
```

Repeat for 1% error .

```
> rand/fsd=.01 q(1)             <== 1% error

> fsd(i)                  <== after fitting examine identifiability
* VALUES MAY NOT RELATE TO CURRENT PARAMETERS
* L ( 2, 1)   8.305E-01    FSD( 1)  2.034E+00
* L ( 1, 2)   2.535E-01    FSD( 2)  1.991E+00
* L ( 0, 2)   1.200E-01    FSD( 3)  4.284E+00
* L ( 0, 1)   9.333E-02    FSD( 4)  1.792E+01
```

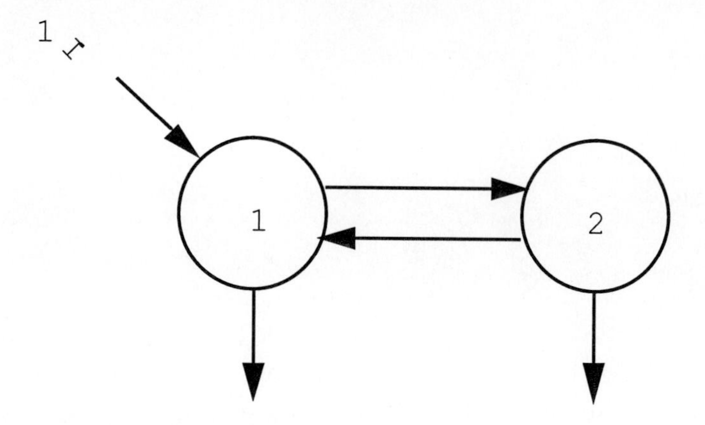

Fig 16.6 Two compartment model

IV. Ranges of parameters

Whereas a particular experiment may not resolve a specific model it may never-the-less allow us to identify a range of models for which the parameters are consistent with certain physical constraints. This topic is treated in Berman and Schoenfeld (1956) and Berman and Van Eerdewegh (1983).

Example 5: Resolving a Range of Models from a Bi-exponential Response

Given a response from an experiment of the form:

$$f_1 = 0.8e^{-t} + 0.2e^{-0.1t} \tag{16.15}$$

We find the range of physically realizable models consistent with this response.

Our response can be written $f = A \cdot e^{at}$ and, from the above,

$$A = \begin{bmatrix} 0.8 & 0.2 \\ -A_{2,1} & A_{2,1} \end{bmatrix} \qquad a = \begin{bmatrix} -1 & 0 \\ 0 & -0.1 \end{bmatrix}$$

From our earlier work we also know that

$$L = A \cdot a \cdot A^{-1}, \quad \text{and hence } L = \begin{bmatrix} -0.82 & 0.144 / A_{2,1} \\ 0.94 A_{2,1} & -0.28 \end{bmatrix}$$

However, since $L = \begin{bmatrix} -L_{1,1} & L_{1,2} \\ L_{2,1} & -L_{2,2} \end{bmatrix}$ we immediately have

$$L_{1,1} = 0.82$$
$$L_{2,2} = 0.28$$
$$L_{2,1} = 0.9 A_{2,1}$$
$$L_{1,2} = 0.144 / A_{2,1}$$

as well as

$$L_{1,1} = L_{2,1} + L_{0,1}$$
$$L_{2,2} = L_{1,2} + L_{0,2}$$

Note: for simplicity, we have abbreviated the notation, in some cases writing, e.g., $L_{1,1}$ for L(1,1). For physical realizability we have

L(1,2) ≥ 0, L(2,1) ≥ 0, L(0,1) ≥ 0, and L(0,2) ≥ 0, i.e. all fractional turnover rates must be greater or equal to zero.

L(1,2) ≥ 0 implies 0.144/A21 ≥ 0 i.e. 0 ≤ A(2,1) ≤ infinity

L(2,1) ≥ 0 implies 0.9 A(2,1) ≥ 0 i.e. 0 ≤ A(2,1) ≤ infinity. Note there are no serious constraints here.

Further,

L(0,1) ≥ 0 implies 0.82 - .9 A(2,1) ≥ 0 or A(2,1) ≤ 0.91.

L(0,2) ≥ 0 implies 0.28 - 0.144/A(2,1) ≥ 0 or A(2,1) ≥ 0.51.

Thus 0.51 ≤ A(2,1) ≤ 0.91

A(2,1) = 0.51 L(1,2) = 0.28 L(2,1) = 0.46
L(0,1) = 0.36 L(0,2) = 0.00

A(2,1) = 0.91 L(1,2) = 0.16 L(2,1) = 0.82

L(0,1) = 0.00 L(0,2) = 0.12 (16.16)

The parameter ranges are thus

0.16 ≥ L(1,2) ≥ 0.28 0.46 ≥ L(2,1) ≥ 0.82

0.00 ≥ L(0,1) ≥ 0.36 0.00 ≥ L(0,2) ≥ 0.12

L(1,1) = 0.82 L(2,2) = 0.28

Through this process a range of models can be identified where the parameters are consistent.

IV. Stability

Once steady state solutions are found for a model it is important to ask the question: 'If small random disturbances are applied at the steady state will they lead to drastic changes or will they die away?' That is, is the steady state stable, or is it unstable, or is somewhere in between? At steady state the effect of a small perturbation can be explored by use of a linearized representation of the response of the model at or near the steady state.

The linearized form of a model is $[F']=[J][F]$ where $J_{i,j}=\dfrac{\partial F_i'}{\partial F_j}$

and $[F']=\left[F_i'\right]$.

For such a system a steady state is stable provided Trace (L) < 0 and Det (L) > 0, where L is the Jacobian evaluated at the steady state, Trace (L) is the sum of the diagonal elements of L, and Det (L) is the determinant of L.

Example 6: Predator and Prey Model

Predator prey models (Roberts et al., 1983) are very important from environmental considerations. Predators prosper to the extent that their food permits, their food is of course their prey. However if the predators are too prosperous they in fact start to encounter food supply problems anyway. They become disproportionately populous compared to their prey and eventually succumb to food (prey) shortage. As the predator population subsequently declines, the prey start to become more populous and the cycle continues. Delicate cycles may build up and, under conditions of temporary imbalance, plagues and infestations may emerge.

We examine such a situation: Lynx and Hares (High Performance Systems, 1996) and make the following assumptions:

Hare birth rate = 1.25 hares per hare
Hare death rate = lynx numbers * the rate of hare kills per lynx
Lynx birth rate = 0.25 lynx per lynx
Lynx death rate = lynx numbers * lynx mortality
Hare kills per lynx = Hare density
Lynx mortality = 0.5 - 0.005 * hare density
Hare density = Hares / Study area
Study area (a) = 1000 hectares
Initial Hares = 50000
Initial Lynx = 1250

Thus

$$h' = 1.25h - l \cdot h / a$$
$$l' = 0.25l - l \cdot (0.5 - 0.005h / a) \qquad (16.17)$$

The linearized form for this system is:

$$h' = (1.25 - \bar{l} / a)h - (\bar{h} / a)l$$
$$l' = (0.005\bar{l} / a)h - (0.25 - (0.5 - .005\bar{h} / a))l \qquad (16.18)$$

where overbars denote steady state values.

The steady state solution to (16.16) yields

l = 1.25 a, and h = 50 a.

Substituting the steady state values for l, and h into the Jacobian yields the following:

$$L = \begin{bmatrix} 0 & -50 \\ 0.00625 & 0 \end{bmatrix} \qquad (16.19)$$

where Trace (L) = 0, and Det (L) = 0.3125.

Based on the above conditions, only one of the stability requirements is satisfied. Solving for the eigenvalues of L we obtain $a = \sqrt{-.3125}$,
which tells us that, the solution will oscillate about the steady state with period: $2\pi / \sqrt{-.3125}$.
The WinSAAM input file representing this model, which we can use to test our results, follows. By dropping the number of lynx by 400 over the period of 3 months, starting at the beginning of the second year, we jolt the system away from its steady state. The plot (Fig 16.7) shows how the two species are expected to respond to this perturbation.

```
 1: A SAAM31
 2: H PAR
 3: C
 4: C COMP. 1 = HARE
 5: C COMP. 2 = LYNX
 6: C G(1) IS THE LYNX MORTALITY FUNCTION
 7: C
 8: C INITIAL HARES AND LYNXES
 9: C
10:     IC(1)     50000
11:     IC(2)     1250
12: C
13: C P(1)  = HARE NATALITY
14: C P(2)  = LYNX MORTALITY
15: C P(3)  = HECTARES OF ENVIRONMENT
16: C
```

```
17:      P(1)      1.25
18:      P(2)      .25
19:      P(3)      1000
20: C
21: C K(2) IS SET TO JUXTAPOSE PLOTS
22: C
23:      K(2)      40
24: C
25: C FOR CONVENIENT PLOT SET YSCA G 40000 60000
26: C
27: H DAT
28: X UF(1)=F(1)*P(1)-F(1)*F(2)/P(3)
29: X UF(2)=F(2)*P(2)-G(1)*F(2)-F(4)
30: X G(1)=0.5-0.005*F(1)/P(3)
31: 101
32:               0
33: 2             .1                          360
34: 102
35:               0
36: 2             .1                          360
37: 104 QO
38:               0            0
39:               2            400
40:               2.25         0
```

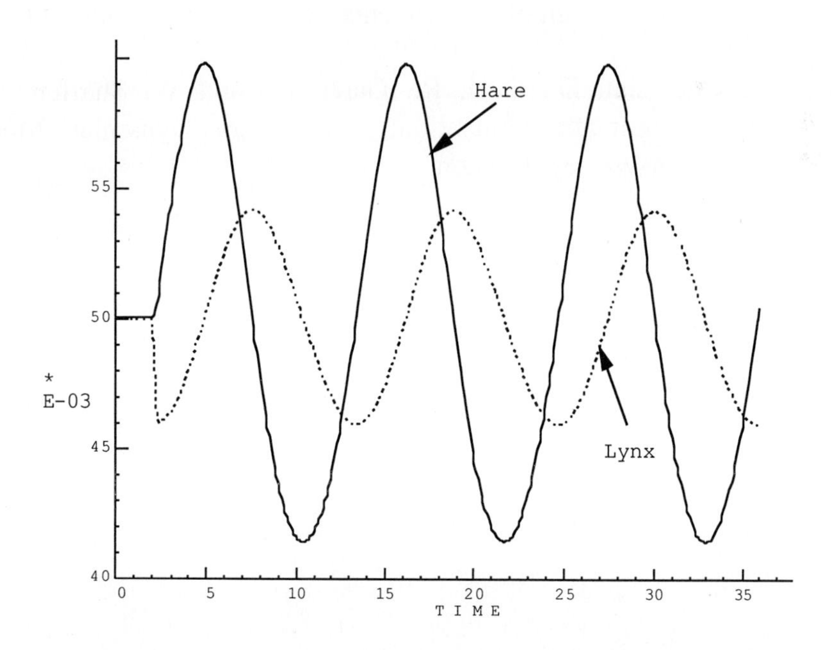

Fig 16.7 Plot of the numbers of hares and lynx predicted by model.

REFERENCES

1. Berry, K. V. 1975. Statistical Methods in Applied Science. J. Wiley and Sons, N.Y
2. Bard, Y. 1974 Nonlinear Parameter Estimation. Academic Press, New York, NY
3. Berman, M. and Schoenfeld, R. 1956. Invariants in experimental data on linear kinetics and the formulation of models. J.Appl. Physics 27: 1361-1370.
4. Berman, M. and Van Eerdewegh, P. 1983. Information content of data with respect to models. Am. J. Physiol. 245: R620-R623.
5. Box, G .E. P. and Draper, N. 1987. Empirical model-building and response surfaces. Wiley and Sons, New York NY
6. Daniels, F. and Alberty, R. A. 1960. Physical Chemistry. J. Wiley and Sons, New York, NY
7. Edelstein-Keshet, L 1987. Mathematical Models in Biology. Random House, New York, NY
8. Frank, P.M. (1978) Introduction to system sensitivity theory Academic Press, New York, NY
9. Godfrey, K. 1983. Compartmental Models and Their Application. Academic Press, New York, NY
10. High Performance Systems, 1996. STELLA Applications, Hanover, NH
11. Jacquez, J. A. 1985. Compartmental Analysis in Biology and Medicine. University of Michigan Press, Ann Arbor, MI
12. Jacquez, J. A. 1998. Design of Experiments Journal of the Franklin Institute, 335b:259-279.
13. Roberts, N., Anderson, D., Deel, R., Garet, M, and W. Shaffer. 1983. Introduction to Computer Simulation: A system Dynamics Modeling Approach, Addison-Wesley, London.

SECTION V

Evaluating and Using Published Models

17

WHY USE A PUBLISHED MODEL?

This chapter discusses the use of published models in research and education. First, some reasons for making use of information in published models are presented. Then, using examples from the literature, the type of information that can be obtained from published models is demonstrated. Thirdly, it is argued that using published models for a system to analyze new data is a more efficient way to progress knowledge than developing models *de novo*. These arguments are based on philosophical issues as well as practical and economic factors. Fourthly, the use of models for experimental design and for making predictions is discussed. Finally, the potential for using models as educational tools for teaching dynamics and quantitative biology is discussed.

I. Why use published models?

There are three reasons why published models should be used in research and education. First, scientific advances are made by building upon what is known and this means building upon current published models. The second reason is economic. Developing models is expensive in terms of the experimental resources that are used in performing the study and analyzing the data. Using information from current models to identify areas where data are lacking, and designing new studies efficiently so that they do not repeat previous studies, is a way to ensure that experimental resources are used efficiently. By using published models, we capitalize on the large investment in research resources that is made to develop a model. Third, models encapsulate or condense a large amount of information about a system. They can therefore be used as a reference for information on a system.

A partial listing of some published compartmental models for metabolism of drugs, enzymes, hormones, the immune system, metabolites, minerals, nonessential metals, proteins, trace elements and vitamins are shown in Table 17.1.

II. Using models to understand and explore systems

Models have a life beyond the initial study for which they were developed. Because models condense a large amount of information on a system, they serve as a resource of information on the system. This information relates to the system structure, kinetic properties, dynamic properties, unexplained features, and areas lacking data. Whether the model is of a cell or metabolism in the whole body, it will show how the compound(s) of interest moves among compartments by how the compartments are interconnected. An example of a complex system is that of copper (6, 7). The model shows not only copper distribution between various tissues, but also the form of copper and the interconversion of copper between those forms. The model therefore not only shows the system structure (number of compartments, relation of compartments to various tissues) but also the kinetic properties of the system (interconnections between compartments, compartment mass, sites of copper input into the system, and sites of loss from the system, and sites where there is a delay in movement within the system).

Table 17.1 Some published compartmental models for metabolism

Substance	Reference
DRUGS	
methotrexate	Reich et al, J Pharm Biopharm 5:421, 1977
warfarin	Covell et al, J Pharm Biopharm 11;127, 1983
ENZYMES	
adenylate cyclase	Rodbell et al, Acta Endocrin S191:11, 1974
BAM	Hensley et al, J Biol Chem 265:15300, 1990
METABOLITES/HORMONES/PROTEINS	
alanine	Hall et al, Fed Proc 36, 239, 1977;Foster et al, Am J Physiol 239:E30, 1980
aldersterone	Ayers et al, J Clin Invest 41:884, 1962
antibody	Eger et al, Cancer Res 47:3328, 1987
apoproteins	Berman et al, J Lipid Res 19:38, 1978;Zech et al, J Lipid Res 24:60, 1983
bicarbonate	Irving et al, Am J Physiol 245:R190,1983;Barstow et al, Am J Phys 259:R163,1990
cholesterol	Avigan et al, Lipid Res 3:216, 1962; Schwartz et al, J Clin Invest 61:408, 1978; Schwartz et al J Clin Invest 70:105, 1982; Magot et al, BBA 921:587, 1987
collagen	Phang et al, BBA 230:146, 1971
glucose	Shames et al, J Clin Invest 50:627,1971;Hall et al, Fed Proc 36:239, 1977 Cobelli et al, Am J Physiol 257:E943, 1989
insulin	Sherwin et al, J Clin Invest 53:1481,1974;Insel et al, J Clin Inv 55:1057, 1975; McGuire et al, Diabetes 28:110, 1979; Berman et al, Diabetes, 29:50, 1980
ketone bodies	Cobelli et al, Am J Physiol 243:R7,1982; Wastney et al, J Lipid Res 25:160, 1984
lactate	McGuire et al, J App Phys 41:565,1976; Foster et al, Am J Phys 239:E30,1980; Hetenyi et al, Am J Phys 239:E39,1980; Stanley & Lehman, Biochem J 256:1035,1988
plasma proteins	Fleck, Ann Clin Biochem 22:33, 1985
FFA	Eaton et al, J Clin Invest 48:1560,1969;Shames et al, J Clin Invest 49:2298,1970
lipoprotein	Phair et al, Fed Proc, 34:2263, 1975; Zech et al, J Clin Invest 63:1262, 1979
thyroxine	McGuire & Berman, Endocrinol 103:567, 1978; McGuire & Hays, J Clin End Metab 53:852, 1981
TRACE ELEMENTS/NON-ESSENTIAL METALS/MINERALS	
cadmium	Kjellstrom et al Environ Res 16:248,1978; Nordberg & Kjellstrom, Env Hth Per 28:211,1979
calcium	Neer et al, J Clin Invest 46:1364,1967; Phang et al, J Clin Invest 48:67, 1969; Birge et al, J Clin Inv 48:1705,1969; Ramberg et al, Am J Physiol 219:1166,1970; Moore et al, Ped Res 19:329,1985; Massaldi et al, Am J Physiol 259:R172,1990
chromium	Onkelinx, Am J Physiol 232:E478, 1977;Lim et al, Am J Physiol 244:R445, 1983
cobalt	Onkelinx, Tox Appl Pharm 38:425, 1976
copper	Weber et al, Aust J Ag Res 31:773,1980; Gooneratne et al, Br J Nutr 61:373,1989; Buckley, Can J Anim Sci 71:155, 1991; Dunn et al, Am J Physiol 261:E115, 1991
iodine	McGuire & Berman, Endocr 103:567,1978;McGuire & Hays, J Endocr Metab 53:852,1981
iron	Vuille, Acta Phys Scand 65:S253,1965;Nathanson & McLaren, J Nutr 117:1067,1987
magnesium	Avioli & Berman, J Appl Physiol 21:1688, 1966
phosphorus	Grace, Br J Nutr 45:367, 1981; Schneider et al, Aust J Ag Res 33:827,1982; Schneider et al, Am J Physiol 252:R720, 1987
nickel	Sunderman et al, PSEBM 191:5, 1989
selenium	Patterson et al, Am J Physiol 257:R556,1989; Swanson et al, Am J Clin Nutr 54:917,1991
vanadium	Patterson et al, Am J Physiol 251:R325, 1986
zinc	Popov & Besel, TEMA3:168, 1977; Foster et al, Am J Physiol 237:R340, 1979; Jain et al, Ann Biomed Eng 9:347, 1981; Babcock et al, Metab 31:335, 1982; Wastney et al, Am J Phys 251:R398,1986;Dunn & Cousins, Am J Phys 256:E420,1989
VITAMINS vitamin A	Green et al, J Lipid Res 26:806, 1985; Lewis et al, J Lipid Res 31:1535, 1990
vitamin B-6	Coburn & Townsend, Prog Food Nutr 12:227, 1988

The dynamic properties of copper metabolism are shown by the model of Buckley et al. for cows (4, 5). In this model, the increase in liver copper, while plasma copper decreases, was modeled by a nonlinear function on release of copper from liver. Other nonlinear pathways were exchange of plasma copper with copper in body tissues, excretion of copper in urine and movement of copper from the gut into feces. Unexplained features of the copper models are the slower pools within tissues. The identity of these pools is not known but they are necessary to fit the experimental data and are thus considered to relate to intracellular metabolism of ceruloplasmin (6). The unexplained features of a model generally equate to areas that lack data. Dunn (6) states that compartment 9 oversimplifies a number of steps in the poorly understood intracellular metabolism of ceruloplasmin copper.

III. Using models to design studies

Models bridge the gap between theory and experiment. They can be used therefore to better utilize experimental resources. Specifically, models can be used to identify areas that are lacking in data and to optimize new studies in terms of number of samples and length of study required. The final choice of a model may be arbitrary (e.g., series versus a quaternary) model. The choice as to which model is more correct can often be determined only by obtaining more data that would allow the models to be distinguished.

A model, as described above, often identifies areas where data are lacking. In this sense, the modeling process identifies critical experiments. It is important that published models be examined prior to experimentation to identify key areas that need to be studied so as to avoid duplication. In this regard it would be helpful if publications of modeling studies included a statement about new areas of investigation that are identified by the current study.

Models can be used to determine length of experiment necessary to estimate parameters of interest. For example, the model for fluid phase endocytosis (3) can be used to predict movement of a compound through various cellular compartments. If a particular compartment is of interest, a simulation can be used to determine how long the system should be loaded before the experiment in order to study a compartment. For example, (Fig 17.1) shows a three compartment system, if compartment 1 is of interest, the system should be loaded for about 5 minutes to maximize the label in this compartment. For compartment 2, the system should be labeled for 25 min, and for compartment 3, the system should be labeled for longer than 50 minutes.

The number of samples, the optimal sampling times, and the length of the study for determining parameters of interest can also be determined using models (8). The number of samples required can be determined by generating values at a number of time points (e.g., 12 points during a study) and adding some error to the values. This can be done in SAAM by using the RAND command (see Tools, Chapter 8). The model can then be fitted to the values with the added error by adding weights and making the parameter values of interest adjustable. If the parameters of interest are not well-determined, increase the number of samples and repeat the simulation. For example, increase the number of samples from 12 to 24 and see the effect on the parameter estimation errors. To determine the timing of samples, use the INFO command in SAAM (see Chapter 8). This command will list the relative information content of each time point (2). Critical points can be identified as those with the highest information content.

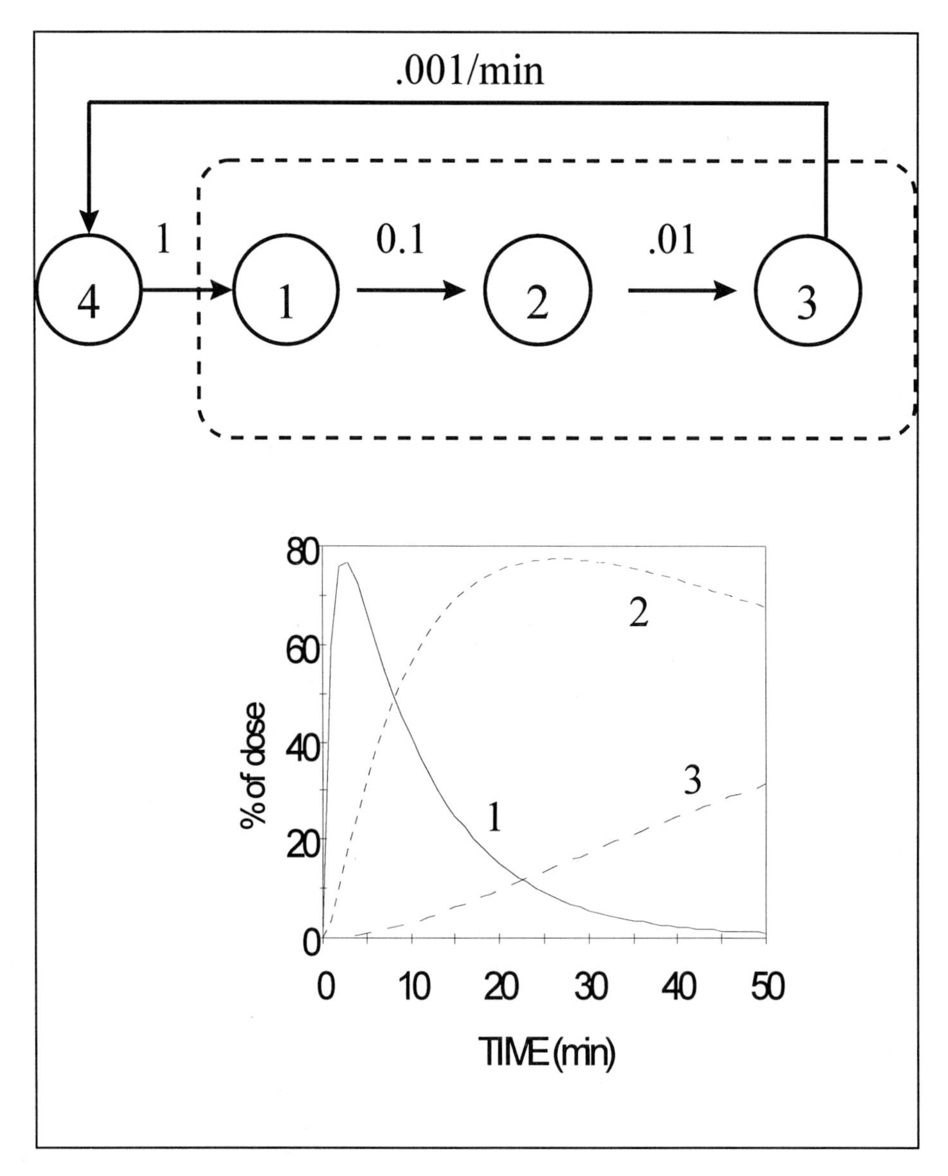

Fig 17.1. A model for cellular uptake. Compartment 4 is extracellular and compartments 1, 2 and 3 are intracellular. The lower figure shows the simulated accumulation of a compound in each intracellular compartment after an initial dose in the extracellular compartment.

IV. Using models to make predictions

Models can be used as laboratories to ask 'what if' questions. For example, a zinc model developed for humans was used to predict the changes in plasma zinc concentration with varying zinc intakes (11). It was predicted that when zinc intake was reduced by 50% (5 mg/d versus a normal 10 mg/d) plasma zinc would remain in the normal range for 6 months, and even on very low intake (0.2 mg/d, or only

2% of normal), it would remain in the normal range for almost 2 months. These results were in agreement with experimental studies. If the results differed, the model could be used to predict which parameters were involved in the homeostatic mechanisms. A model for selenium metabolism in humans has been used to make predictions about blood selenium levels during high selenium intake (10). Models are useful then, for predicting beyond the range of the experimental study used to develop the model. Models are also useful for exploring the system under investigation and developing scenarios for further study.

V. Using published models to analyze data from new studies

Published models should be used to analyze data from new studies. It is often said that 'there are an infinity of models that fit a particular set of data and an infinity of models that do not fit the data'. Because of this, there is a high probability that a new model, although it may have a slightly different structure (e.g., series vs. quaternary) will only confirm the current model and not necessarily contribute any new information to our understanding about the system. Proposing a new model that differs modestly from a previous model does not contribute new information about a system. In fact, it is difficult to compare models with structural differences that do not affect the fit to the data. Knowledge is obtained by refining or expanding current models and only discarding them if they are inadequate to fit the new data. To identify sites of change in a system, it is particularly important to use the same model for comparing data from control and experimentally or physiologically perturbed conditions. However, there maybe justifications for developing a new model to analyze data; no current model exists, the current model contains an error, or the current model does not fit the new data.

VI. Using models for educational purposes

The biomedical sciences are traditionally taught by a didactic approach using lectures or by a hands-on approach in laboratories. While labs have more appeal to students, they have the draw back of expense in terms of materials and time of preparation. An alternative method is to use simulation where students 'perform' experiments on a computer. This approach addresses real-world problems but allows students to learn through interaction and inquiry, and allows them to ask questions and seek answers in creative ways. Properties of a system such as saturation kinetics, feed back loops, or interactions can also be demonstrated visually by using models.

Once a system is represented by a model, equations can be solved, and the results mimic the behavior or response of a system. The model can be used to explore the system and to answer 'what if' questions by varying parameter values, solving the model, and comparing the results. In this way, models can be used to represent systems in a healthy state and to determine the effects when one or more processes are deficient or excessive, such as during a disease. Biomodels are developed and solved on computers. They are therefore a ready-made tool for educating students about complex systems such as those encountered in biomedicine.

The benefits of using modeling and visual simulation in science education are manifold, and to a large extent mirror the benefits of this technique in science research; namely, these tools enhance the ability to observe, think about, experiment with, and discover patterns and relations that are inaccessible because, for example of spatio-temporal scale such as microscopic nanoscale or large planetary-scale.

Biomodels of various biological systems can be used for teaching: i) that biological systems are dynamic with interrelated processes, ii) that modeling is a tool for representing systems in a logical and quantitative manner, and iii) that systems can be investigated by varying parameter values of biomodels and predicting their effect. In this way, biomodels can be used to compare and assess systems, for example during health and disease. Modeling software allows students to interactively run the model, manipulate its parameters, and even modify it by rearranging or adding parameters. By allowing this type of exploration the model engages students and draws them into an understanding of the modeled system by active learning (9). An added impetus for the use of models in education, is the proposed Physiome project, where modeling will be used to integrate and interpret data from various biological levels for understanding physiology in the whole organism (1).

REFERENCES

1. Bassingthwaighte, J. B. 1995. Toward modeling the human physionome. *Adv Exp Med Biol*. 382:331-339.
2. Berman, M., and P. Van Eerdewegh. 1983. Information content of data with respect to models. *Am. J. Physiol.* 245:R620-R623.
3. Blomhoff, R., M. S. Nenseter, M. H. Green, and T. Berg. 1989. A multicompartmental model of fluid-phase endocytosis in rabbit liver parenchymal cells. *Biochem J.* 262:605-610.
4. Buckley, W. T. 1991. A kinetic model of copper metabolism in lactating dairy cows. *Can. J. Anim. Sci.* 71:155-166.
5. Buckley, W. T. 1995. Copper metabolism in dairy cows: development of a model based on a stable isotope tracer. *In* Kinetic Models of Trace Element and Mineral Metabolism During Development. K. N. Siva Subramanian and M. E. Wastney, editors. CRC Press, Boca Raton, FL. 37-52.
6. Dunn, M. A. 1995. Historical overview of copper kinetics. *In* Kinetic Models of Trace Element and Mineral Metabolism During Development. K. N. Siva Subramanian and M. E. Wastney, editors. CRC Press, Boca Raton, FL. 171-186.
7. Dunn, M. A., M. H. Green, and R. M. Leach, Jr. 1991. Kinetics of copper metabolism in rats: a compartmental model. *Am. J. Physiol.* 261 (Endocrinol. tab. 24)::E115-E125.
8. Jacquez, J. A. 1998. Design of experiments. *J. Franklin Institute*. 335B:259-279.
9. Modell, H. I. 1996. Preparing students to participate in an active learning environment. *Adv Physiol Educat*. 15:S69-S77.
10. Patterson, B. H., L. A. Zech, C. A. Swanson, and O. A. Levander. 1995. An overview of selenium kinetics in humans. *In* Kinetic Models of Trace Element and Mineral Metabolism During Development. K. N. Siva Subramanian and M. E. Wastney, editors. CRC Press, Boca Raton.
11. Wastney, M. E., K. N. Siva Subramanian, N. Broering, and R. Boston. 1997. Using models to explore whole-body metabolism and accessing models through a model library. *Metabolism*. 46:330-332.

18

REVIEWING AND SUMMARIZING PUBLISHED MODELS

As discussed in Chapter 17, published models contain a large amount of information about a system. While some information can be obtained from the publication describing the model, it is often necessary to reconstruct the model to obtain other information of interest. For example, solving the model using different parameter values or solving the model for a longer period in order to make predictions. To use models for their intended purpose (exploring systems) it is necessary to obtain a working version of the model. In some cases, this can be accomplished by simply contacting the authors. If the authors are unable or unwilling to supply the model, it may be possible to download the model from a model library (Chapter 22). If the model is not available in a library, it will be necessary to reconstruct the model. This chapter describes the first steps in this process, i.e., identify the purpose of the model, determine the model type, identify the model assumptions, ascertain the source of the data, and evaluate the model. In this chapter, and the following chapters, examples will be used to demonstrate how to interpret published models. These examples were obtained from various sources (1, 6, 7) and were translated into WinSAAM text input files (3, 4) to probe the systems, verify their robustness, and to make predictions (8).

I. Identifying the purpose of the model

There may be a number of models developed for a system for different purposes. The purpose of a model is the question about the system that the model was developed to answer. The purpose determines the scope, type, and detail of the model. A purpose can be general, for example, to integrate knowledge about a system, or specific, to determine a particular parameter. The purpose of the model is identified at the beginning of the paper or in the abstract. For example, an epidemic model for acquired immmunodeficiency syndrome (AIDS) first presented by Anderson et. al. (1) and reviewed by Murray (7), stated that the purpose was to investigate the spreading rates of early stage of human immunodeficiency virus (HIV) and the influence of an incubation period on the spread rate of AIDS, in a homosexual population (This model will be discussed in Chapter 21). In a model of ruminant digestion, the purpose was to simulate the biochemical processes in the rumen so that animal nutrition could be related eventually to growth (6).

II. Identifying the model type

The model type determines how the model will be described mathematically, and the type depends, in turn, on the purpose of the model. For example, Fox et. al. (5) modeled ruminant metabolism by using a *static model*, composed of a set of algebraic equations, to partition the flow of metabolites and nutrients through the subsystems. By contrast, *kinetic models* are composed of a system of linear differential equations, and these can be used to examine the responses of the system to a small perturbation. Our representative example is a *dynamic model* that uses nonlinear differential equations to explore a global response of the system to a large perturbation.

Models can also be classified by the influence of random or unpredictable variables on the system. *Deterministic models* ignore this randomness, assuming it to be unimportant to the decision, while *stochastic models* include random components of the system. By considering the changes of state within the system, models can also be identified as discrete, continuous, or a combination of these. If the state of the modeled system changes only at isolated times, it is called a *discrete model*, whereas if a model treats the variation in state as a continuous process and uses sets of algebraic and/or differential equations to represent it, it is called a *continuous model*.

III. Identifying the model assumptions

The aim of modeling is to simplify the nature of the actual physical system under investigation and this involves making assumptions about the system. The model assumptions are usually found in the methodological account of modeling papers. Some common assumptions involve i) identifying which aspects or compartments of a system are controlled, intermediate, and responsive ii) determining the constituents in each compartment and the flow between compartments, and iii) specifying the parameters pertaining to a particular compartment and to the interactions between the compartments. Assumptions of tracer studies are that the tracer mixes rapidly (compared to the sampling schedule), that it does not perturb processes in the system, and that the tracer behaves (i.e., is metabolized) in the same way as the tracee (or unlabelled material).

IV. Identifying the sources of data and information

The data used to develop a model determines to some extent the model's reliability and validity. To test a published model, we need to retrieve the data and information used for developing the model. Data usually come from the following sources: experimental observations collected during a study, previous publications, or generation of data based upon assumptions and theoretical considerations.

The source of data should include details of the subjects studied, the experimental conditions, protocols, the length of the study, the times and sites where observations were made, a description of the analytical procedures, and units. The identified data source may have limitations due to its nature, such as the number of observations, applicable range of the data, and measurement errors. By understanding how the data were obtained experimentally, we are more likely to understand how the model operates. Data, for example for the sheep ruminant digestive model of France et al. (6), that is described in Chapter 19, are based on the experiments reported previously by Beever et al (2).

V. Evaluating the model with respect to its intended application

A model can be evaluated by how well it represents the system under study, how comprehensively it includes current knowledge on a system, and by how easily it can be modified to incorporate new knowledge on the system. A user can evaluate how well a model meets its purpose by comparing the results of the model with data in the paper and with other known information on the system. The results include parameter values, pathways, and the statistics on the parameters. Conclusions of any modeling study also need to be evaluated with respect to assumptions underlying the model.

REFERENCES

1. Anderson, R. M., G. F. Medley, R. M. May, and A. M. Johnson. 1986. A preliminary study of the transmission dynamics of the human immunodeficiency virus (HIV), the causative agent of AIDS. *IMA J Math Appl Med Biol.* 3:229-263.
2. Beever, D. E., D. F. Osbourn, S. B. Cammell, and R. A. Terry. 1981. The effect of grinding and pelleting on the digestion of Italian ryegrass and timothy by sheep. *Br J Nutr*. 46:357-370.
3. Boston, R. C., P. C. Greif, and M. Berman. 1981. Conversational SAAM - an interactive program for kinetic analysis of biological systems. *Computer Programs in Biomedicine*. 13:111-119.
4. Foster, D. M., and R. C. Boston. 1983. The use of computers in compartmental analysis: the SAAM and CONSAM programs. *In* Compartmental Distribution of Radiotracers. J. Robertson, editor. CRC Press, Boca Raton. 73-142.
5. Fox, D. G., C. J. Sniffen, J. D. O'Connor, J. B. Russell, P. J. Van Soest, and W. Chalupa. 1991. Using the cornell net carbohydrate system for evaluating dairy cattle rations. *In* Large Dairy Herd Management Conference, Syracuse, NY.
6. France, J., H. M. Thornley, and D. E. Beever. 1982. A mathematical model of the rumen. *J Agric Sci Camb.* 99:343-353.
7. Murray, J. D. 1991. Epidemic models and the dynamics of infectious diseases. *In* Mathematical Biology. Vol. 19. S. Levin, editor. Springer-Verlag, New York. 610-650.
8. Wastney, M.E., X.Q. Wang, and R.C. Boston. 1998. Publishing, interpreting and acessing models. *J. Franklin Instit*. 335B: 281-301

19

THE MODEL TRANSLATION PROCESS

There are a number of steps to follow to set up published models so they can be used interactively for investigating a system. The initial steps were described in Chapter 18, as identifying the model purpose, type, assumptions and the data source. This Chapter describes the next steps, namely to identify the initial conditions of the model, the parameters, and the mechanisms, and to translate these into equations. The example to be used was introduced in Chapter 18, and is on ruminant digestion.

I. Identify the initial conditions and state variables of the model

To start rebuilding a model, we have to understand all the essential substates, or subsystems of the model, the variables that characterize each subsystem, and the initial conditions. These are listed, for the dynamic model of ruminant digestion developed by France et al. (1) in Table 19.1. Although four combinations of dietary inputs are presented in the paper, we use only one input (chopped Italian ryegrass) to demonstrate the model robustness. Other combinations of inputs can be tested by using different sets of steady state input parameters given in France et al. (1) that describe the starting state of the system.

Table 19.1 lists the name, the biological meaning, and the initial condition of each substate, or compartment. Initial condition refers to the amount of substance present in the compartment at the beginning of the study and it is important to note the units for each compartment for mass considerations. In addition to the amount present in each compartment at the beginning of the study, various components are added in the feed. The amounts are calculated from analysis of the feed, and the quantity of feed ingested. Notice that all intake units are expressed as Kg/d or cubic meters (m^3). Keeping units consistent is important for determining rates of conversion of compounds between compartments.

II. Identify the parameters in the model

Once the compartments have been identified, it is necessary to identify the constant parameters of the model. The names of the parameters, and a description of the biological process they relate to, are listed in Table 19.2. Again, it is important to indicate the units (those without units are fractions), and values.

III. Understand the mechanisms

To understand the mechanisms or process occurring in the system, the model can be drawn, showing each subsystem or compartment, with pathways showing all the interactions between the compartments (Fig 19.1). The type of interactions between compartments can then be identified, together with the transient state, and the input/output relations. The diagram (Fig 19.1) summarizes the organization and the structure of the model.

Table 19.1 State variables and initial conditions of the model of France et al. (**1**).

Name	Meaning	Initial Condition		Steady State Input *	
		Unit	Value	Unit	Value
C	Soluble CHOH	kg/ m^3	7.0	kg/day	215×10^{-3}
M	Microbial Mass	kg/ m^3	26.5		
N	Non protein Nitrogen	kg/ m^3	2.0	kg/day	42.2×10^{-3}
V	Rumen Fluid	m^3	0.005	m^3/day	4.2×10^{-3}
A	a-Hexose	kg/ m^3	5.0	kg/day	26.6×10^{-3}
Bn	Non Degradable b-Hexose	kg/ m^3	36.0	kg/day	67.3×10^{-3}
Br	Degradable b-Hexose	kg/ m^3	30.0	kg/day	493.7×10^{-3}
Pn	Non Degradable Protein	kg/ m^3	4.0	kg/day	24.3×10^{-3}
Pr	Degradable Protein	kg/ m^3	1.5	kg/day	72.9×10^{-3}

*for Chopped Italian Ryegrass

Table 19.2 Parameters used in the model of France et al (1)

Parameter	Purpose	Units	Value
K_{AM}	a-Hexose utilization for microbes	$m^3\ kg^{-1}$	1.5
K_{BrM}	b-Hexose utilization for microbes	$m^3\ kg^{-1}$	2.0
K_{PrM}	Degradable protein utilization for microbes	$m^3\ kg^{-1}$	2.0
m_M	Asymptotic microbial growth rate	d^{-1}	5.0
K_C	Microbial growth dependence on C		0.2
K_N	Microbial growth dependence on N		0.2
K_{CN}	Microbial growth dependence on interaction of C and N		0.0
l_M	Maximum specific rate of microbial catabolism	d^{-1}	0.2
K_m	Dependence of microbial catabolism on growth rate		5.0
f_C	Microbial C composition		0.3
f_N	Microbial N composition		0.5
Y_M	Conversion efficiency of substrate CHOH into microbial CHOH		0.1

The names of the compartments in the model diagram correspond to the names used in the table of state variables (Table 19.1) and pathway labels correspond to those in the parameter table (Table 19.2). In addition, the model diagram identifies the types of processes occurring (utilization, dietary input, salivary input, out flow and recycling).

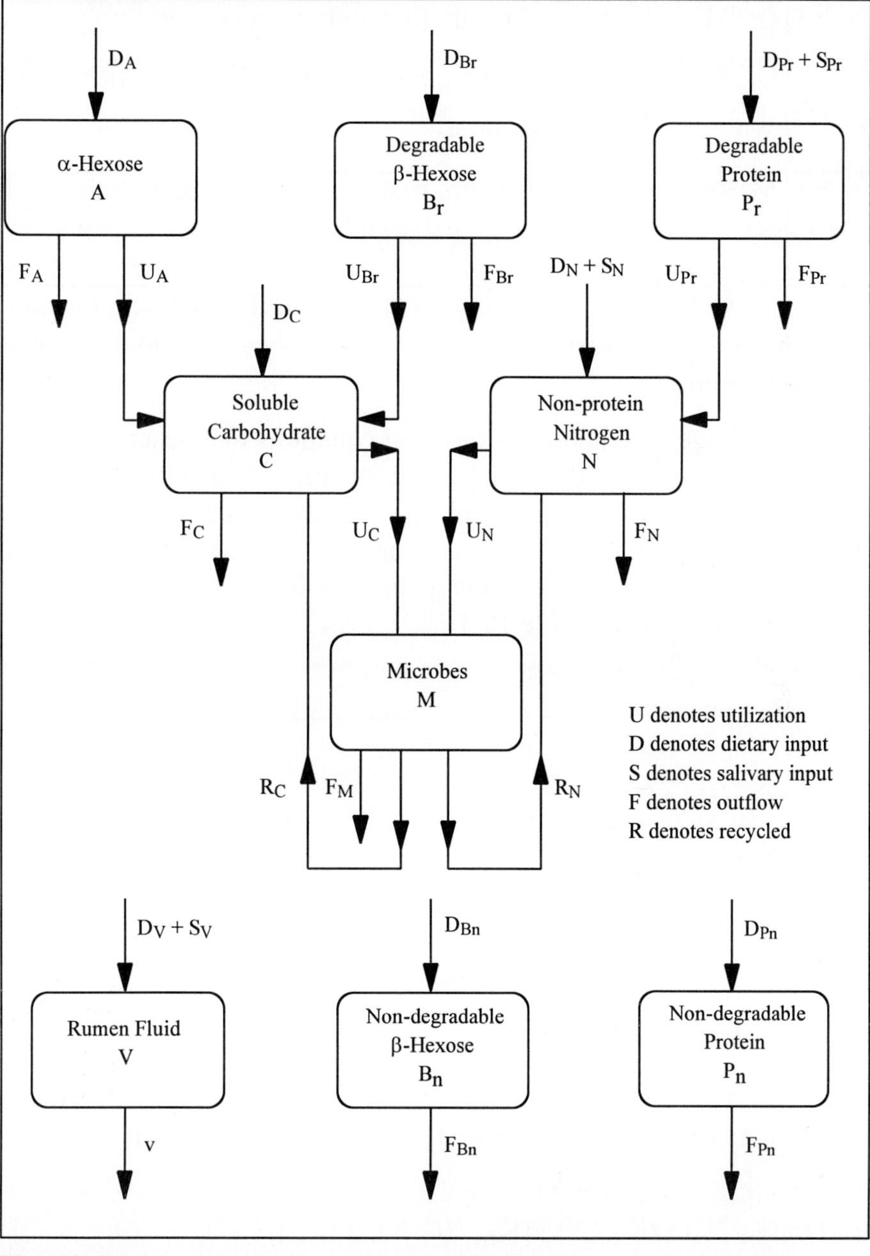

Fig. 19.1 Model diagram for ruminant metabolism based on France et al.(1)

Table 19.3. France's Model (1): State equations.

State Variable	Dependent on State Variables	Type of Equation
V	0	Zeroeth order
Bn	Bn	Zeroeth and first order
Pn	Pn	Zeroeth and first order
A	A, M	Zeroeth, first and second order
Br	Br, M	Zeroeth, first and second order
C	C, M, A, Br	Zeroeth, first, second order and Michaelis-Menten process
N	M, Pr	Zeroeth, first, second order and Michaelis-Menten process
M	M, C, N	First order and Michaelis-Menten process

IV. Governing equations

The system of algebraic and/or differential equations that describe the changes in each compartment are called the governing equations. The state variables, their dependence on other state variables, and the type of each equation, i.e., whether the processes are zeroeth-, first-, second-order or mixed or if they involve Michaelis-Menten processes are shown in Table 19.3. The equations are expressed mathematically in Table 19.4.

Table 19.4 State Equations for the model.

$$V' = D_V + S_V - v$$

$$B_n' = (D_{B_n} - B_n \cdot (D_V + S_V)) / V$$

$$P_n' = (D_{P_n} - P_n \cdot (D_V + S_V)) / V$$

$$A' = (D_A - A \cdot (D_V + S_V)) / V - K_{AM} \cdot A \cdot M$$

$$B_r' = (D_{B_r} - B_r \cdot (D_V + S_V)) / V - K_{B_rM} \cdot B_r \cdot M$$

$$P_r' = ((D_{P_r} + S_{P_r}) - P_r \cdot (D_V + S_V)) / V - K_{P_rM} \cdot P_r \cdot M$$

$$C' = (D_C - C \cdot (D_V + S_V)) / V + \lambda \cdot f_C \cdot M - \mu \cdot f_C \cdot M / Y_M + K_{AM} \cdot A \cdot M + K_{B_rM} \cdot B_r \cdot M$$

$$N' = ((D_N + S_N) - N \cdot (D_V + S_V)) / V + \lambda \cdot f_N \cdot M - \mu \cdot f_N \cdot M + K_{P_rM} \cdot P_r \cdot M$$

$$M' = M \cdot (\mu - \lambda - (D_V + S_V) / V)$$

$$\mu = \mu_m / (1 + (K_C / C) + (K_N / N) + (K_{CN} / (C \cdot N)))$$

$$\lambda = \lambda_\mu / (1 + K_\mu \cdot \mu)$$

V. Model Solution

Solution of the model and further description of the translation process is provided in Wastney et al., (2). A working version of the model can be accessed in the model library (See Chapter 22) at http://biomodel.georgetown.edu/model.

REFERENCES

1. France, J., H. M. Thornley, and D. E. Beever. 1982. A mathematical model of the rumen. *J Agric Sci Camb.* 99:343-353.
2. Wastney, M. E., X. Q. Wang, and R. C. Boston. 1998. Publishing, interpreting and accessing models. *J. Franklin Instit.* 335B:281-301.

20

VERIFICATION AND VALIDATION

Once a published model has been translated into a mathematical form, it is necessary to verify that the model has been built correctly. This means the translated model must be verified with respect to the published model. Validation is the process by which the model is confirmed to be a true representation (approximation) of the system that it is designed to represent. This chapter will demonstrate the steps associated with the translation of a moderately complex model for use with the WinSAAM software, how to verify a model, and then some issues relating to model validation are discussed.

I. Verifying a translated model

Example 1. The minimal model of glucose disappearance

Glucose metabolism is controlled in the body at a number of sites. After a meal, or infusion of glucose, insulin is released. Insulin suppresses synthesis of glucose and stimulates uptake of glucose by tissues. Bergman and colleagues (1, 4) have developed a model to describe these processes (Fig 20.1). The purpose of the model is to provide a basis for estimating S_I (a measure of insulin sensitivity in the tissues) and S_G (sensitivity to glucose) based on data obtained during an intravenous glucose tolerance test (IVGTT). This test involves the administration of a standard amount of glucose followed by blood sampling over a period of 3 hours. The blood samples are analyzed for glucose and insulin. The purpose of the model is to describe the relationship between the amount of insulin released in response to the administered glucose, and the rate at which glucose is utilized. The model assumptions are: a) insulin action is achieved after it is delivered to a remote site, b) insulin in this remote site causes glucose to be taken up by tissues, c) the effect of insulin action and tissue uptake of glucose can be resolved, and d) glucose synthesis and disposal can be resolved from a single assault to the system.

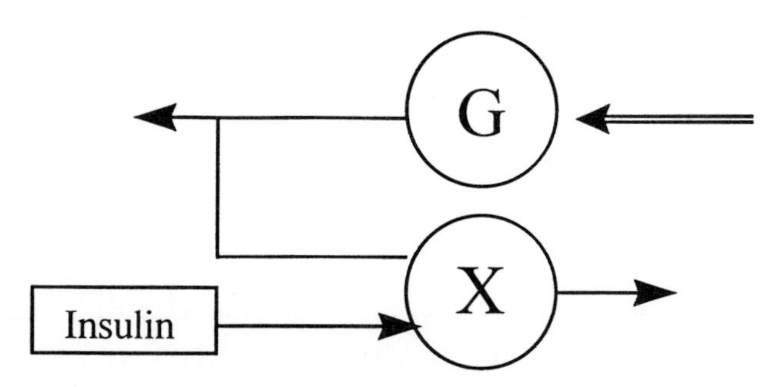

Fig 20.1 Minimal model for glucose. G -glucose, X -remote insulin, I -insulin

The equations for the model are;

$$
\begin{aligned}
G' &= -(p_1 + X)\, G + p_1\, G_b \\
X' &= -p_2\, X + p_3\, (I - I_b) \\
S_I &= p_3/p_2 \\
S_G &= p_1
\end{aligned}
\qquad (20.1)
$$

The symbols are defined in Table 20.1 and the WinSAAM input file for this model is shown in Fig 20.2.

Table 20.1 WinSAAM constructs used in translating this model

Name	Construct	Equation	Description
Compartments:	F(6)	G	Glucose
	F(7)	X	Remote insulin
Fractional Transfers	L(0,7)	-P(2).X	Loss of insulin
Input Rates	UF(6)	$-(P(1)+X)G + P(1).G_b$	Input of glucose
	UF(7)	$p_3.(I - I_b)$	Input of insulin
G Functions:	G(8)	$(I(t) - I_b)$	Increase in insulin
QL Functions:	FF(8)	I	Insulin
Parameters	P(1)	S_G	Glucose sensitivity
	P(2)/P(3)	Si	Insulin sensitivity
	P(4)	Gb	Basal glucose
	P(5)	Ib	Basal insulin
	P(6)	G0	Glucose at time 0
	P(7)	Si	Insulin sensitivity
Initial conditions	IC(6)	G_0	Initial glucose
Statistical Description:	FSD	$s(t) = k.G(t)$	

```
 1: A SAAM31 BERGMAN GLUCOSE DISAPPEARANCE MODEL USING SAAM
 2: 2         25
 3: H PAR
 4: C
 5: C P(1)=SG
 6: C P(2)/P(3)=SI
 7: C P(4)=GB
 8: C P(5)=IB
 9: C P(6)=G0
10: C P(7)=SI
11: C
12:    P(1)       2.586522E-02  0.000000E+00   1.000000E+02
13:    P(2)       3.704002E-02  0.000000E+00   1.000000E+02
14:    P(3)       1.399679E-05  0.000000E+00   1.000000E+02
15:    P(4)      82
16:    P(5)      9.2
17:    P(6)       2.864497E+02  0.000000E+00   1.000000E+04
18:    L(0,7)=P(2)
19:    UF(7)=P(3)                                       8G08
20:    P(7)=P(3)/P(2)
21: C
22: C COMPT. 6 = GLUCOSE
23: C COMPT. 7 = REMOTE INSULIN
24: C COMPT. 8 = PLASMA INSULIN
25: C
26:  IC(6)=P(6)
27: C
28: H DAT
29: X UF(6)=-(P(1)+F(7))*F(6)+P(1)*P(4)
30: X G(8)=FF(8)-P(5)
31: X FF(8)=F(8)
32:    P(7)
glucose data
33: 106                                           WT=0.0
34:             0             92
35:             2             350
36: 106                                           FSD=.01
37:             4             287
38:             6             251
57:             162           85
58:             182           90
insulin data
59: 108QL
60:             0             11
61:             2             26
```

Fig 20.2 WinSAAM Input for Minimal Model

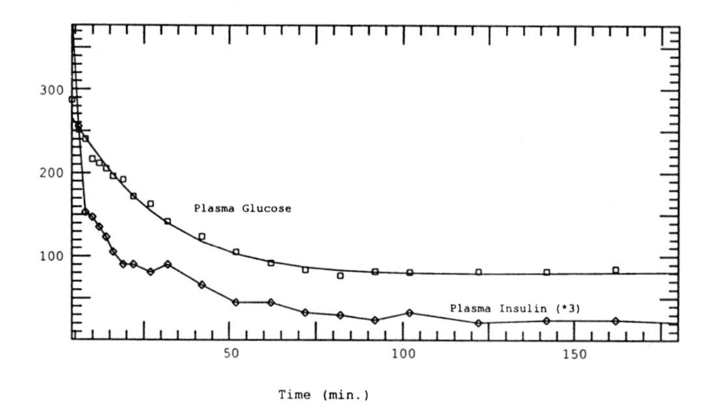

Fig 20.3 Fit of model in Fig 20.1 to glucose and insulin IVGTT data. Note that the insulin data were fitted using a QL function, to drive the input into the remote insulin compartment.

Parameters estimates obtained from fitting the data in Fig 20.3 are;

PARAMETER	VALUE	ERROR	FSD
P(1)	2.587E-02	6.130E-03	2.370E-01
P(2)	3.704E-02	1.182E-02	3.192E-01
P(3)	1.400E-05	7.726E-06	5.520E-01
P(6)	2.864E+02	6.463E+00	2.256E-02

Example 2. The minimal model for insulin kinetics

After glucose is administered iv there is a biphasic response of insulin release indicated by two peaks in the insulin data curve.
Purpose: To enable assessment of pancreatic responsiveness (i.e., 1st and 2nd phases)
Assumptions are that a) insulin is released from the pancreas in direct response to plasma glucose above a threshold level and b) insulin is lost at a rate proportional to insulin level. The model is shown in Fig 20.4.

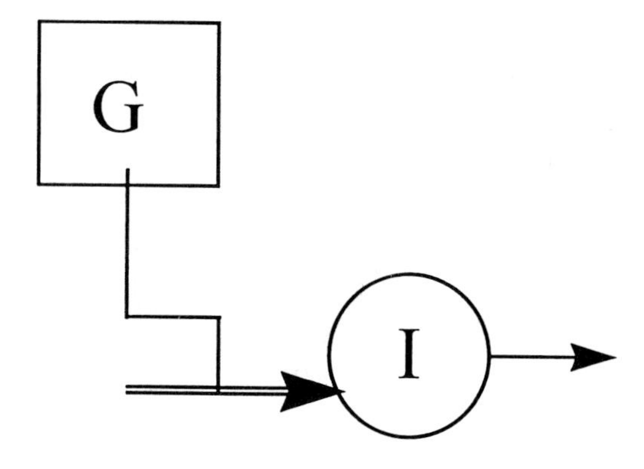

Fig 20.4 A minimal model for insulin kinetics.

The equations for the minimal model are;

$$
\begin{aligned}
I' &= -N \bullet I + \gamma(G-h)t \\
\Phi_1 &= (I_{max} - I_b) / [n \bullet (G_0 - G_b)] \\
\Phi_2 &= \gamma \bullet 10^4
\end{aligned}
\qquad (20.2)
$$

where I' is change in plasma insulin, N is the fractional insulin clearance, γ is the proportionality factor between glucose concentration and rate of increase of second phase insulin secretion, G is plasma glucose, h is threshold glucose level, I_{max} is the first peak insulin concentration, I_b is baseline insulin, Φ_1 and Φ_2 are two parameters that describe the biphasic pancreatic response to rapid elevation of plasma glucose. Symbols, described in Table 20.2, were used to model the system in WinSAAM (Fig 20.5).

Table 20.2 WinSAAM Constructs Used for Insulin Model

Name	Construct	Equation
Compartments:	F(9)	I
Fractional Transfers:	L(0,9)	- n.I
Input Rates:	UF(9)	γ.(G - h) t
QL Functions	FF(6)	G(t)
G Functions	G(10)	G(t) - h)t
	G(11)	I = I(t) + I_b
Parameters	P(9)	n
	P(10)	G
	P(11)	h
	P(12)	I_0
	P(13)	I_b
	P(14)	ϕ_1
	P(15)	ϕ_2

The equations are translated into a WinSAAM text input file in Fig 20.5. In this example the glucose data are used as the forcing function to drive insulin release. The results from fitting the model are shown in terms of the parameter values that were calculated by the fitting process and the values of P(14) and P(15).

```
A SAAM31 BERGMANS INSULIN KINETICS MODEL USING SAAM
2        25
H PAR
C
C P(9)  = N = .3
C P(10) = G = .0033
C P(11) = H =  83.5
C P(12) = I(0)
C P(13) = IB
C
C P(14) = PHI-1
C P(15) = PHI-2
C
   P(9)       1.417502E-01  0.000000E+00   1.000000E+02
   P(11)      8.990284E+01  0.000000E+00   1.000000E+02
   P(12)      1.572435E+02  0.000000E+00   1.000000E+04
   P(13)      8.126204E+00  0.000000E+00   1.000000E+02
   P(14)=(132.5-P(13))/(P(9)*(287-92))
   P(15)=P(10)*10000
C
   P(10)      1.210761E-03  0.000000E+00   1.000000E+02
   L(0,9)=P(9)
   UF(9)=P(10)                                             8G10
C
   IC(9)=P(12)
C
C COMP. 6 = GLUCOSE
C COMP. 9 = CENTRAL INSULIN
C
H DAT
X FF(6)=F(6)
X G(10)=(FF(6)-P(11))*T
X G(11)=F(9)+P(13)
110
   P(14)
   P(15)
.. .. glucose data
106 QL
C          0          92
           2         350
           4         287
           6         251
           8         240
.. .. insulin data
109 G(11)                                 FSD=.02
C          0          11
C          2          26
           4         130
           6         85.
           8         51.
```

Fig 20.5 Text input file for determining insulin response

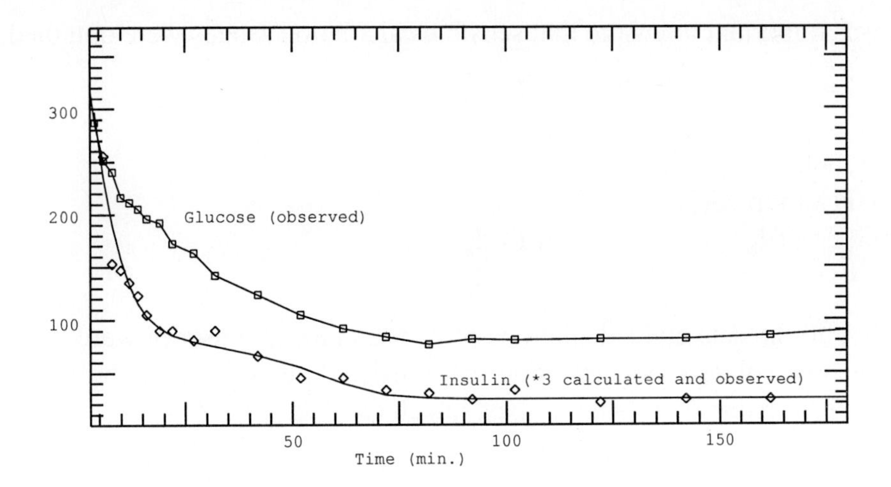

Fig 20.6 Result of fitting the model in Fig 20.5. Note that the glucose data were fitted using a forcing function.

Parameters estimates from solving the model in Fig 20.5 and fitting the data in Fig 20.6 are;

PARAMETER	VALUE	ERROR	FSD
P(11)	8.990E+01	6.895E-01	7.670E-03
P(13)	8.126E+00	6.198E-02	7.628E-03
P(9)	1.418E-01	3.762E-03	2.654E-02
P(12)	1.572E+02	5.226E+00	3.324E-02
P(10)	1.211E-03	5.595E-05	4.621E-02

Results show that the model fitted the observed data and determined parameters for the model, including the rates of biphasic release of insulin from the pancreas.

Tabular Output:

CAT--	T	QC	QO	QO-QC	WT	FSD
P 14	0.00E+00	4.49957E+00	0.00000E+00	-4.500E+00	0.00E+00	2.635E-02
P 15	0.00E+00	1.21076E+01	0.00000E+00	-1.211E+01	0.00E+00	4.621E-02
QL 6	2.00E+00	0.00000E+00	3.50000E+02	3.500E+02	0.00E+00	0.000E+00
QL 6	4.00E+00	0.00000E+00	2.87000E+02	2.870E+02	0.00E+00	0.000E+00
QL 6	6.00E+00	o.00000E+00	2.51000E+02	2.510E+02	0.00E+00	0.000E+00
QL 6	8.00E+00	0.00000E+00	2.40000E+02	2.400E+02	0.00E+00	0.000E+00

Important aspects associated with re-creating these models were as follows. Firstly, it was necessary to confirm that adequate information was provided in the manuscript to enable precise reconstruction of the model. Secondly, it was important to confirm that parameter values were provided along with their units and, preferably, uncertainties. Thirdly, the most efficient modeling language structures were used in the translation process. In WinSAAM, for example, the rate of change of remote insulin was written as an equation;

$$X' = -p_2 X + p_3 (I - I_b)$$

and this was modeled by using the observed data as a forcing function:

QL(Insulin)=>FF(8)

Next, it was important to ensure that sensible calculation ranges were obtained;

$$X' = -p_2.X + p_3.(I - I_b) \quad \text{if } I > I_b$$
$$X' = -p_2.X \quad \text{if } I \leq I_b$$

or, in modeling syntax

$$UF(X) = p_3.G(I)$$
$$G(I) = I - I_b \quad \text{if } I > I_b$$
$$G(I) = 0 \quad \text{if } I \leq I_b$$

Finally, all the data and calculations provided in the manuscript were carefully checked against values produced using the translated model. It may be necessary to refine 'mean' parameter values to best match the data extracted.

II. Model Validation

Validating published models refers to whether the right model has been used to represent a system given the purpose for the model. In the examples above, the purpose was to analyze the insulin response data to provide a tool for clinically assessing diabetes. Though the models are simplifications of all the processes that occur in glucose homeostasis, they have been validated as predictive tools for identifying differences between normals and subjects who develop diabetes (3).

Berman (2) described validation as an effort to support a model through alternate, independent approaches. This means testing how well predictions of the model can be fitted experimentally. Confidence in the model increases when it is supported by additional data. A model will need to be expanded and revised, however, when the model is unable to fit new data on the system.

References

1. Bergman, R. N., and C. R. Bowden. 1981. The minimal model approach to quantification of factors controlling glucose disposal in man. *In* Carbohydrate Metabolism. C. Cobelli and R. N. Bergman, editors. Wiley. 269-296.
2. Berman, M. 1982. Kinetic analysis and modeling: Theory and applications to lipoproteins. *In* Lipoprotein Kinetics and Modeling. M. Berman, S. M. Grundy, and B. V. Howard, editors. Academic Press, NY. 3-36.
3. Martin, B. C., J. H. Warram, A. S. Krolewski, R. N. Bergman, J. S. Soeldner, and C. R. Kahn. 1992. Role of glucose and insulin resistance in development of type 2 diabetes mellitus: results of a 25-year follow-up study. *Lancet.* 340:925-929.
4. Pacini, G., and R. N. Bergman. 1986. MINMOD: a computer program to calculate insulin sensitivity and pancreatic responsivity from the frequently sampled intravenous glucose tolerance test. *Comp. Meth. and Prog. In Biomed.* 23:113-122.

21

USING THE MODEL

This chapter shows how a published model can be used to investigate a system and make predictions. The model to be discussed is a simple homogeneous mixing model by Anderson *et al.* (1) for the acquired immunodeficiency syndrome (AIDS). The model will be described, depicted graphically, and then, it will be used to make some predictions. Further discussion on this model is available in Wastney et al. (10) and a working version of the model can be obtained through the model library (see Chapter 22).

I. Model Description: AIDS model

The AIDS model of Anderson et al. (1) describes the relationship between four populations; Susceptible (S), Infectives (Y), AIDS (A), and Seropositive (Z) (Fig. 21.1). It is a dynamic (or nonlinear) model and is based on a number of assumptions. The model assumes that the natural death rate of each population is the same, but that AIDS subjects have, in addition, disease-induced death. Other assumptions are listed in Table 21.1 together with the source of the data used to develop the model and the application of the model for predicting the course of an epidemic. The descriptive table (Table 12.1) also lists the initial conditions, parameter values, and the model equations. Despite the simplicity of the model, it gives a clear picture of the development of an epidemic after the introduction of HIV into a susceptible homosexual population. The art of Anderson's AIDS model is not to mimic the reality precisely, but to capture the essence of the AIDS epidemic without including unnecessary details.

The equations for the model are shown in Table 21.2. A graphic of the model is shown in Fig 21.1 and the WinSAAM text input file for the model is shown in Fig 21.2. The model has been used to investigate the effect of number of partners on the predicted size of the infective population (Fig 21.3) and the seropositive population (Fig 21.4). The use of the model to predict the size of the different populations over time is shown in Fig 21.5.

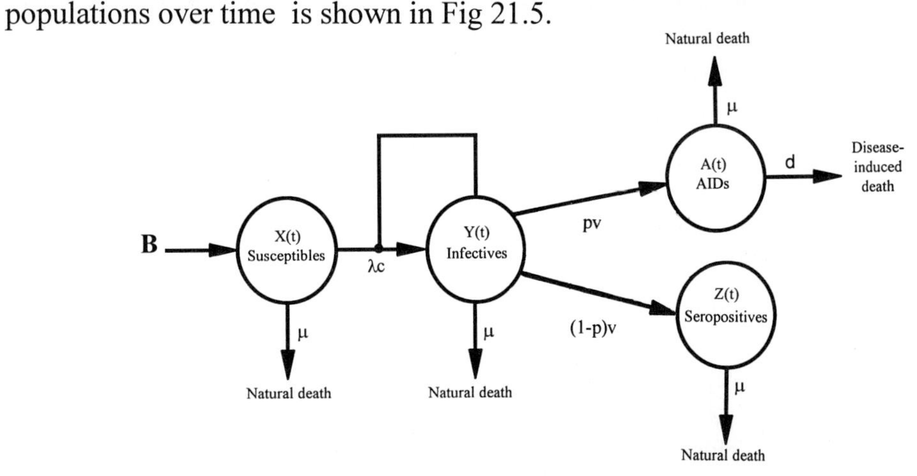

Fig. 21.1 The flow diagram of Anderson's AIDS model (8). For definitions of symbols, see Table 21.1.

Table 21.1. AIDS epidemic model in a homosexual population (1, 8).

Purpose	Modeling the AIDS transmission dynamics.
Type	Dynamic model
Assumptions	• There are four classifications under the total modeled population: Susceptible X, Infectives Y, AIDS A, and Seropositive Z; • The immigration rate of susceptible males into a population is constant, B; • Susceptibles acquire infection via sexual contact with infectious people; • Patients with AIDS are effectively withdrawn from circulation in the population such that they do not generate new cases of HIV infection; • Infected people are infectious for a period $1/v$ time units, after which a proportion p proceeds to develop AIDS while the remaining $1-p$ passes into a seropositive but non-infective class; • The incubation period is depicted as a linear function of the time that a person has been incubating the infection, i. e. $v(\tau)=\alpha\tau$.
Data sources:	Epidemiological data from various sources: • Peterman et al.(9) • CDC (3-6) • Carne et.al., (2) • McKusick et al., (7) • Weller & Carne (unpub)
Applications	Assess how various processes influence the course of initial epidemic following the introduction of the virus, based on a survey of the available epidemiological data on HIV infection and the incidence of AIDS.
Model diagram	Fig. 21.1
State variables	Susceptibles, X(t), Infectives, Y(t), AIDS Patients, A(t), and Seropositive, Z(t), Total homosexual population investigated, N(t).
Initial conditions	X(0) = N(0) = 999,995 Y(0) = N(0)-X(0) = 5 A(0) = Z(0) = 0.
Parameters	B: the recruitment rate of susceptibles (13333.3 yr^{-1}). μ: natural death rate (1/32 yr^{-1}). d: the death rate of AIDS patients (1 yr^{-1}). λc: the rate of transferal from the susceptible to the infectives class. Where λ is the ability of acquiring infection from a randomly chosen partner, $\lambda=\beta Y(t)/N(t)$, here β is the transmission probability (0.26).

	c is the number of sexual partners (2 ~ 6). p: is the proportion of seropositives who are infectious (0.3). v: is the rate of conversion from infection to AIDS (0.2 yr^{-1}). N(t) = X(t)+Y(t)+A(t)+Z(t), total population size (1,000,000 at t=0).
Equations	The Model Equations: $\frac{dX}{dt} = B - \mu X - \lambda c X, \quad \lambda = \frac{\beta Y}{N},$ $\frac{dY}{dt} = \lambda c X - (v + \mu) Y,$ $\frac{dA}{dt} = pvY - (d + \mu) A,$ $\frac{dZ}{dt} = (1 - p) vY - \mu Z,$ and $N(t) = X(t) + Y(t) + A(t) + Z(t),$ Initial conditions: $X(0) = 999{,}995 \approx N(0),$ $Y(0) = N(0) - X(0) = 5,$ $A(0) = Z(0) = 0.$
WinSAAM model	Fig 21.2
Results	Fig. 21. 3 The effect of the number of homosexual partners per year on the development of AIDS population. Increasing the partner number significantly accelerates the maximum occurrence of the AIDS incidence. For the partner number, c=4, the AIDS incidence reaches a maximum around 17 to 18 years, whereas for c=8,the maximum occurs around 8 years. Fig. 21.4. The influence of the partner number per year on the sero-positive development. Increasing the partner number per year accelerates the maximum occurrence of the sero-positive incidence. For c=4, the sero-positive incidence reaches a maximum around 25 to 27 years, whereas for c=8, the maximum occurs around 16 to 17 years. Fig. 21.5 Comparison of the heterogeneous development of different populations: infective males, sero-positives, and AIDS patients, given the partner number per year, c=4.

```
A SAAM31              AIDS.S
C  REF: ANDERSON RM AND MAY RM. PHIL. TRANS. R. SOC. LOND. 314:
533-570,
C 1986
C DEMONSTRATION OF AIDS SPREAD WITH TIME
C
H PAR
C P(1)=B=13333.3 YR**-1
C P(2)=V=0.2 YR**-1
C P(3)=U=.034 YR**-1
C P(4)=D=1.0 YR**-1
C P(5)=P=.3
C P(6)=C=4 PARTNERS
C P(7)=R=0.88 YR**-1
C P(8)=BETA=.26
C
C A0=0
C Z0=0
C X0+Y0=N0=100000
C
C X=SUSCEPTIBLES
C Y=INFECTIOUS MALES
C A=AIDS PATIENTS
C Z=SERO-POSITIVE NON INFECTIOUS
C
C COMP 1 = X
C COMP 2 = Y
C COMP 3 = A
C COMP 4 = Z
C
   P(1)      13333.3
   P(2)      .2
   P(3)      .034
   P(4)      1
   P(5)      .3
   P(6)      4
   P(7)      .88
   P(8)      .26
C
   IC(1)     999995
   IC(2)     5.
   IC(3)     0
   IC(4)     0
H DAT
X G(1)=P(8)*F(2)/G(2)
X G(2)=F(1)+F(2)+F(3)+F(4)
X UF(1)=P(1)-P(3)*F(1)-G(1)*P(6)*F(1)
X UF(2)=G(1)*P(6)*F(1)-(P(2)+P(3))*F(2)
X UF(3)=P(5)*P(2)*F(2)-(P(3)+P(4))*F(3)
X UF(4)=(1-P(5))*P(2)*F(2)-P(3)*F(4)
103
             0
2            .5                              60
104
             0
2            .5                              60
```

Fig 21.2 Text Input file for fitting the AIDS model.

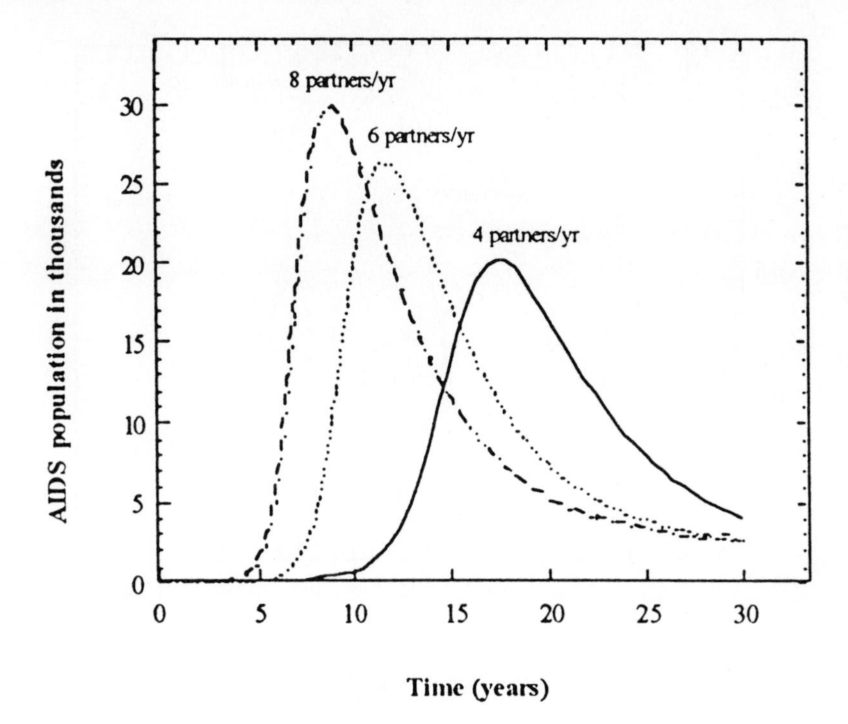

Fig 21.3. The influence of partner number, c, on the development of AIDS patient population (A).

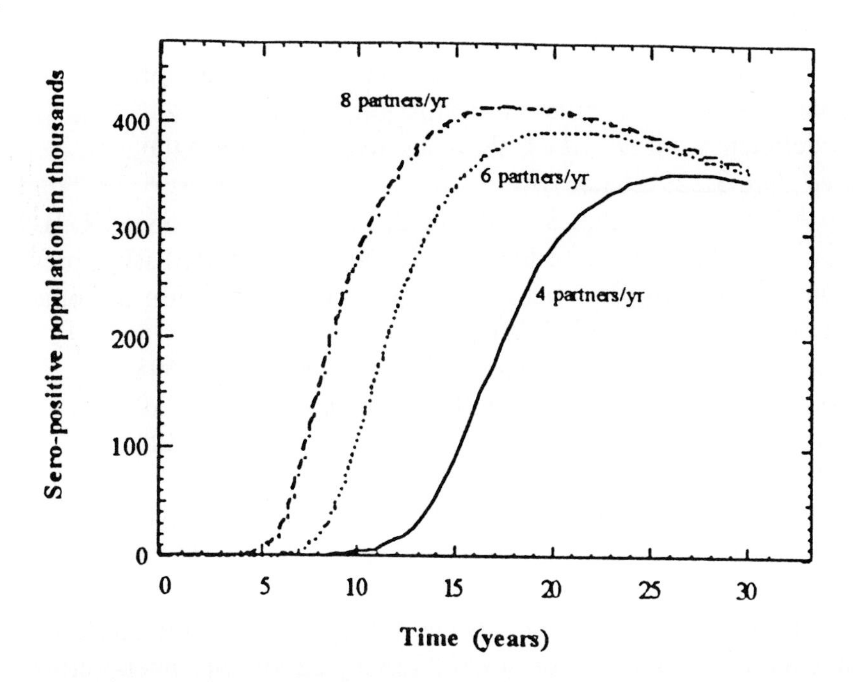

Fig 21.4. The influence of partner number, c, on the development of sero-positive population (Z).

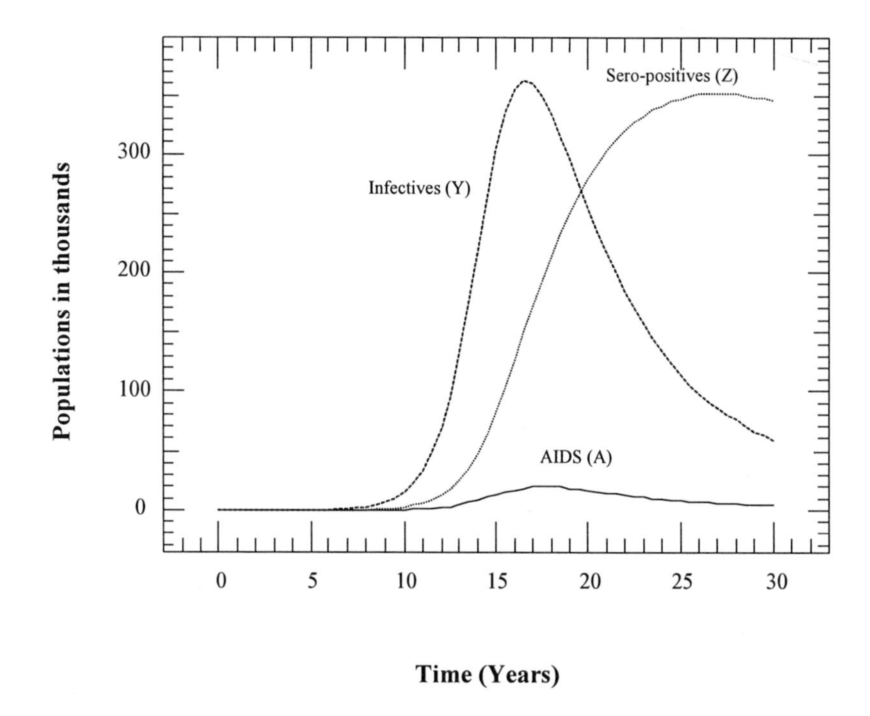

Fig 21.5. The development of different populations: infectives (Y), sero-positives (Z) and AIDS patients (A).

REFERENCES

1. Anderson, R. M., G. F. Medley, R. M. May, and A. M. Johnson. 1986. A preliminary study of the transmission dynamics of the human immunodeficiency virus (HIV), the causative agent of AIDS. *IMA J Math Appl Med Biol*. 3:229-263.
2. Carne, C. A., I. V. D. Weller, S. Sutherland, R. Cheinsong-Popov, R. B. Ferns, P. Williams, A. Mindel, R. Tedder, and M. S. Adler. 1985. Rising prevalence of human T-lymphotropic virus type-III (HTLV-III) infection in homosexual men in London. *Lancet*. I:1261-1262.
3. CDC. 1985. *Morbidity and Mortality Weekly Report*. 34:681-695.
4. CDC. 1985. *Morbidity and Mortality Weekly Report*. 34:583-589.
5. CDC. 1985. *Morbidity and Mortality Weekly Report*. 34:573-583.
6. CDC. 1986. *Morbidity and Mortality Weekly Report*. 35:17-21.
7. McKusick, M. L., W. Horstman, and T. Coates. 1985. AIDS and sexual behaviour reported by gay men in San Francisco. *Am J Public Health*. 75:493-496.
8. Murray, J. D. 1991. Epidemic models and the dynamics of infectious diseases. *In* Mathematical Biology. Vol. 19. S. Levin, editor. Springer-Verlag, New York. 610-650.
9. Peterman, T. A., D. P. Drotman, and J. W. Curran. 1985. Epidemiology of the acquired immunodeficiency syndrome (AIDS). *Epidemiol Rev*. 7:1-21.
10. Wastney, M. E., X. Q. Wang, and R. C. Boston. 1998. Publishing, interpreting and accessing models. *J. Franklin Instit.* 335B:281-301.

22

A LIBRARY OF MODELS

Mathematical models are used in biomedicine, engineering, meteorology, and other applied sciences to describe and analyze the behavior of complex systems. Models are used because a large amount of information pertaining to a system can be condensed and summarized using mathematical equations. Because of the complex nature of biological reactions *in vivo* and the rate at which scientific knowledge is increasing, modeling may be the most expedient way to maintain and integrate information on biological systems. For biomedical scientists in the next decade, understanding modeling may be as important as understanding molecular biology has been in this decade. Published models on biological systems (biomodels) therefore need to be accessible to all biomedical investigators. This chapter discusses the need to provide on-line access to published biomodels; a facility called a Model Library that is being developed to meet this need. Then future developments of modeling on the Internet are discussed.

I. Need for an on-line library of published models

There are a number of scientific and economic reasons why models should be communicated among the scientific community. Scientifically, there is a need to make use of information inherent in models to design and analyze new studies (3); for example, to indicate what data should be collected, when it should be collected during an experiment, and the number of experiments required for a study. Models can be used in this way to optimize use of experimental resources. There is also a need to communicate models to allow integration of studies at different levels; for example, to relate metabolism at the cellular level to that at the organ or whole body level. One broad aim of the recently proposed Physiome Project is to integrate and interpret information obtained through the genome project to hierarchically higher levels of organization (4). Educationally, models provide a hands-on method for learning and understanding biological systems (1).

Models are expensive to develop and there is a large investment of research funds in modeling; several resources have been funded to develop modeling software (2, 5, 8) and a review of grants funded by NIH in 1996 included 40 with 'kinetics' in the title and 128 with 'dynamics'. At an average of $200,000/grant, over $33 million were invested in studies which have the potential to result in the development or use of a model. A review of MEDLINE showed that over 80 papers per month include the terms biological model and computer simulation. To increase the return on the investment of research funds in studies and software development, models and data from modeling studies need to be made available for use by other investigators.

A standardized well-organized and user-accessible biomathematical model library would allow users from biological, physiological, epidemiological, pharmaceutical, agricultural, and psychological fields to identify and easily access published model in their area of interest. They would then be able to evaluate the model, translate the characteristics of the model, rebuild the model, and explore new model features based on their own requirements. Although such an idea has been employed in other areas (6), it has not yet been applied to managing published biomathematical models.

There are a number of problems and barriers to communicating biomodels. The problems can be worked around but the barriers will need to be removed before seamless communication is achieved. Some of the problems are philosophical differences between investigators who develop models in terms of the approach used, and whether models should be shared. A second problem relates to the mathematical complexity of models and a third problem relates to the biological complexity; models often require interdisciplinary knowledge to be fully understood. These problems can be addressed by presenting multiple versions of models of varying detail and complexity to promote general understanding of the model, and also by presenting models in formats that other investigators can use. The Internet will be a valuable medium for communicating these facets of modeling.

One barrier to communicating models is lack of accessibility to working versions of models. Published models, in general, are underutilized because they are not accessible in a useable form. Models are published descriptively although they are developed and used electronically on computers. To use a model, an investigator must either obtain a copy from the author, which assumes that the author is available and has the time to supply a copy, or reconstruct the model from the paper. Reconstruction is time-consuming and error-prone because published descriptions are limited by space and editorial policies and are often incomplete. We have initiated the development of a biomodel library on the Internet to address this issue and provide access to working versions of published models (9). Other libraries of physiological models have been developed for educational purposes (1, 7) (http://www.phys-main.umsmed.edu/).

A modeling environment must be developed that will accommodate the dissemination of user data, models, experimental designs, and analytic methods across different structures and platforms. The utility of biomodels resides in their use for simulation and to manipulate them to test theories (13). The Internet is a valuable medium for communicating models because it allows working interactive forms of models to be obtained, and dynamic demonstration of models.

II. Accessing the on-line library

Because reconstructing models from publications is time-consuming and error-prone, an alternative approach is to make working versions of published models available over the Internet. This project called a 'model library' has been started as a collaboration between Georgetown University and the University of Pennsylvania (9-12). The project is funded by the National Science Foundation. The purpose of the library is two-fold; to make published models available, thereby encouraging their use as a tool for probing systems, and as an educational facility for students to learn about modeling biological systems through a hands-on approach. Criteria for inclusion are that the model has been published, peer-reviewed, and has been fitted to data.

Users may access the library at the following address;

http://biomodel.georgetown.edu/model/

The library can be used to locate information on modeling sites, conferences, recent modeling publications, and to find working versions of models (Fig 22.1). From the home page users may link to pages describing the purpose of the project, listing the personnel on the project, a Search page for locating a model, recent Publications, Modeling resources, and conferences and courses. Published models can be submitted to the library through an on-line menu.

WELCOME TO THE LIBRARY OF
MATHEMATICAL MODELS OF BIOLOGICAL SYSTEMS

Georgetown University Medical Center, the University of Pennsylvania
and Vanderbilt University
(NSF Grant No. BIR-950-3872)

This project is under development for accessing working versions of published models:

Purpose
Personnel
Search for a model
Submit a model
Publications
Who's Who
Suggestions
Modeling Resources
Conferences, Courses, and Symposia

Copyright 1997, Grantholders of NSF Grant No. BIR-950-3872, Disclaimer
Date last modified: January 27, 1998

Fig 22.1 Home page of the Library of Mathematical Models of Biological Systems

From the search page, users may locate a model by typing in a search term, or list models in the library by model type, model software, species, or hierarchy (Fig 22.2). In addition, users may access a list of commonly used chemical and biochemical models (Fig 22.3). These models are shown graphically, and using WinSAAM formats. Models can be viewed in terms of a description, graphic, equations, and plots of the model solution. A model can be downloaded, using the browser, in formats compatible with several modeling packages for the user to run with his/her own modeling software.

LEVEL	NAME
Molecular	
	Restriction Endonuclease
Cell	
	Endocytosis
Tissue/Organ	
	Rumen Digestion
	Glucose/Insulin
	Sheep Rumen
Whole Body	
	Lithium in rats
	Pharmacokinetic model
	Zinc in humans
	Zinc in rats
	Calcium in humans
	Calcium in pregnant sheep
	Selenium in humans
	Magnesium in humans
	Mg kinetics in Adolescent Girls
Population	
	AIDS

Fig 22.2 Listing of models currently in the library, by hierarchy.

Making models available through a model library allows users to locate models for a particular system. This is important because models for similar systems are often published in different fields, (e.g., zinc kinetics is studied in nutrition and toxicology). By searching the library by subject, species, model type, and modeling software, users can locate different models for the same system, for comparison purposes, and similar models for different systems, for educational purposes. By having models collated, cross-referenced, and available within a single facility, they can be readily located and then used for their intended purpose, to investigate systems through simulation.

Future plans for the library involve developing a system for models to be automatically deposited in the library at the time of publication. We plan to add models to the site as they are published. By linking a journal article to a site containing a working version of the model, the use of published models for research and understanding of systems in health and disease should increase.

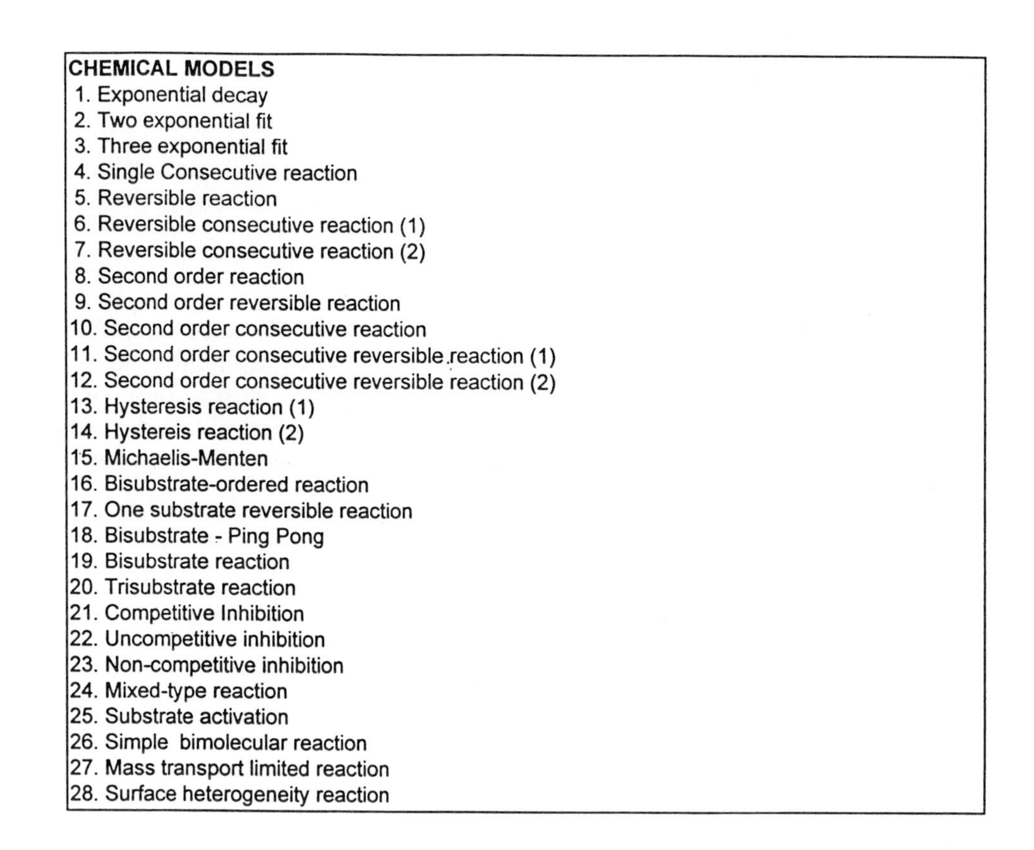
CHEMICAL MODELS
1. Exponential decay
2. Two exponential fit
3. Three exponential fit
4. Single Consecutive reaction
5. Reversible reaction
6. Reversible consecutive reaction (1)
7. Reversible consecutive reaction (2)
8. Second order reaction
9. Second order reversible reaction
10. Second order consecutive reaction
11. Second order consecutive reversible reaction (1)
12. Second order consecutive reversible reaction (2)
13. Hysteresis reaction (1)
14. Hystereis reaction (2)
15. Michaelis-Menten
16. Bisubstrate-ordered reaction
17. One substrate reversible reaction
18. Bisubstrate - Ping Pong
19. Bisubstrate reaction
20. Trisubstrate reaction
21. Competitive Inhibition
22. Uncompetitive inhibition
23. Non-competitive inhibition
24. Mixed-type reaction
25. Substrate activation
26. Simple bimolecular reaction
27. Mass transport limited reaction
28. Surface heterogeneity reaction

Fig 22.3 Listing of chemical and biochemical models available in the Model Library.

III. The future of modeling on the Internet

Plans are being developed to make large databases of information from the genomic to the whole body levels available on the Internet. This information will form the basis for developing models to integrate physiological systems from cells, tissues, organs, to whole body and populations (4).

References

1. Aarts, J., D. Moller, and R. Van Wijk van Brievingh. 1992. Modeling and simulation in biomedicine. *In* Symposium on Computer Applications in Medical Care. P. Clayton, editor. McGraw Hill, NY. 900-902.
2. Anon. 1991. Specialized resources for research. *J NIH Research*. 3:108-110.
3. Bassingthwaighte, J. B. 1987. Strides in the technology of systems physiology and the art of testing complex hypotheses. *Fed Proc*. 46:2473-2476.
4. Bassingthwaighte, J. B. 1995. Toward modeling the human physionome. *Adv Exp Med Biol*. 382:331-339.
5. Collins, J. C. 1992. Resources for getting started in modeling. *J Nutr*. 122:695-700.
6. Morton, R. F., J. R. Hebel, and R. J. McCarter. 1996. Keys to understanding articles on epidemiologic studies. *In* A Study Guide to Epidemiology and Biostatistics. ASPEN Publ., MD. 161-173.
7. Summers, R. L., and J.-P. Montani. 1994. Interface for the documentation and compilation of a library of computer models in physiology. *JAMIA*. Symposium Supplement:86-89.
8. van Milgen, J., R. Boston, R. Kohn, and J. Ferguson. 1996. Comparison of available software for dynamic modeling. *Annales de Zootechnie*. 45:257-273.
9. Wastney, M. E., N. Broering, C. F. R. Ramberg, Jr. , L. A. Zech, N. Canolty, and R. C. Boston . 1995. World-wide access to computer models of biological systems. *Info Serv & Use*. 15:185-191.
10. Wastney, M. E., B. N, B. J, and R. Boston. 1997. On-line access to published models of biological systems. *In* Trace Elements in Man and Animals - 9. Proceedings of the Ninth International Symposium on Trace Elements in Man and Animals. P. Fischer, M. L'Abbe, K. Cockell, and R. Gibson, editors. National Research Council of Canada, Ottawa. 146-147.
11. Wastney, M. E., K. N. Siva Subramanian, N. Broering, and R. Boston. 1997. Using models to explore whole-body metabolism and accessing models through a model library. *Metabolism*. 46:330-332.
12. Wastney, M. E., D. C. Yang, D. F. Andretta, J. Blumenthal, J. Hylton, N. Canolty, J. C. Collins, and R. C. Boston. 1998. A mechanism for distributing working versions of published models via the Internet. *Adv. Exp. Med. Biol.*:(In press).
13. Yamamato, W. S. 1985. Converting mathematical models of physiological systems to relational database schemes for analysis and comparison. *IEEE Trans Biomed Engineer*. 32:273-276.

Appendix 1: Glossary based on reference (1).

MODELING TERM	DEFINITION	UNITS
Compartment	An anatomical, physiological, chemical or physical subdivision of a system.	Same as the initial condition
Compartmental analysis	Analysis of a system using ordinary differential equations	
Clearance	Net transport, as volume containing the amount transferred per unit time, from one part of a system (e.g., blood) to another part (e.g., urine)	Mass/time
Convolution integral	Defines response of a system to an input based on its response to a unit input	
Cycles	Number of time a particle returns to a compartment. Is equal to (residence time/transient time) minus 1.	
Deconvolution	Determining an input from a response	
Delay	Time lag between input and output	Time
Dynamics	Study of non-linear system	
Initial condition	Amount of material in a compartment at the start of the study	Amount
Input	Entry of material from the outside	Amount/time
Fractional catabolic rate (FCR)	Fraction of material cleared from a compartment per unit time	1/time
Fractional transfer coefficient	see Rate constant	
Function	Equation	
Kinetic analysis	Interpretation of tracer behavior in terms of a mathematical model	
Kinetics	Study of linear systems	
Linear system	System with a linear response, i.e., if response to input I_1 is R_1, and to I_2 is R2 then response to I_1+I_2 is R_1+R_2	
L-inverse, LI(I,I)	Inverse of the L(I,J) matrix and equals the residence time	Time
Model	Mathematical or physical construct used to simulate behavior of a system	
Modeling	Interpretation of behavior of a system in terms of a model	
Rate constant	Fraction leaving a compartment per unit time in a compartmental system. For a chemical system, rate constant is related to the order of the system	1/time

MODELING TERM	DEFINITION	UNITS
Residence time	Total time a particle spends in a compartment, LI(I,I), or 1/FCR	Time
Specific activity	Ratio of tracer to tracee.	
Steady state	Amount of tracee in a system is constant during the study	Mass
System	Part of the universe of interest	
Transit time	Time for a particle to be transferred between two parts of a system	Time
Transient time	Turnover time	
Turnover rate	Fraction leaving a compartment per unit time	1/time
Turnover time	Reciprocal of transfer out of a compartment, 1/L(I,I). Time a particle spends in a compartment	Time
Tracee	Material of interest that is being studied	Mass
Tracer	Labeled particle used to tag the movement of the compound of interest through a system.	E.g., Activity (radioisotope) Enrichment (stable isotope)
Transport	Amount of material transferred from a compartment per unit time	Mass/time
Volume (or space) of distribution	The apparent volume tracer distributes in, by dividing the amount of tracer by the concentration of tracer.	Volume

1. Brownell, G. L., M. Berman, and J. S. Robertson. 1968. Nomenclature for Tracer Kinetics. *Int J Appl Radiat Isotop*. 19:249-262.

Appendix 2: WinSAAM Dictionary*

SYMBOL	MEANING
A SAAM31	Text on first line of a WinSAAM text input file.
ADJU	Command that lists the adjustable parameters.
C	C in the first character space of a line in the text input file denotes a comment.
CALC	Command to open an on-line calculator.
CALC LI	Command to calculate the inverse matrix.
CALC TR	Command to calculate the transfer matrix.
Category	Solution stored at an address in WinSAAM (e.g., F(I), solution of compartment I)
Component	Address of a solution in WinSAAM; Symbol is Q.
COR(I,J)	Command to request correlation matrix for the adjustable parameters.
DECK	Command that translates the model in the working file for solving.
DEPE	Command that calculates the dependent parameter values.
DN(n)	Resolution (or number of divisions) in a delay.
DT(I)	Delay duration. The time material is delayed before appearing in a compartment
EXCLude	Freeze an adjustable parameter at its current solution.
FCR(I,J)	Fractional catabolic rates for all compartments (I) of the model with inputs in any compartment (J). (See Chapter 14 for additional information).
F(I,T)	A function where I is a compartment number and T is an independent variable, usually time. A compartment contains material that behaves in the same way kinetically. It may be a chemical form or a compound, a tissue, or state.
FF(I)	Forcing function.
FIX L(I,J)	Command that changes an adjustable parameter to a fixed parameter.
FREE L(I,J) a b	Command that makes a parameter adjustable by setting the lower limit of L(I,J) to 'a', upper limit to 'b' (Default is one-third the current value and three times the current value respectively).
FSD	Fractional standard deviation (standard deviation/value) for adjustable parameters. Command lists the FSD's for the parameters.
G(n)	A function or equation.
H DAT	Line in WinSAAM text input file that precedes data.
H PAR	Line in WinSAAM text input file the precedes parameters.
H PCC…TC(n)	Line in WinSAAM Text input file that precedes parameters that are to change in value in Time block n.
H STE	Line in WinSAAM that precedes steady state information.
IC(I)	Initial conditions for the solution of a set of differential equations. The amount of material present in a compartment at the start of a model solution.

SYMBOL	MEANING
INCLude	Command to unfreeze an adjustable parameter that had been frozen at its current value using the EXCLude command.
INF	Infinity. Sets value of T to a large number.
INFO	Command to display informational analysis for current model.
ITERate	Command that adjusts parameter values to obtain the best fit to the data.
K(I)	A proportionality constant the represents the fraction of a compartment that is sampled.
KEEP	Command to store the current solution.
L(I,I)	Total loss from a compartment. [1/turnover].
L(I,J)	Fractional rate constant, or fraction of material transported from compartment J to compartment I per unit time. Also represents the probability that a particle will move into compartment I from compartment J. When L(I,J) is typed as a command, all L(I,J) of a model are listed.
LI(I,I)	Residence time [1/FCR]. Total time a particle spends in a compartment (including recycling).
MAX L(I,J) n	Command that sets the upper limit of L(I,J) to value n
MODE=n	Command to select an integrator (see Table 5.1 for options)
M(I)	The steady state solution for compartment I. The mass, or size of compartment I.
MIN L(I,J) m	Command to set the lower limit of L(I,J) to value m
NEW	Command to open a new device for storing solutions. Stored solutions are accessed by using the RELI or REST commands.
OLD	Command for opening an existing facility containing stored solutions.
PART	Command to calculate the partial derivatives of all components with respect to all adjustable parameters.
P(n)	A general parameter that may or may not have physical meaning. When P(I) is entered as a command all general parameters in a model are listed.
PLOT Q(I)	Command to plot the results of component I. Other options for plotting QO (observed values) and QC (calculated values)
PLOT P <parameter list>	Plot partials of the parameters specified in the list. Another option is S for sensitivities of the listed parameters.
PRINt	Command to list results on the screen. E.g., PRIN Q(1), to list the results of component 1 or PRIN W(1) to list the weights assigned to data in component 1.
Q(I)	Calculated and observed values for component I.
QC(I)	Calculated values for component I.
QO(I)	Observed values for component I.

Appendix 2 cont.

SYMBOL	MEANING
QO(I)	Observed values for compartment I. Measured quantity in a compartment
QO-function	Replaces quantity of material in compartment with observed amount
QF-function	Replaces amount of material in a compartment by the value defined by a forcing function, FF(I).
QL-function	Replaces quantity of material in a compartment with a value determined by linear interpolation between observed values.
RAND	Command to calculate new 'observed' values, by adding random error to the current QC values. The error can be normal Gaussian, exponential, or uniform and the type can be SD, FSD or RQO with any specified value.
R(I)	Ratio of QC/QO.
RESCue	Recalls the latest saved Text Input file into the Working file area.
RES(I)	Residuals
RESI	Command to calculate the correction vector RES(I) based on the partials matrix.
REST n	Restore saved solution number 'n'. The saved parameter values will overwrite the current values in the translated model.
R(I,J)	Transport rate, or movement of tracee from compartment J into compartment I per unit time
RQO	Weighting function
SA	Specific activity. Calculates F(I)/M(I).
Saam	Command to print the output to a separate window.
SD	Standard deviation
SE(I)	Standard error of compartment I.
S(I,J)	Summing coefficient.
SOLVe	Command to solve the model
SS	Sum of squares. SS(I) lists sum of squares for each component.
SWIT CONN ON	Command to connect the observed data symbols in a plot.
T	An independent variable. Usually time
TC(n)	A signal to interrupt the solution to allow for changes in initial conditions parameter or steady state values. Allows condition of an experiment to be simulated, e.g., washout period, or addition of a 2nd tracer.
TH	Theta, a second independent variable
TR(I,J)	Transfer coefficient. Fraction of compartment J transported to compartment I. (See Chapter 14 for additional information).
U(I)	Steady state input.
UF(I)	Input function.
UPDA	Command to update values of model from the translated model to the

	working file.
W(I)	Weights for compartment I.
Y	When Y is entered in the first character space on a line, lines following it are ignored by WinSAAM. A Y- in the first space will cause the program to begin recognizing lines again. Y/Y- are therefore used to exclude sections of the Text input file.

*See also Fig 6.1 and Tables 6.3 and 6.4. I implies a compartment, n is not associated with a compartment. Only the first four letters of a command need to be entered, as indicated by capitals (e.g., ITER is the command for iterate).

REFERENCES

1. Berman, M., W. F. Beltz, P. C. Greif, R. Chabay, and R. C. Boston. 1983. CONSAM User's Guide. In DHEW Publication No 1983-421-123:3279. US Govt Printing Office, Washington, DC.
2. Berman, M., and M. F. Weiss. 1978. SAAM Manual. DHEW Publication No. NIH 78-180. US Printing Office, Washington, DC.
3. Foster, D. M., and R. C. Boston. 1983. The use of computers in compartmental analysis: the SAAM and CONSAM programs. In Compartmental Distribution of Radiotracers. J. Robertson, editor. CRC Press, Boca Raton. 73-142.

Appendix 3: Abbreviations

ACSL	Automated Continuous Simulation Language
AIDS	Acquired ImmunoDeficiency Syndrome
AUC	Area under curve
CONSAM	CONversational SAAM
Cp	Drug concentration
CSMP	Continuous Simulation Modeling Program
CV	Coefficient of variation
DET	Determinant
E0	Baseline drug concentration
EC50	Drug concentration for 50% effect
Emax	Maximum drug effect
EMSA	Extended Multiple Studies Analysis
FCR	Fractional catabolic rate
G_0	Glucose concentration at Time=0
G_b	Basal glucose concentration
GITS	Global iterative two-stage
GTS	Global two-stage
I	Insulin
I_b	Basal insulin concentration
ITS	Iterative two-stage
IV	Intravenous
IVGTT	Intravenous glucose tolerance test
Km	Michaelis-Menten constant
MLAB	Modeling LABoratory
NE	Norepinephrine
NONMEM	NONlinear Mixed Effects Models
NPML	Non-parametric maximum likelihood
OLE	Object Linking and Embedding
PR	Production rate
SAAM	Simulation, Analysis And Modeling
S_g	Glucose sensitivity
S_i	Insulin sensitivity
Vmax	Maximum rate of a reaction
WinSAAM	Windows version of SAAM

Index

A

B

M

T

U

V

W